AXIALLY COMPRESSED STRUCTURES

Stability and Strength

AXIALLY COMPRESSED STRUCTURES

Stability and Strength

Edited by

R. NARAYANAN

M.Sc.(Eng.), Ph.D., F.I.Struct.E., M.I.C.E., F.I.E.
Senior Lecturer, Department of Civil and Structural Engineering,
University College of Cardiff, United Kingdom

APPLIED SCIENCE PUBLISHERS
LONDON and NEW YORK

APPLIED SCIENCE PUBLISHERS LTD
Ripple Road, Barking, Essex, England

Sole Distributor in the USA and Canada
ELSEVIER SCIENCE PUBLISHING CO., INC.
52 Vanderbilt Avenue, New York, NY 10017, USA

British Library Cataloguing in Publication Data

Axially compressed structures.
1. Structural stability
I. Narayanan, R.
624.1'71 TA656

ISBN 0-85334-139-7

WITH 9 TABLES AND 180 ILLUSTRATIONS

Photoset in Malta by Interprint Limited

Printed in Great Britain by Galliard (Printers) Ltd, Great Yarmouth

PREFACE

It is my privilege to write a short preface to this volume on *Axially Compressed Structures*, the first of a planned set of volumes on the stability and strength of structures. The inspiration for these books comes from the recognition of the significant advances made, as a result of research during the last decade, in our understanding of the behaviour of structures; these, in turn, have set new trends and caused major changes in the design codes in many countries in recent times. Even the philosophy of design has seen a major shift from the permissible stress basis to the concepts of limit state.

Much research effort continues to be directed towards a better understanding of the complex behaviour of structures in the post-elastic, post-buckling, and ultimate ranges. Nevertheless, the ultimate benefit to be derived from the substantially improved knowledge which continues to accrue as a result of research depends on its effective implementation. Much needs to be done to bridge the gap between the results of research and their effective use by designers and practitioners of the art. The purpose of this book (and its planned companion volumes) is to present material which has been (or will be) influential in the generation of specifications concerned with design.

Each volume is dedicated to a central theme and will contain a number of chapters of the state-of-the-art type on selected topics. In addition, a few chapters describing current developments in the field are also included. Each topic is presented with sufficient introductory material, so that an engineering graduate who is familiar with basic concepts in structural analysis and structural stability is able to follow it without having to undertake any substantial background reading. An effort has been made to avoid lengthy theoretical discussions and to concentrate on advances of practical significance.

This volume contains nine chapters, written by well-known experts who have made significant contributions in their relevent fields. The first chapter provides an up-to-date state-of-the-art report on the analysis and design of columns, while the second one reviews the concepts of probabilistic treatment of safety, which is the basis for limit state design. The remainder of the book is dedicated to the discussion of recent developments in axially compressed members. Design practices prevalent in North America, Europe, Japan, and the United Kingdom have been compared, so that a truly international exchange of ideas will result.

I am grateful to all the contributors for their willing participation in the venture and for the co-operation they have extended to me in producing this volume in a short time. It is hoped that the book will prove stimulating both to the practising engineer and to the researcher.

R. NARAYANAN

CONTENTS

LIST OF CONTRIBUTORS

W. F. CHEN
Professor and Head of Structural Engineering, School of Civil Engineering, Purdue University, West Lafayette, Indiana 47907, USA.

P. J. DOWLING
Professor of Steel Structures, Department of Civil Engineering, Imperial College of Science and Technology, London SW7 2BU, UK.

I. H. G. DUNCAN
Partner, Buro Happold (South West) Consulting Engineers, 17 Portland Square, Bristol BS2 8SJ, UK.

PIERRE DUBAS
Professor of Statics and Steel Construction, Swiss Federal Institute of Technology, ETH-Honggerberg, 8092 Zürich, Switzerland.

BEN KATO
Professor of Steel Structures, Faculty of Engineering, Department of Architecture, University of Tokyo, 7-3-1 Hongo, Bunkyo-ku, 113 Tokyo. Japan.

J. B. KENNEDY
Professor, Department of Civil Engineering, University of Windsor, Windsor, Ontario, Canada N9B 3P4.

W. I. LIDDELL
Partner, Buro Happold Consulting Engineers, 14 Gay Street, Bath BA1 2PH, UK.

M. K. S. MADUGULA
Lecturer, Department of Civil Engineering, University of Windsor, Windsor, Ontario, Canada N9B 3P4.

D. M. PORTER
Senior Lecturer, Department of Civil and Structural Engineering, University College, Cardiff CF2 1TA, UK.

W. RAMM
Professor, Fachgebiet Massivbau und Baukonstruktion, University of Kaiserslautern, PO Box 3049, 6750 Kaiserslautern, West Germany.

LAMBERT TALL
Professor and Dean of School of Technology, Florida International University, Tamiami Campus, Miami, Florida 33199, USA.

W. UHLMANN
Professor of Steel Construction, Department of Construction Engineering, Technische Hochschule, Darmstadt, Alexanderstrasse 7, 6100 Darmstadt, West Germany.

K. S. VIRDI
Lecturer, Department of Civil Engineering, The City University, London EC1V 0HB, UK.

C. J. K. WILLIAMS
Lecturer, School of Architecture and Building Engineering, University of Bath, Claverton Down, Bath BA2 7AY, UK.

Chapter 1

CENTRALLY COMPRESSED MEMBERS

LAMBERT TALL
Florida International University, Miami, USA

SUMMARY

This chapter considers the strength and behaviour under load of centrally compressed steel members which are axially loaded. The development of column curves for design is described through consideration of theoretical analyses and a comparison of the results of these with experimental test results. The theoretical analyses consider both elastic and inelastic buckling, and include the presence of residual stresses which, together with out-of-straightness, are the prime factors which influence strength. Both the large number of test results presented, and the theoretical column curves prepared, exhibit wide scatter that is not a random phenomenon, but rather is a function of manufacture, fabrication, and geometry. A knowledge of this scatter has led to the introduction of the concept of multiple column curves for design; however, there has been considerable debate on the concept since the wide scatter in the column curves is reduced to a negligible amount for practical compression members with end restraint.

1.1 INTRODUCTION

Compression members are the key elements of almost all structures, and the study of their behaviour and their design is based on the strength of centrally compressed members.

Compressed members, or columns, may be defined as members carrying a compressive load, and whose length is considerably greater than the cross-sectional dimensions. Such a member may carry other loadings, and may have end conditions and end moments of any type. This definition of compressed members includes beam-columns, plates, component parts of frames, and, for instance, the compression flange of beams or plate girders.

This book is concerned with compressed members that have been axially loaded; this chapter is restricted to compressed members under centrally applied loads with no end moments, as shown in Fig. 1. Also

FIG. 1. Simple column under centrally applied load.

called a simple column, such a member is perfectly straight with pin-ends through which the load is applied. A simple column, therefore, is an idealised member not found in practice.

The simple column, and its behaviour, is the basis for the study of all other columns and beam-columns.

1.2 THEORIES FOR STRENGTH AND BUCKLING

1.2.1 Historical Review

The metal column has been in use for centuries, yet the study of its strength is comparatively recent. Van Musschenbroek[47] published the first paper on column strength in 1729, and his empirical results were expressed in a form very similar to those in use today, even though his work was not directed at metal columns. Metal columns became relatively common in the late 18th century when wrought-iron shapes became available; this period followed the publication of Euler's famous treatise[47,61] on the buckling of columns in 1759. Euler was the first to realise that column strength could also be a problem of stability and not merely a matter of crushing, and his study is a mathematical exposition of the natural phenomenon of buckling. Euler investigated the purely elastic phenomenon of buckling, which implies that the elastic limit is not exceeded in any fibre in the cross-section of the compressed member.

The use of wrought-iron shapes was followed by the introduction of steel and its employment in structures, and particularly in railroads as they developed in the early 19th century—railroads in turn needed bridges of substantial load-carrying capacity, and thus the knowledge needed to design them gave impetus to the study of compression members.

Application of the Euler formula was not possible for practical columns which exhibit inelastic instability, and so empirical or semi-empirical methods were used and it was not until the latter part of the 19th century that theories were introduced to define column strength in the inelastic range. The tangent modulus and reduced modulus theories were presented by Engesser in 1889 and 1895[18,47,61] and today still provide a basis for the modern concepts of buckling and strength.

A tremendous effort went into the testing of columns and the development of theories and of empirical design methods in the two decades prior to 1900 and the two decades thereafter, and much of this is summarised in Salmon's treatise.[47] Figure 2 summarises a scatter band

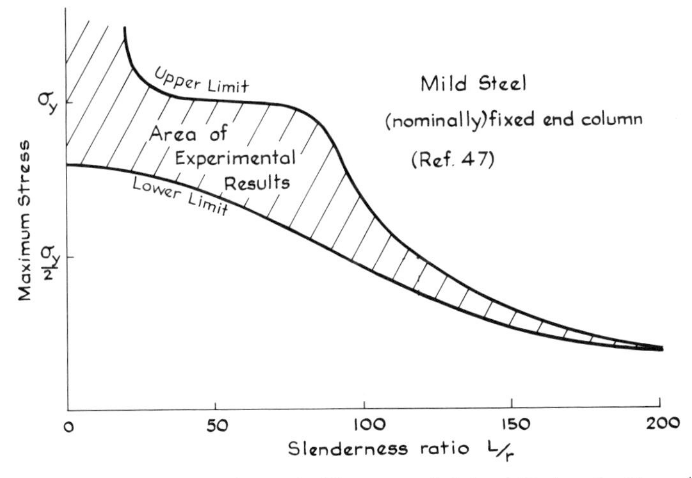

FIG. 2. Column strength in 1915. (Courtesy of Oxford University Press.[47])

of test results reported in Ref. 47—the results are immediately recognisable as being essentially identical to those of today.

The reduced modulus had been accepted as the correct buckling theory for columns in the inelastic range until 1947 when Shanley published a paper[48] giving the buckling load of a centrally loaded column as the tangent modulus load.

Residual stresses were recognised in the late 1940s and the 1950s as being the major influence in the strength and behaviour of steel columns,[7,57] and much of the more recent research has proceeded around this fact. The more recent studies have been supported by experimentation on full-scale members and structures, and by the use of the computer, both of which have expanded knowledge in the topic to the point where many feel that there are no more unknowns.

1.2.2 Strength

The strength of an actual compression member is the load it can support under the conditions of straightness, load eccentricity, and end restraint that exist. The design of the member requires a knowledge of the strength, and this is usually related to the buckling strength of the compression member in the configuration of a simple column. The buckling strength of the simple column may be defined as either the buckling load or the maximum load, where these definitions apply to global column failure, and not to local failure such as local buckling. The buckling load, also called the critical load, is the load corresponding to bifurcation, that is, the lowest load at which the theoretically straight column can assume a deflected position. The maximum load is the ultimate load the member can carry; it marks the boundary between stable and unstable deflected positions of the column and is reached gradually, unlike the buckling load which is an instantaneous phenomenon. These definitions are illustrated in Fig. 3 which shows a plot of load versus central deflection of a compressed member. (It should be noted that the attainment of the maximum load in a compressed member which is a part of a frame does not necessarily mean that the frame cannot carry more load.)

Bifurcation is a phenomenon that will occur only for a perfectly centrally loaded, perfectly straight column, and, as such, will not normally be seen in practice. The strength of a member in practice depends on the out-of-straightness, eccentricity of load, the end fixity, existence of transverse load, local and/or lateral buckling, and residual stress. Tests on columns give the maximum stress attained and do not isolate these

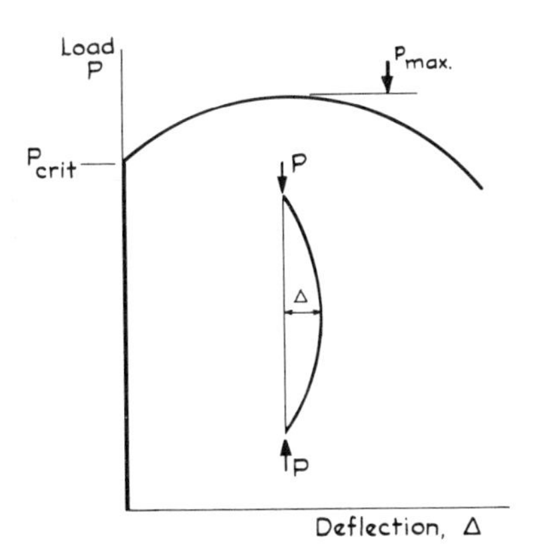

FIG. 3. Load vs. deflection of compressed member.

effects, so that a scatter band of results will be observed, Fig. 2. The strength of a practical column may be expressed in terms of the buckling load, or of the maximum load, of the simple column, and this is considered later in this chapter.

1.2.3 The Column Curve

The strength of a column is defined by the column curve, a plot of the load (or stress) versus slenderness ratio, Fig. 4. Today, as a century ago, the column curve is taken as the line of best fit through the scatter band of test results. The column curve is limited by the yield stress (σ_y), and the Euler curve which defines elastic buckling. The transition in the column curve between σ_y and the Euler curve played an important part in the evolution of design criteria, through the consideration of estimated eccentricities or initial deflections. More recently, this transition was shown to be due entirely to the presence of residual stresses in the cross-section for the hypothetical case of a straight, centrally loaded, pinned-end column; in other words, for such a column without residual stresses the strength is defined by either the yield stress or the Euler curve, depending on the slenderness ratio.

The stress in the column curve normally represents the maximum stress attained—under some definitions used in the past, it may also be the stress corresponding to the first yield of an extreme fibre in the cross-section.

A number of column curves were developed in the 19th century,[47] and variations of these are still in use today. These are empirical or semi-

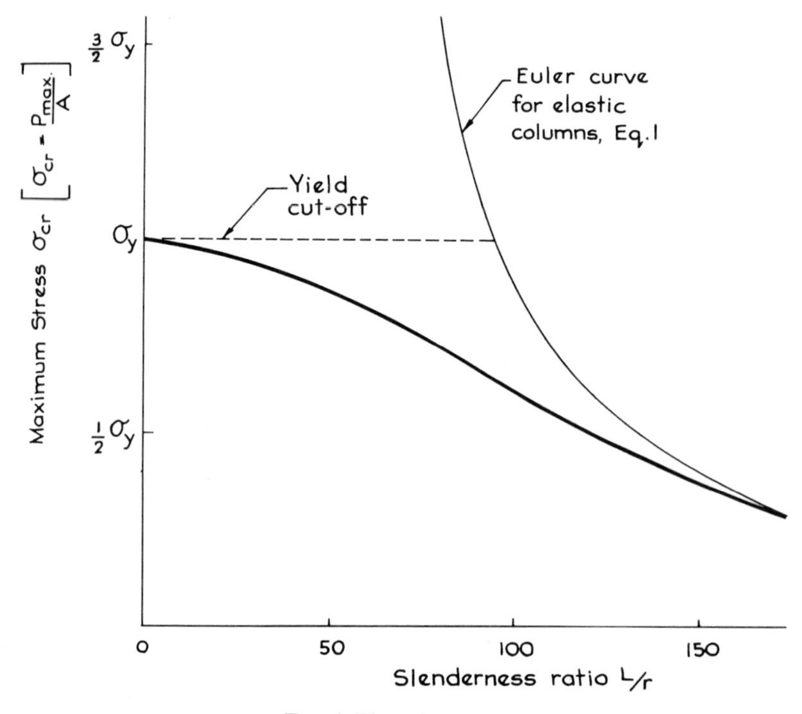

FIG. 4. The column curve.

empirical formulas, of which the following might be regarded as the most significant:

1. Rankine–Gordon (which becomes the secant formula when eccentricity is introduced)
2. Straight-line
3. Johnson parabola.

Figure 5 shows these curves; the Johnson parabola and Rankine–Gordon are almost coincident.

In recent times column curves have been presented as tabulated values of stress vs. slenderness ratio, but in the past an inordinate amount of time was given to the equation defining the column curve. Thus, the straight line assumed an importance because of simplicity, even though it did not represent a line of best fit to the scatter band of test results. The line of best fit is obviously parabolic in form, leading to the empirical Johnson parabola, or to the semi-empirical group represented by the

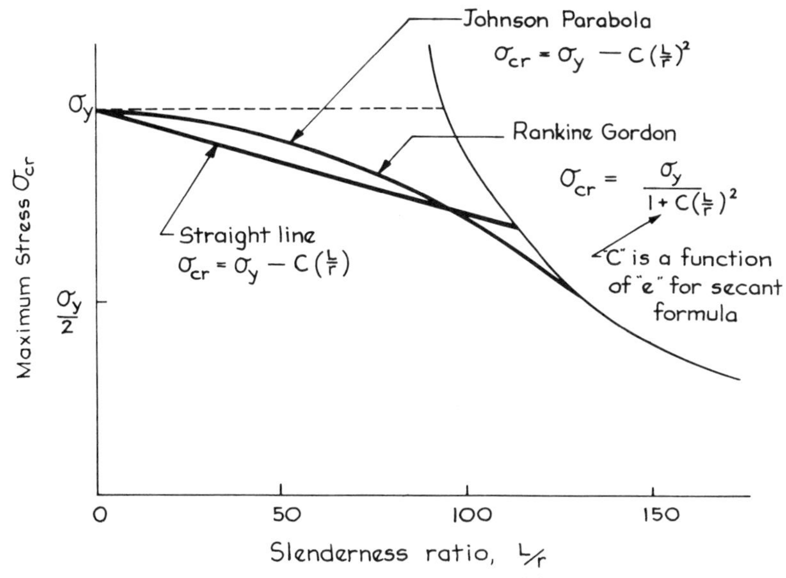

FIG. 5. Basic column curves of the 19th century.

secant, Rankine–Gordon, and Perry–Robertson formulas. The semi-empirical group is based on consideration of elastic behaviour of columns, see below, with the introduction of constants to represent unknown eccentricities or initial curvatures. In effect, the elastic formulas were modified to agree with observed test results, and bore no relationship to actuality; the column was assumed to have failed when any fibre in the cross-section attained the yield stress. Some column curves were a combination of a straight line and one of the parabolic formulas.

Finally, it should be realised that it is not necessary for a column curve to have any other basis than good correlation with experimental and practical results. Additional information on column curves is presented in section 1.5, together with some details on modern usage.

1.2.4 Elastic Behaviour

The elastic behaviour of centrally compressed members is summarised below and includes simple beam behaviour and the bifurcation behaviour of a buckled column. The latter is known as the Euler buckling load and is the most important reference point in column behaviour. The former are essentially formulas for bending behaviour, and they formed the basis for the semi-empirical group of column curves described above.

Euler Load

For a pinned-end member, perfectly elastic, perfectly straight, and loaded centrally, the buckling load, or critical load, is known as the Euler load,

$$P_e = \frac{\pi^2 EI}{L^2} \tag{1}$$

where P_e is the Euler load, E the modulus of elasticity, I the moment of inertia of the cross-section, and L the length between the pin ends. In terms of stress, and introducing the effective length factor (see section 1.4), the average stress corresponding to the Euler load is

$$\sigma_e = \frac{P_e}{A} = \frac{\pi^2 E}{(K\ L/r)^2} \tag{2}$$

where r is the radius of gyration of the cross-section.

Equation (1) is derived from a consideration of the equilibrium of the external and internal moments at any cross-section along the member, when it is in its deflected position after bifurcation.

Initial Curvature

For a pin-ended member, perfectly elastic, with an initial curvature assumed as sinusoidal, and loaded at the pin ends, Fig. 6, it may be

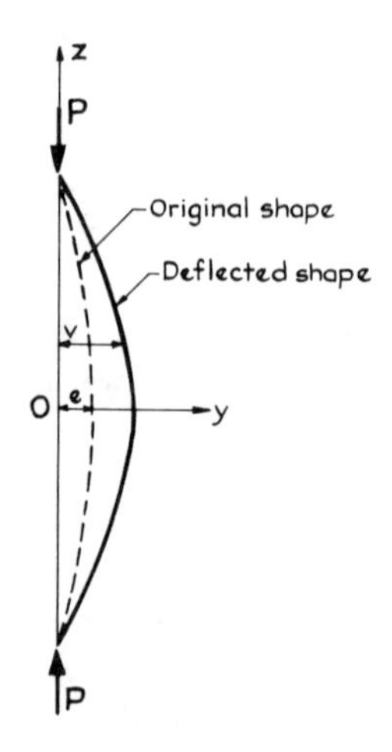

FIG. 6. Column with initial curvature.

shown that

$$\sigma_{max} = \frac{P}{A}\left(1 + \frac{ec}{r^2} \cdot \frac{P_e}{P_e - P}\right) \tag{3}$$

where e is the initial mid-height deflection of the column, and c the distance from the neutral axis of bending to the extreme fibre. If the maximum stress is assumed to be the yield stress σ_y, then it may be shown that

$$\sigma_{ave} = \frac{1}{2}\left[\sigma_y + \sigma_e\left(1 + \frac{ec}{r^2}\right)\right] - \sqrt{\left[\sigma_y + \sigma_e\left(1 + \frac{ec}{r^2}\right)\right]^2 - 4\sigma_y\sigma_e} \tag{4}$$

where σ_{ave} is the average stress in the cross-section corresponding to attainment of first yield. Equation (4) defines column strength for the criterion of failure being defined by first yield—it is the Perry–Robertson formula, and formed the basis for design column curves in a number of countries in the past.

Eccentrically Applied Load
The maximum compressive stress in a pin-ended member with an eccentricity of load, e, Fig. 7, may be shown to be[57]

$$\sigma_{max} = \frac{P}{A}\left[1 + \frac{ec}{r^2} \sec\left(\frac{L}{2r}\sqrt{\frac{P}{AE}}\right)\right] \tag{5}$$

Equation (5) is known as the secant formula, and was the basis of many rational attempts at column design.

Many other formulas, such as for initial curvature combined with eccentrically applied load, have been prepared and may be found in the literature. When such formulas became the basis for column curves, the initial eccentricities were used to take account of all factors which could not be evaluated at the time—in effect, values were assigned to the initial eccentricity to cause the column curve to correspond to observed experimental behaviour.[54]

1.2.5 Inelastic Buckling

The elastic behaviour of columns is mainly of academic interest. Elastic buckling occurs only for very slender columns. The use of allowable stress design formulas like the secant formula, that is, the definition of load-carrying capacity as that corresponding to first yield, neglects the additional load-carrying capacity up to the maximum load. Most practi-

FIG. 7. Column with eccentrically applied load.

cal columns fail in the inelastic range. The development of modern column curves has been based on inelastic buckling analyses, and two theories have evolved:

1. Tangent modulus theory
2. Reduced modulus theory.

Both have been defined in terms of the Euler load, and reflect the fact that the early tests on steel columns never attained the Euler load. Thus, the inelastic buckling strength is

$$\sigma_{cr}=\frac{P_{cr}}{A}=\frac{\pi^2 E_m}{(K\ L/r)^2} \tag{6}$$

where E_m is a modified modulus of elasticity; $E_m < E$.

In 1889, Engesser introduced the tangent modulus theory in its original form. He assumed that the column 'remained straight up to the moment of failure, and the (tangent) modulus of elasticity remained constant right across the cross-section.'[47] He modified Euler's buckling formula by replacing E by E_t, the tangent modulus at the stress corresponding to the buckling load:

$$\sigma_t=\frac{P_t}{A}=\frac{\pi^2 E_t}{(K\ L/r)^2} \tag{7}$$

where σ_t is the critical buckling load.

Figure 8 illustrates the assumptions made. The generalised stress–strain relationship assumed implies that stress is not directly proportional to strain, as is the case with elastic buckling.

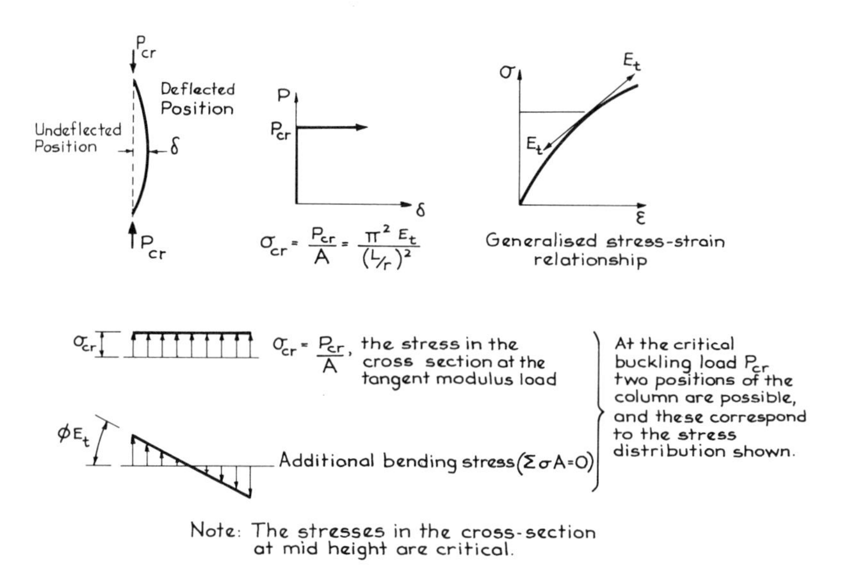

FIG. 8. Original Engesser theory.

Reduced Modulus Theory

The original Engesser theory presented a contradiction in that the tangent modulus does not remain constant at the moment of bifurcation. Actually, with the slightest deflection, the material on the convex side would continue loading. Engesser recognised this, and in 1895 replaced his original tangent modulus theory by the reduced modulus theory. The reduced modulus theory assumes that strain reversal of fibres takes place on the convex side of the bent compression member when it passes from the straight to the deflected configuration, see Fig. 9. Thus, the critical buckling stress by the reduced modulus load is

$$\sigma_r = \frac{P_r}{A} = \frac{\pi^2 E_r}{(K\ L/r)^2} \tag{8}$$

where E_r is the reduced modulus load. This expression may be derived[57] from the equation of internal and external moments on the cross-section of the member in its deflected shape. The reduced modulus, E_r, depends on the shape of the cross-section, and is a complicated expression for

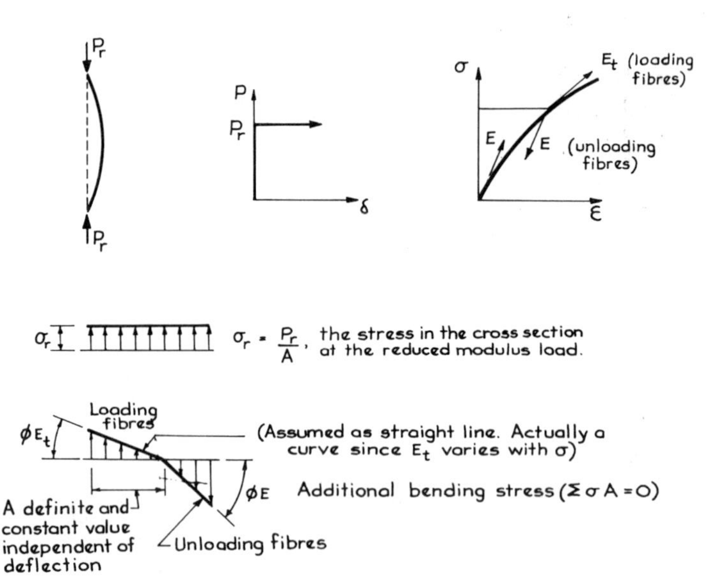

FIG. 9. Reduced modulus concept.

most shapes. For a rectangular shape[57]

$$E_r = \frac{4E_t E}{(E^{1/2} + E_t^{1/2})^2} \tag{9}$$

The reduced modulus concept appeared logical because of the consideration of the simultaneous loading and unloading of fibres—and so there seemed to be no explanation for the fact that experimental results tended to be approximated by the tangent modulus theory.

The Shanley Contribution

It was not until 1947 that the correct relationship between the tangent modulus and reduced modulus theories was shown. Shanley drew attention to the assumption in the reduced modulus theory of the column remaining perfectly straight up to the reduced modulus load. (The criterion of perfect straightness is an assumption for any stability problem.) He showed[48] that an initially straight column will buckle at the tangent modulus load, and then will continue to bend with increasing axial load—this was the introduction of a completely new concept of column behaviour, which other investigators later illustrated.[24,36,51,52]

The behaviour of a centrally loaded column is typified by the load–

deflection curve in Fig. 10. With the Shanley concept, the tangent modulus load is the lower bound for column strength; it is the load at which an initially straight column will start to bend. The upper bound is the reduced modulus load since it is the maximum load a column will sustain if it is temporarily supported up to that load. The maximum load

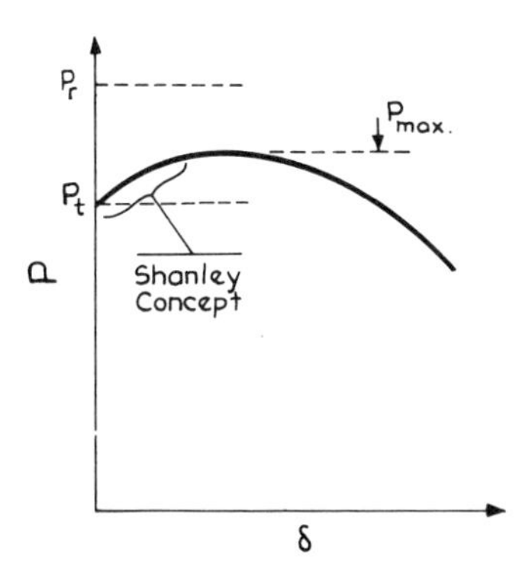

FIG. 10. Deflection of initially straight centrally compressed member.

of a compressed member will lie between these two limits, although generally, test results will tend to approximate the tangent modulus load. As it is a lower bound, the tangent modulus load has been used as a column strength formula,[21] since the effect of the usual out-of-straightness and load eccentricity approximates the difference between the maximum load and the tangent modulus load.

Tangent Modulus Theory

After consideration of the Shanley contribution, the tangent modulus concept was reformulated to take account of the fact that 'no strain reversal takes place on the convex side of the bent column when it passes from the straight form to the adjacent deflected configuration.'[18] Equation (7) holds true, where E_t and the stress distribution are defined as shown in Fig. 11.

As noted above, eqn. (7) has been used as the basis for column strength, and the Column Research Council in 1952 issued a memorandum[21] stating that 'the tangent modulus formula for the buckling strength affords a proper basis for the establishment of working load formulas.'

It is necessary to realise that the tangent modulus load is an in-

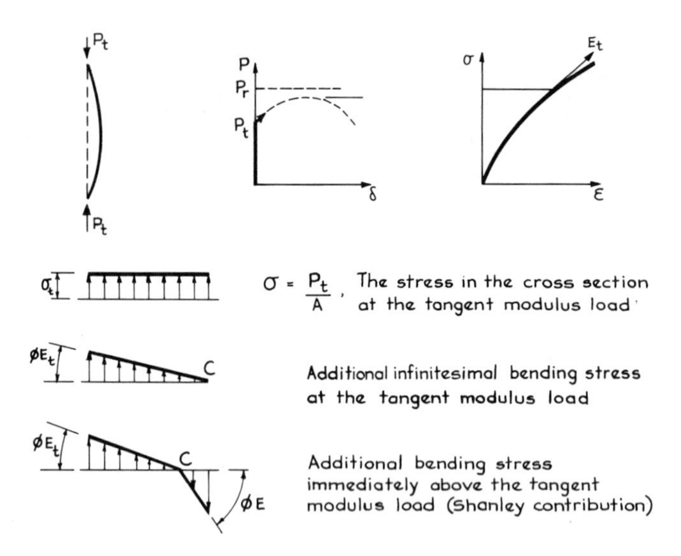

FIG. 11. Tangent modulus concept.

stantaneous phase only[54]—as soon as the compressed member buckles, unloading of some fibres takes place, and the member load increases with increase in member deflection. It is not necessary to understand the phenomenon involved in the concept to be able to use eqn. (7); however, an insight into this concept helps in the application of the equation. In Fig. 11, consider the position of the point C. (C is the position of the fibre which has a zero increase in strain between any two adjacent deflected positions of the member.) When the member is straight, the point C is at infinity. At the tangent modulus buckling load, the point C may be regarded as moving from infinity to the edge of the cross-section, instantaneously. In Fig. 11 this limiting condition is shown by the infinitesimal stress distribution. As soon as the member is no longer straight, the tangent modulus load has been exceeded and point C is inside the cross-section. As soon as this happens, unloading of some fibres occurs with loading of other fibres of the cross-section.

The load–deflection curve of Fig. 10 may be prepared from theoretical considerations of the equation of internal and external stresses and moments. For practical columns containing residual stresses, the variation of stress distribution with deflection is not simple, see Refs. 51 and 54.

1.3 INFLUENCE OF RESIDUAL STRESS

1.3.1 Residual Stress

Residual stresses have been studied extensively for many decades. However, it was only in the 1950s that it was shown conclusively that residual stresses are a major influence in the strength of compression members.[7,54,57] The influence of residual stresses had been suspected, and the first recorded instance of this is believed to be in 1908 when they are noted as the probable cause of lower column strengths than expected in a test series.[31] The first systematic study of the influence of residual stresses on the strength of compression members began in the late 1940s at Lehigh University under the guidance of the Column Research Council, and this led to major studies there which continued through the 1950s and 1960s, and well into the 1970s.

It has been shown[7] that residual stresses are the cause of the hitherto unexplained transition curve in the column curve, and that variation in their distribution and magnitude exerts comparatively great influence on the strength of compression members and on the strength of plates in compression. Residual stresses exist in rolled, welded, and cold-straightened shapes—in fact, residual stresses exist in all shapes. Their removal by annealing is costly and sometimes impossible, but a control over their influence is possible.

Residual stresses are formed in a structural member as a result of plastic deformations; they are stresses which exist in the cross-section even before the application of an external load.[51,54,57] These plastic deformations may be due to cooling after hot-rolling or welding, or due to fabrication operations such as flame-cutting, cold-bending, or cambering. In rolled shapes, these deformations always occur during the process of cooling from the rolling temperature to air temperature; the plastic deformations result from the fact that some parts of the shape cool more rapidly than others, causing inelastic deformations in the slower-cooling portions. (The flange tips of a wide-flange shape, for example, would cool more rapidly than the juncture of flange and web.) Residual stresses are also introduced during the welding operation as a result of the localised heat input and resultant plastic deformation. Reference 59 compares a number of measuring techniques, and Ref. 58 summarises the theoretical methods available for their prediction by calculation.

Residual stress patterns for a wide variety of shapes and fabrication procedures, and consideration of their influence on column strength, are given in Refs. 1, 2, 3, 8, 22, 25, 28, 29, 32, 33, 35, 38, 44, 46, 49, 51, 54, 55, 56, 60, and 62.

1.3.2 Column Strength

The discussion of strength theories above applies also to members with residual stresses—the presence of residual stresses implies that yielding is reached earlier than otherwise in some fibres, and the inelastic theories of tangent modulus and reduced modulus apply. It has been shown[7,38,54,57] that the tangent modulus theory describes the strength of compression members of rolled H-shapes, but that this is not realistic for welded shapes for which a maximum strength analysis is needed. Further, welded columns tend to have greater out-of-straightness than rolled shapes, and the effect of this on welded columns is so great that it cannot be neglected.

An important tool in the prediction of column strength from experimental data is the stub column,[50] from which is obtained a stress–strain curve for the complete cross-section. Such a stress–strain curve reflects the presence of residual stress—and the tangent modulus is applied to this curve and not to the stress–strain relationship obtained from the tension test on a small coupon. The tangent modulus concept applied to cross-sections containing residual stresses results in equations for column strength which are functions of E_t, rather than utilising E_t directly as in Fig. 8. For cross-sections containing residual stresses, the tangent modulus and reduced modulus theories for column buckling define loads differing from those for the same members free of residual stress.[49]

For an idealised stress–strain relationship, Fig. 12, it may be shown that, for a compression member of symmetrical cross-section containing residual stresses in a symmetrical distribution, the critical stress at the

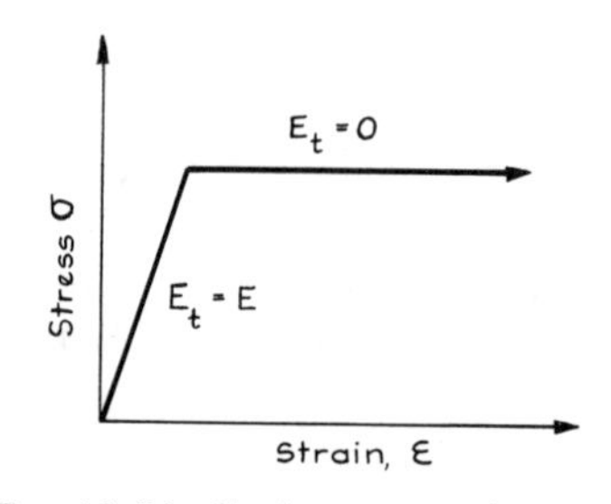

FIG. 12. Idealised stress–strain curve.

buckling load as defined by the tangent modulus concept is given by[32]

$$\sigma_{cr} = \frac{P_{cr}}{A} = \pi^2 \frac{E(I_e/I)}{(L/r)^2} \tag{10}$$

where EI_e is the effective bending rigidity of the column, derived through the realisation that the yielded portion of a structural shape offers no additional resistance to bending, and the buckling strength is a function of the moment of inertia of the elastic part.[62]

The solution of eqn. (10) requires the function relating I_e to σ_{cr}, which is accomplished as indicated in Refs. 32 and 62. For a small rolled H-shape, and using the E_t determined from the stress–strain relationship of a stub column, the following holds true

$$\left.\begin{aligned} E\frac{I_{ex}}{I_x} &= \frac{AE_t - \frac{2}{3}A_w E}{2A_f + A_w/3} \left[\approx E\frac{E_t}{E}\right] \\ E\frac{I_{ey}}{I_y} &= E\left(\frac{AE_t}{A_w E} - \frac{A_w}{2A_f}\right)^3 \left[\approx E\left(\frac{E_t}{E}\right)^3\right] \end{aligned}\right\} \tag{11}$$

where A_w = area of the web
and A_f = area of the flange

Figure 13 summarises the relationship between the column curve and the residual stress distribution in the cross-section. Thus, the column curve, Fig. 13(d), results from the use of the stub column stress–strain relationship, Fig. 13(b), and the tangent modulus curve, Fig. 13(c), in eqns. (10) and (11). The use of the stub column curve to predict column strength is described further in Ref. 66.

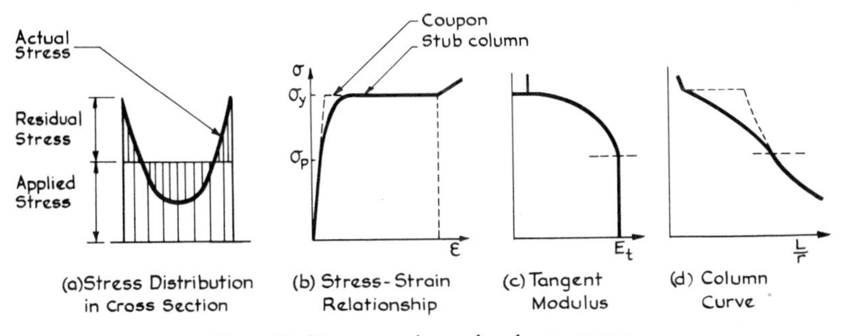

FIG. 13. Stress–strain and column curve.

Equations (11) apply only to small and medium-size rolled shapes, and to values of E_t obtained from a stub column stress–strain curve. For all other shapes and fabrication processes, the relationship is more complex and requires a mathematical formulation of the stress–strain relationship and the residual stress distribution.

When column curves based on the tangent modulus concept are prepared for small rolled H-shapes, then these curves may be approximated by straight line and parabolic curves for the weak and strong axes, respectively. Such curves also correspond satisfactorily to test results, since the loads carried by such columns do not exceed the tangent modulus load enough to warrant the use of a post-buckling maximum load analysis. Some test results are compared with the straight-line and parabolic assumptions in Fig. 14 for rolled H-shapes.[7] The column curves are cut off at $L/r = 20$, to take account of the effect of strain-hardening. The CRC* Basic Column Curve[23] is an average parabolic curve used for bending about both axes; the curve is a compromise, being the average of test results for bending about both axes. It is the first

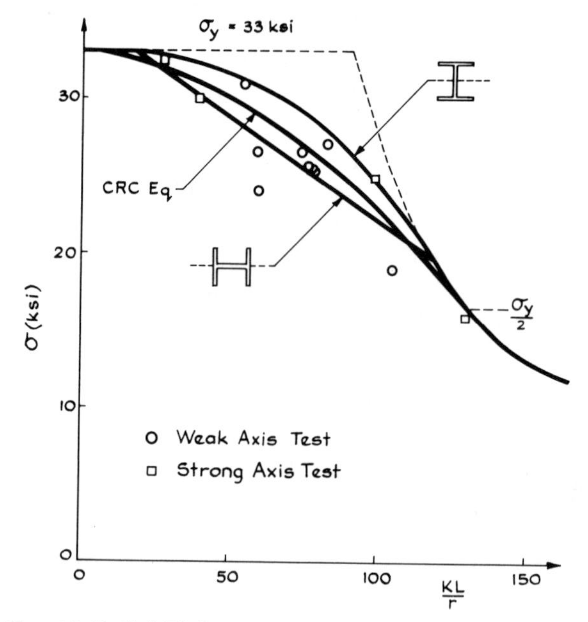

FIG. 14. Rolled H-shapes: test results and column curves.

*Column Research Council (presently the Structural Stability Research Council)

column curve based on a theoretical study reflecting actual conditions. The CRC curve was adopted in 1960 by the American Institute of Steel Construction[5] as the maximum strength curve used as the basis for the design curves for compression members of steel with yield points of 33 ksi to 50 ksi. Over the years, steels with higher yield strengths have been added; today (1982) the same curve applies to heat-treated steels with yield points up to 100 ksi.[6] Even though it was based on the strength of small rolled shapes, the AISC column design curve currently is used for all shapes, materials, and fabrication processes. (This leads to a discussion of multiple column curves, see section 1.5.)

1.3.3 Manufacturing Process

Structural steel members may be rolled, or they may be fabricated from plates by welding; they may be straightened, they may be small or heavy, of high-strength steel or of mild steel, they may be annealed. There are many manufacturing processes, all of which have an influence on the residual stress distribution and magnitude, and thus on column strength.

The residual stress distribution set up in a cross-section due to welding may be vastly different from that set up in a rolled shape due to cooling, as seen from the comparison of the residual stress distributions in Fig. 15. The fact that welding induces a different distribution of residual stress implies that welding may induce different column strength properties. Welded columns have high residual stresses, and, at the weld there is a tensile residual stress with a magnitude equal to the yield point of the weld metal. The magnitude and distribution of residual stresses in welded

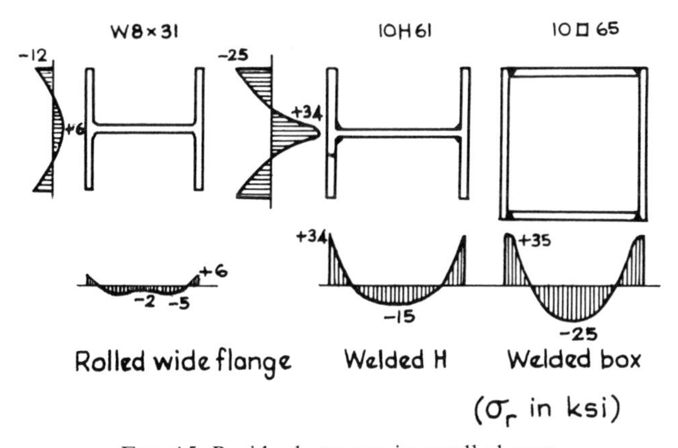

FIG. 15. Residual stresses in small shapes.

shapes are markedly influenced by the geometry of the cross-sectional shape.[3,20,51]

The strength of axially compressed welded members, or members which have undergone other manufacturing processes, can be predicted by the same techniques as for rolled shapes. However, it has been shown that the use of the tangent modulus concept is not realistic for the prediction of the strength of welded columns.[3,29,51,67] Because of the large magnitudes of residual stress, a maximum strength analysis is necessary. Welded columns tend to have a greater out-of-straightness than rolled columns,[8] and even though all columns are straightened to minimum tolerances, the combined effect of the out-of-straightness and the large magnitudes of residual stress is so great that out-of-straightness must be considered with welded columns, where it is normally neglected with rolled columns.[51] Thus, welded columns need a maximum strength study, whereas the tangent modulus buckling load presents a realistic figure for rolled columns. (A maximum strength study, either with or without the inclusion of out-of-straightness, is a complicated consideration of behaviour in the inelastic range. References 51 and 54 demonstrate the steps and assumptions needed to carry out such an analysis.)

Welded columns have lower strengths than corresponding rolled columns. This is illustrated by the test results in Fig. 16 for small welded

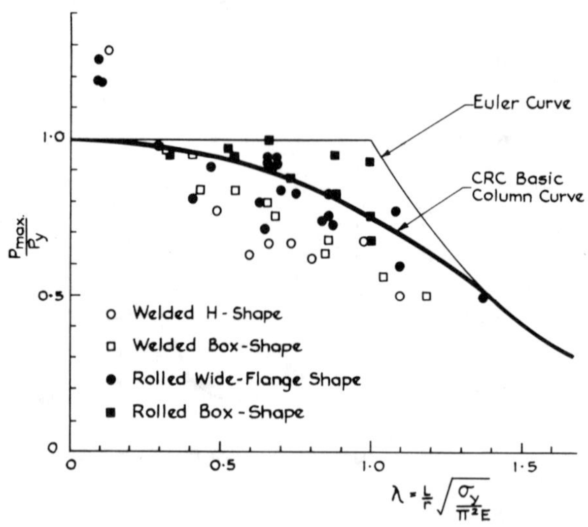

FIG. 16. Test results for small to medium rolled and welded shapes.

shapes (H-shapes and box-shapes), compared with the CRC curve which is a good average for small rolled shapes.[8,51] The reason for the somewhat lower column strengths of welded members of small and medium cross-section is two-fold: the effect of residual stresses due to welding, and the effect of initial out-of-straightness. For the more practical slenderness ratios, $L/r < 60$, welded box-shapes tend to be stronger than welded H-shapes, since the box-shapes retain the corners in the elastic condition throughout the bending history of the columns, owing to the favourable tensile residual stresses there as a result of the weld. Similarly, box-shapes are able to sustain the maximum load for much larger deflections than H-shapes.[51] H-shapes, with the compressive residual stress at the flange tips, lose a major part of their rigidity very early under load since the flange tips yield first. References 9 and 67 summarise the available information on box-shapes.

The above has considered welded shapes built up from plates with mill-rolled edges, also called universal mill (UM) plates. Actually, a great proportion of welded shapes are built up from oxygen-cut plates, also called flame-cut (FC) plates, since UM plates do not normally satisfy edge straightness requirements. For H-shapes, the use of FC plates generally leads to improved column strength because of the tensile residual stresses set up on the flange tips by the operation of flame-cutting. For this reason, the use of FC plates results in a 'favourable' residual stress distribution, favourable in that improved column strength results.

Figure 17 shows residual stress distributions in typical FC plates, together with the residual stress distribution in a welded H-shape fabricated from these same plates,[41] and Fig. 18 compares test results with strength predictions of columns. Of interest in Fig. 18 is the fact that the tangent modulus prediction estimates the column strength of the flame-cut welded shapes fairly well. This means that the post-buckling reserve above the tangent modulus load of a fictitious perfectly straight column is of approximately the same magnitude as the reduction in strength due to unintentional out-of-straightness of a practical column. Thus, the tangent modulus concept may be used for the design of such members, that is, small FC welded shapes, including the effect of residual stresses. For the welded shapes of UM plates in Fig. 18, the post-buckling reserve is considerable, and an accurate maximum strength analysis is necessary to obtain close correlation with the data. It has been concluded[41] that the strength of columns of small-to-medium size FC welded shapes is much the same as that of similar rolled shapes, for mild steel.

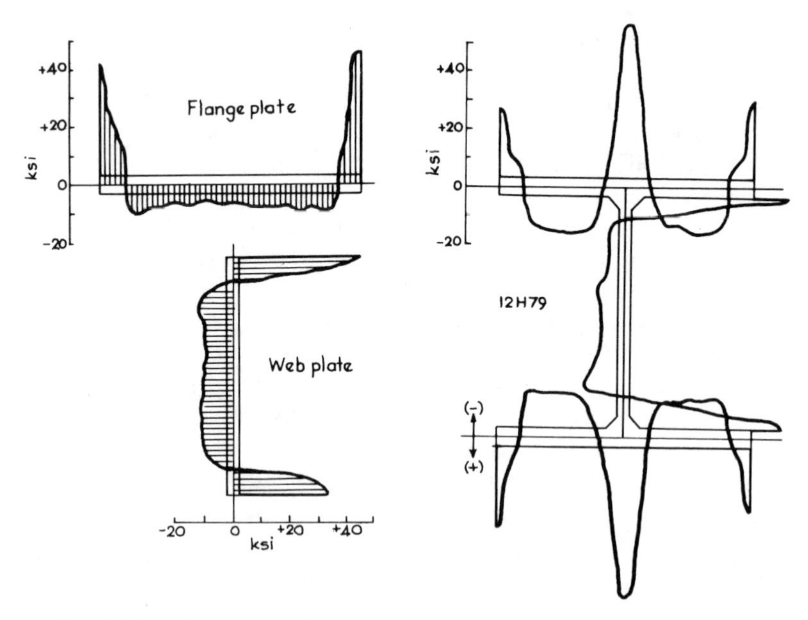

FIG. 17. Residual stresses in component FC plates and in welded shape.

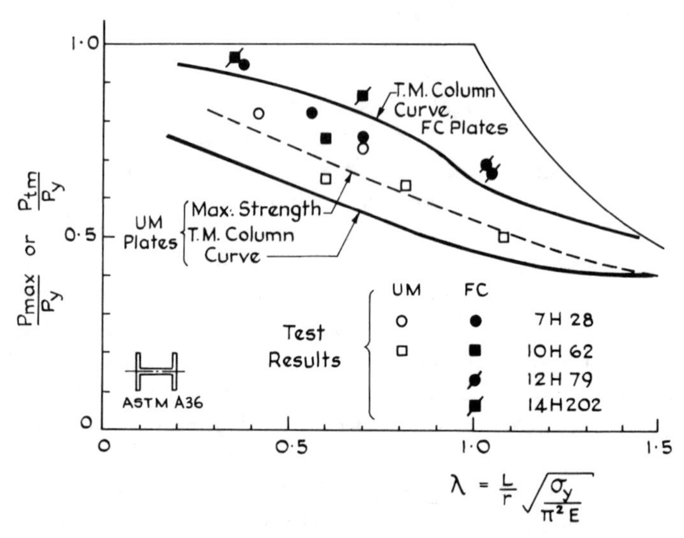

FIG. 18. Theoretical and experimental column strength of welded H-shapes of FC and UM plates.

Some early studies[30,42] showed similar results—improved column strength resulted from both the laying of a weld bead on the flange tips of a rolled shape, and the reinforcement of a rolled shape by the welding of cover plates. Both of these methods are methods of reinforcement since they lead to increased strength capacity, useful when loading conditions have changed. (Note that both methods may require the support of the compression member and its loads during welding.) Figure 19 compares the strength of columns before and after reinforcement by welding.[30]

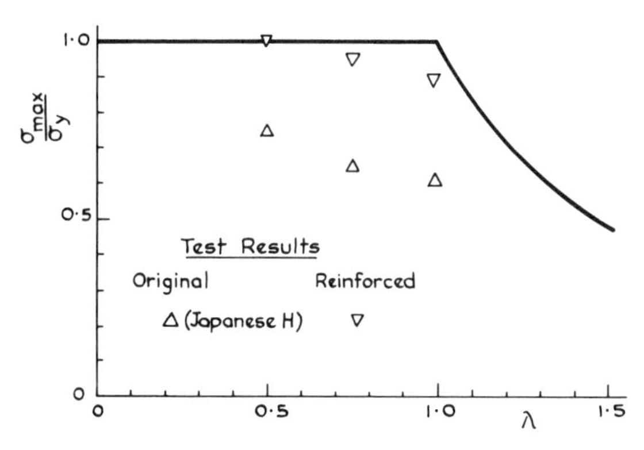

FIG. 19. Reinforcement of columns.

1.3.4 Size, Mechanical Properties, and Other Factors

Structural shapes used as compression members may vary in size from the very smallest with a depth of a few inches, to huge rolled 'jumbo' shapes weighing almost 800 lb per foot, (1200 kg per metre) to even heavier shapes fabricated by welding.[12,57] Different steels are available for these structural shapes; the most important mechanical property for structural purposes is the yield point. The ready availability of different yield points extends from 36 ksi up to 100 ksi in North America (250 to 700 N/mm^2) with some geometrical limitation of size for certain yield points, see Table 2.6 of Ref. 57. Both size and yield point influence the strength of a compressed member.

Most structural shapes used in practice may be classified as small to medium in size, but there are many situations requiring the use of 'heavy' shapes. A heavy shape may be defined as one in which the thinnest component plate exceeds $1\frac{1}{2}$ in (40 mm) in thickness. Clearly, this definition is arbitrary; it is related to the fact that up to approximately 1 in (25 mm) thickness, residual stresses are fairly constant through the thickness, and that there may be substantial variation, up to 10 ksi (70 N/mm^2) in residual stresses through the thickness from above about $1\frac{1}{2}$ in (40 mm) thickness.[15] It has been shown[19] that a realistic prediction of the strength of columns of heavy shapes requires consideration of the variation of residual stress through the thickness.

The process of cooling in heavy shapes is such that the residual stress magnitudes are considerably larger than in small-to-medium shapes.[1] This is illustrated in Figs. 20 and 21[60] for a heavy rolled and a heavy

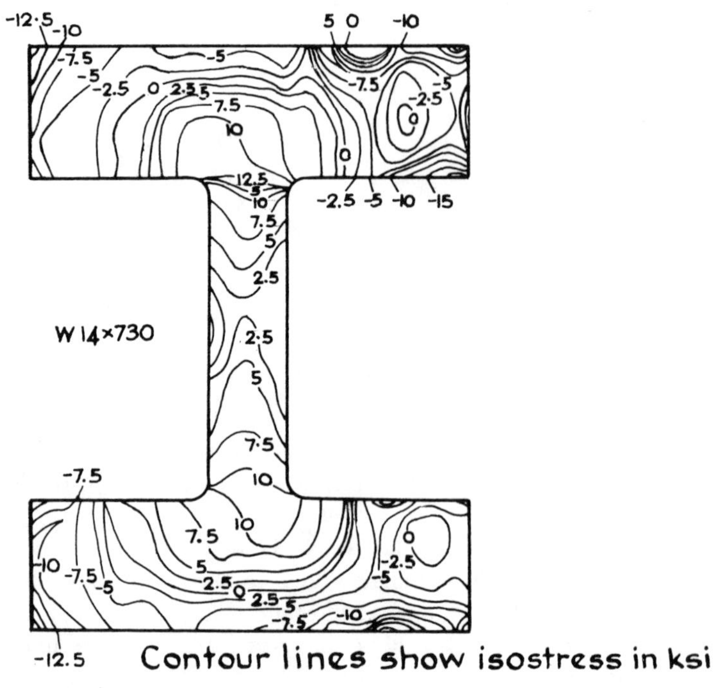

FIG. 20. Residual stresses in heavy rolled shape, W14 × 730.

welded shape respectively. The residual stresses in heavy shapes are influenced very little by welding; the individual plates are so large that the effect of the welding is localised around the weld. Thus, most of the residual stress distribution is created during the operations of cooling after rolling and cooling after flame-cutting.[3]

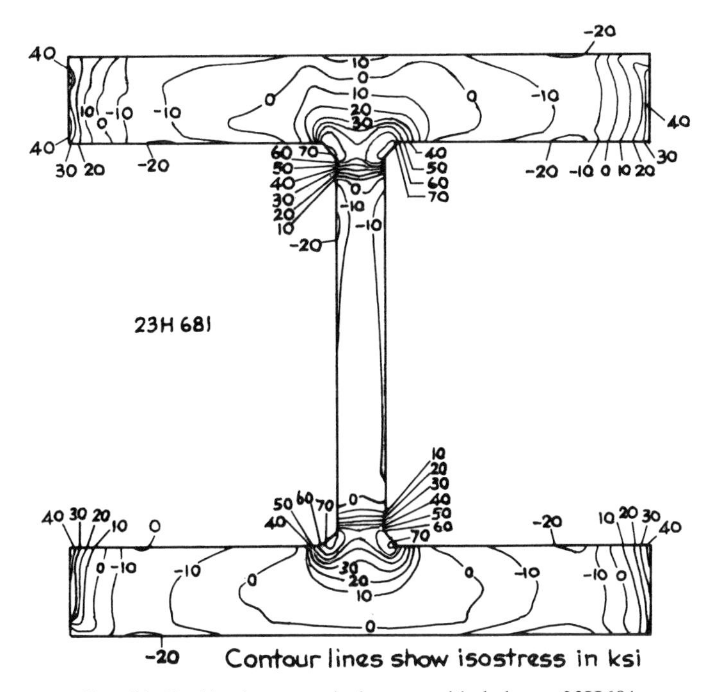

FIG. 21. Residual stresses in heavy welded shape, 23H681.

Heavy shapes, whether rolled or welded, have reduced column strengths due to the large magnitudes of residual stress, combined with the variation through the thickness.[1,2,26] This is shown in Fig. 22 in comparison to the CRC curve; the least reduction in strength is shown for the shorter columns, while the larger slenderness ratios show great reductions.

In general, higher column strengths are obtained most simply by using steel of a higher yield strength. The results of tests have indicated that the residual stresses arising in shapes of high-strength steels are of the same order of magnitude as in mild steels—residual stresses are mainly a function of geometry.[53] Hence, the effect of residual stresses becomes comparatively smaller for steels of higher yield strengths.[8,28,39,43]

The strengths of rolled and welded columns are compared for the same shape, for low yield point and high yield point steels, in Figs. 23,[54] 24, and 25.[39,65] The comparison is not quite complete in the case of rolled A 514 steel shapes, since A 514 steel is quenched and tempered, and thus contains only very small magnitudes of residual stress owing to the tempering operation.[45,65] Reference 63 gives some information on the

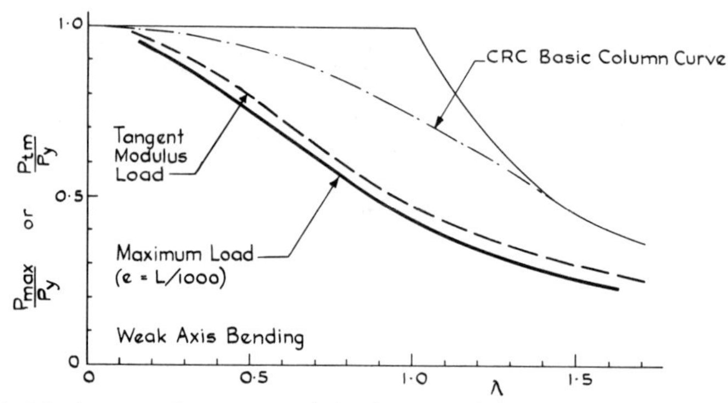

FIG. 22. Maximum column strength for heavy rolled shape W14 × 730 predicted from assumed residual stress distribution.

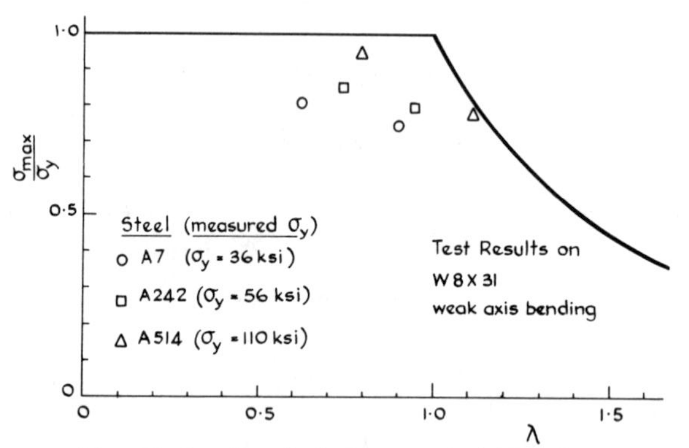

FIG. 23. Small rolled shape and yield point.

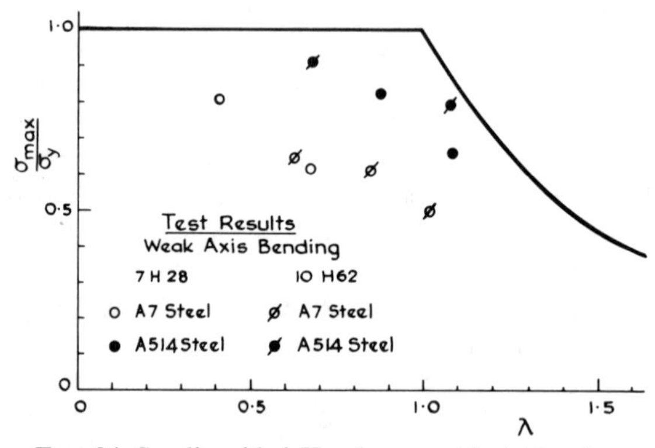

FIG. 24. Small welded H-columns and yield point.

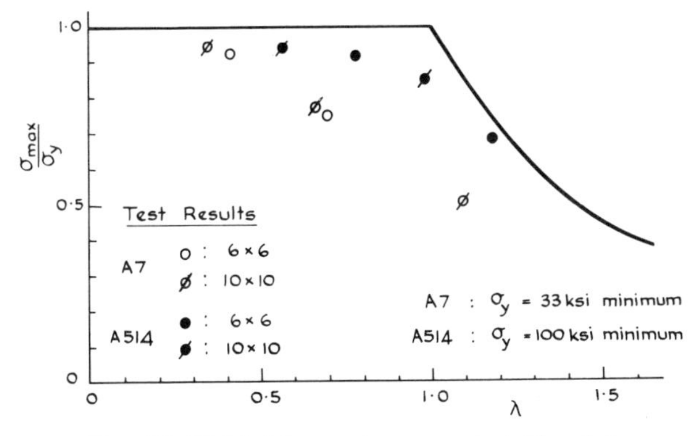

FIG. 25. Welded box columns and yield point.

behaviour of a rolled column of a 130 ksi yield quenched and tempered steel. The results of column tests for mild steel (yield strengths of 33 and 36 ksi), and for A 514 steel (quenched and tempered steel, yield strength 100 ksi) are compared in Fig. 26[65] for both rolled and welded shapes.

Manufacturing and geometrical conditions, together with the type of steel, are the most important variables in the strength of a compressed steel member. Some of these factors need to be considered, as well as other factors—the shape of the cross-section, annealing, out-of-straightness, cold-straightening, and effective length. Out-of-straightness and effective length are considered in section 1.4.

No particular form of cross-sectional shape can be regarded as being

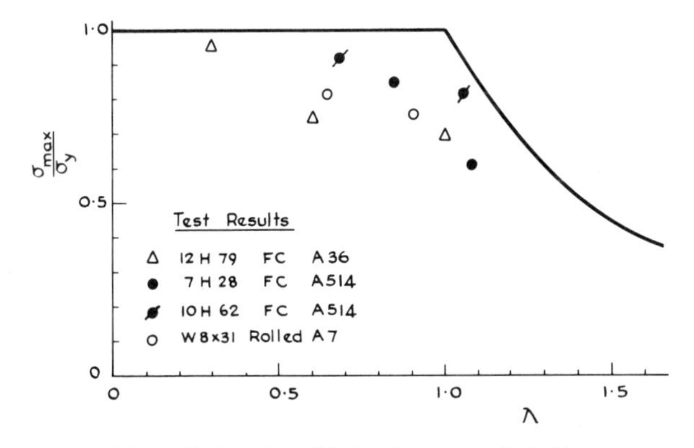

FIG. 26. Rolled and welded columns and yield point.

the best for use as a compressed member—every situation will require its own evaluation. However, all other conditions being the same, box-shapes[9,67] are stronger than H-shapes. This is because of the favourable residual stress distribution, discussed above. For the low slenderness ratios (up to 60) where out-of-straightness is not an important factor, columns with favourable residual stress distribution will be stronger than columns with unfavourable distribution. If the material farthest from the axis of bending is in a state of residual compression, then this material will yield first under load, leading to column failure at a lower load than would otherwise be expected.

Thus, riveted columns are very similar in behaviour to rolled columns,[29] because the process of riveting does not change the cooling residual stresses of the component rolled parts. On the other hand, annealing reduces the residual stress magnitude to very small values, and increases in column strength may be expected, as shown in Fig. 27.[56]

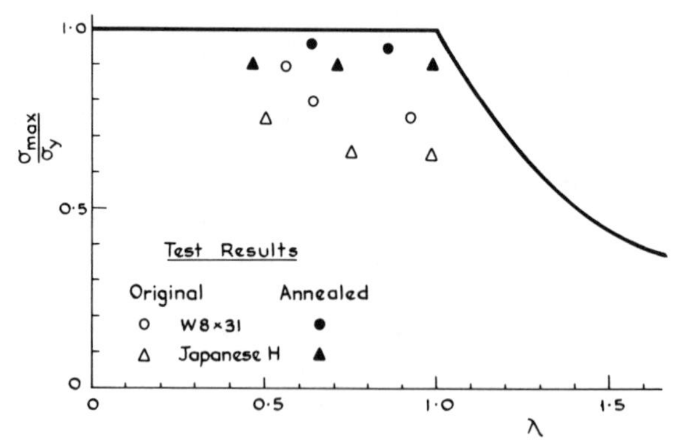

FIG. 27. Strength of annealed columns.

Every structural member, whether welded or rolled, normally will be cold-straightened to specification tolerances, causing a reduction and redistribution of residual stresses.[33] Thus, the compressive strength of straightened structural members will be at least as high as in such members before straightening. Since some structural members after rolling or welding are straight within specification tolerances, it may be expected that they will not be cold-straightened. Thus, the compressive strength is based on the cooling or welding residual stresses since there is no assurance that these residual stresses will be changed to lower magnitudes or to a more favourable distribution. If there is a guarantee that certain shapes would be rotorised (roller-straightened), then such columns could be assigned a more favourable column strength.[4]

1.4 OUT-OF-STRAIGHTNESS AND EFFECTIVE LENGTH

Out-of-straightness refers to the crookedness of the structural member—it may also refer to eccentricity of load and to non-symmetry of residual stresses, that is, deviations which result in an eccentrically loaded column. Out-of-straightness is unavoidable and results in lowered column strengths; it may be regarded as a situation between those of the simple column and the beam-column.

Most compression members are either straightened, or else are framed columns, and have a deflected shape other than the simple single-wave deflection curve, so that the problem of out-of-straightness is normally not a factor to be checked in design. The strength of a compression member is dependent to a great extent on its effective length, rather than on its actual length, and it is the effective length which is used in design.

The discussion of this chapter has considered pin-ended columns. Such members do not usually exist in structures, although pinned-joint trusses and other structures were used often in earlier years, which simplified design even though a perfect pin-joint was never obtained. Most compression members are parts of frames, and the effect of adjacent members (called end fixity or end restraint) needs to be considered.

The question may be asked: why consider a pin-ended column? The reason is that a pin-ended column may be regarded as a basic or limiting condition, and a knowledge of its behaviour under load is necessary in the study of compression members in general. That is, the strength of the pin-ended column is the reference or anchor point for the strength of beam-columns.

1.4.1 Beam-Columns

A beam-column is a member carrying both an axial load and a bending moment. Generally, to differentiate a beam-column from a column with out-of-straightness, a beam-column is described as having significant amounts of bending and compression.

The secant formula, eqn. (5), has served as the basis of some design specifications for beam-columns,[5,7] with specified initial eccentricities and effective lengths to give a conservative design approach. Interaction equations are commonly used for the design of beam-columns. While the concept originally was based on empirical reasoning and experimental correlation, today the interaction equations have been modified extensively to reflect actual behaviour.[38,57] Thus, Fig. 28 compares the simple straight-line interaction formula with the analytical exact solutions, while Fig. 29 compares the analytical solutions with the interaction formula

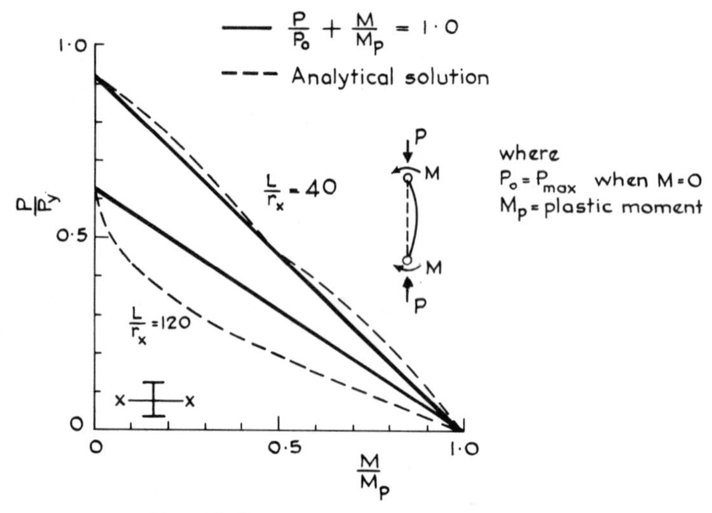

FIG. 28. Straight-line interaction formula.

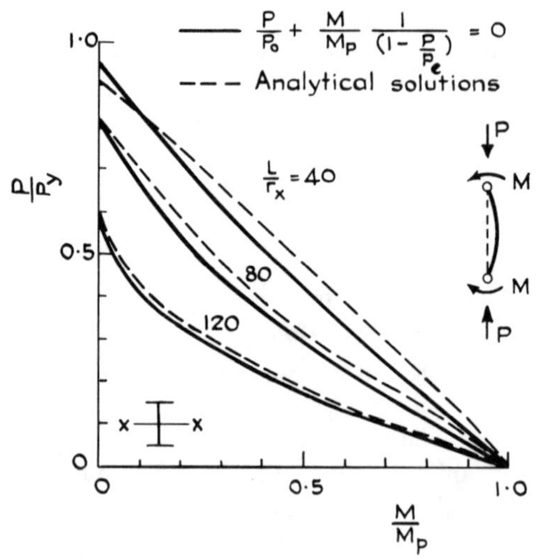

FIG. 29. Interaction formula with amplification factor.

modified with an amplification factor. (Additional modifications, not shown in the equation in Fig. 29, have been introduced to represent the variation of bending moment.[57])

The vertical ordinate of Fig. 29 contains the column curve for simple pin-ended members. Thus, the beam-column strength is defined in terms of the strength of the simple column, and this illustrates the critical importance of a knowledge of the simple column despite the fact that it does not exist.

1.4.2 Effective Length

In design use, the length L of the column is modified by the effective length factor K, to take account of the fact that columns do not have pinned ends. For the case of flexural buckling alone, KL is defined as the distance between inflection points, Fig. 30. In an actual framed structure,

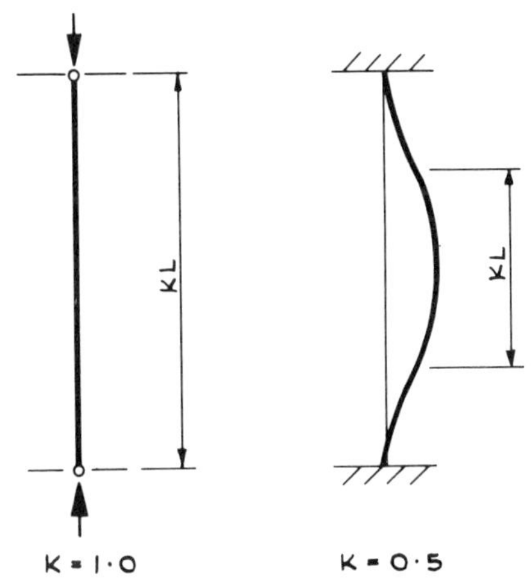

FIG. 30. Effective length factor.

the determination of K may be quite complex, involving end restraint, translation, and the overall stability of the structure. Estimations of the value of K vary from the simple to the complex, and further information on this may be found in Ref. 38.

1.5 COLUMN TESTS, COLUMN CURVES, AND DESIGN

Figure 31 presents test data for columns of different shapes, yield strength, and manufacturing and fabrication methods.[2,54] The scatter of

A7 and A36
- o W, Weak Axis
- ⊙ W, Strong Axis
- • Annealed
- Δ Cold Straightened
- I Welded H
- II Riveted H
- □ Welded Box
- B Round Bar
- P Perforated
- ■ Rolled Box

A242
- + W, Weak Axis

A572 ($\sigma_y = 50$ ksi):
- Welded H, Weak Axis

HS Steel ($\sigma_y = 50$ ksi):
- d Perforated

Japanese Tests: ($\sigma_y = 45$ ksi)
- J Welded H
- Annealed
- r Reversed σ_r

5 Ni–Cr–Mo–V ($\sigma_y = 130$ ksi)
- ▲ W, Weak Axis

A514:
- V Round Bar
- Welded H, Weak Axis
- X W, Weak Axis

Hybrid Columns:
- Welded H, A514 Flange-A 36 Web
- Welded H, A514 Flange-A441 Web
- ▽ Welded H, A441 Flange-A36 Web

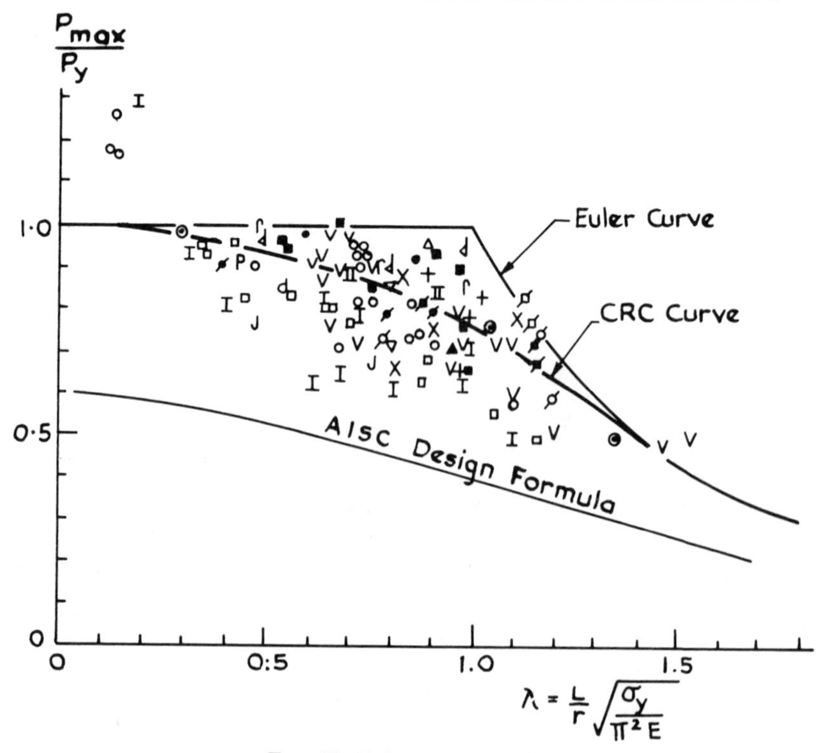

FIG. 31. Column test results.

test results is wide, in fact much wider than that of the earlier exhaustive work of Salmon at the beginning of the century, Fig. 2. Today's availability of steels of different yield strengths, and a vastly greater variety of cross-sections and sizes, are the major reasons.

This scatter is not a test phenomenon. Figure 32 shows the envelope of 112 column curves computed from measured residual stress distributions using the maximum strength concept.[13,16,64]

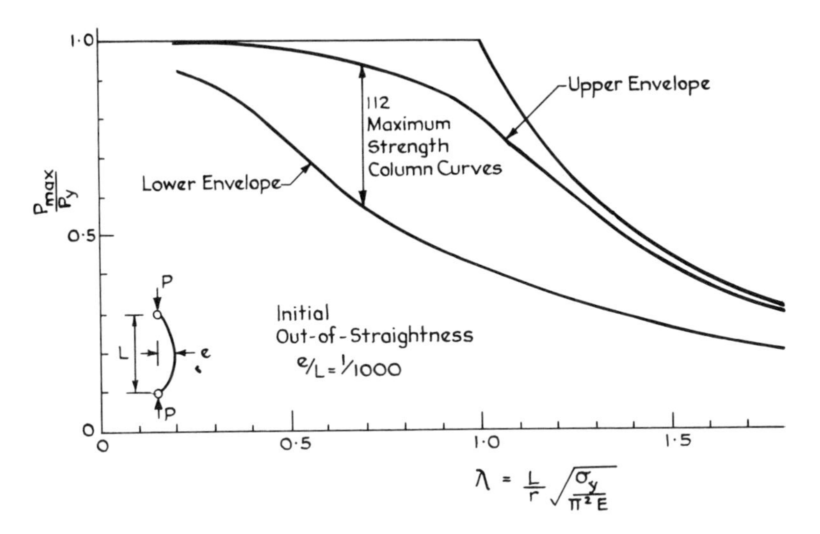

FIG. 32. Envelope of 112 maximum strength column curves.

The CRC Column Curve is shown in Fig. 31, and the immediate reaction is to question whether one single design curve is appropriate or correct to use with all the different types of columns. With one design curve, some columns will be overdesigned, some underdesigned, and the variation could be as much as 20% in each direction from an 'average' design curve. This variation in strength is not random, but rather is a function of manufacture, fabrication, and geometry; when these parameters are known, column strength can be predicted accurately from theoretical considerations.[16,64]

Multiple column curves have been under consideration at Lehigh University[54] since the late 1950s when the first results of the major investigation there into residual stresses and column strength indicated the scatter that was later obtained in great detail. The need for multiple column curves appeared logical and definite, and this was reinforced by the 1959 introduction of the German Standard, DIN 4114, which specified a special column curve for tubes which had higher allowable stresses than the basic column curve used for all other columns.

Within the scatter band, in general, the lower bound corresponds to very heavy column shapes and to welded shapes, and the upper bound to annealed columns and columns of high-strength steel. Other variations within the scatter reflect flame-cutting, light or heavy shapes, axis of bending, and geometrical shape.

There was heavy debate in the early 1960s in the United States in the interested committees of the CRC (later SSRC) and the AISC when the concept of the multiple column curve was introduced by the Lehigh University research team, and this debate has continued essentially unabated to the present time. Thus, the topic was not noted in the second edition of the CRC Guide[37] in 1966, although it had been introduced in Ref. 54 in 1964. The debate was not concerned with the validity of the scatter of results, but rather with the practical need for the complications of having more than one column curve. It was recognised that the scatter would be drastically reduced for compressed members in actual structures where the slenderness ratios tend to be small and where end restraint has an important influence. The scatter becomes even less pronounced with beam-columns—with increased moment, the effect of load on the interaction curve becomes less, so that at a certain level of applied moment there is no significant difference between interaction curves for different slenderness ratios. It may be concluded that any advantage in using the multiple column concept is purely academic.

The first definitive studies on the multiple column curve concept were reported in 1968 by the Lehigh University team.[40] These studies were under the review of CRC Task Group 1 and were based on the tangent modulus loads of a large number of columns taking into account their actual residual stress distributions, and a set of three column curves was proposed.[11,13,40] Based on a refined computer program for maximum strength,[64] the next development was the preparation of multiple column curves based on the maximum column strength using actual measured values of all the column strength parameters for 112 columns.[14,16,17] The column curves for these 112 columns are shown in Fig. 32, and the multiple column curves proposed are shown in Fig. 33. The curves in Fig. 33 were prepared from a deterministic approach—a probabilistic approach was also used which resulted in similar curves with slightly higher column strength predictions. The curves in Fig. 33 were adopted

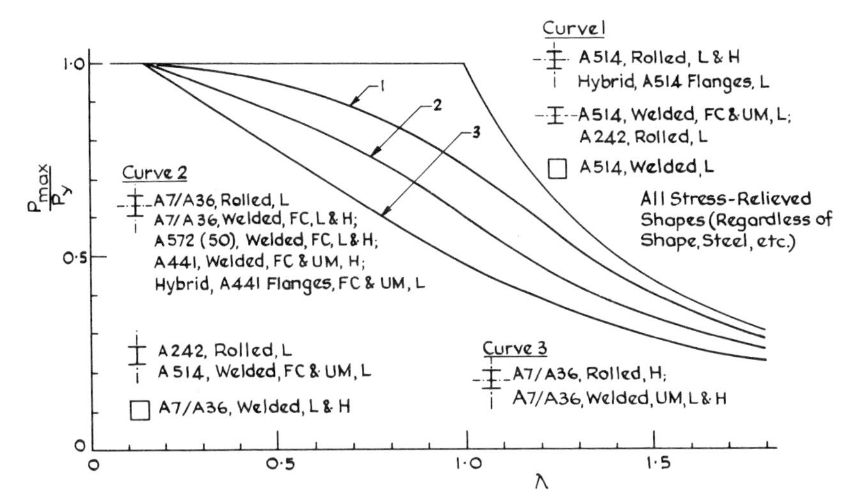

FIG. 33. Proposed US multiple column curves (where L = light, H = heavy).

by the SSRC for consideration, and were introduced in the third edition of the SSRC Guide in 1976.[38] A Column Selection Table was prepared to facilitate the use of these curves.[13]

In the mid-1960s, Commission 8 of the European Convention for Constructional Steelwork (ECCS) commenced a major study into the preparation of specifications for column design. Part of this study resulted in the introduction of a set of three multiple column curves[10] which was later modified by adding a curve at the top for high-strength steels and one at the bottom for heavy shapes, Fig. 34. It is of interest that the American and European multiple column curves correlate very well, Fig. 35, despite the different approaches used. The American study used actual measured values, while the European studies used theoretical data as the basis for computations which were then compared with test results.

The European column curves were adopted by the ECCS in 1978[27] for design practice, and a number of European countries are considering their incorporation into their design codes. On the other hand, US designers have not adopted the SSRC multiple column curves and it seems doubtful that there will be a movement to do this because of the negligible savings such a complicated effort will bring. The noticeable differences between the multiple column curves for the simple columns do not exist for the practical restrained columns typically used in structures.

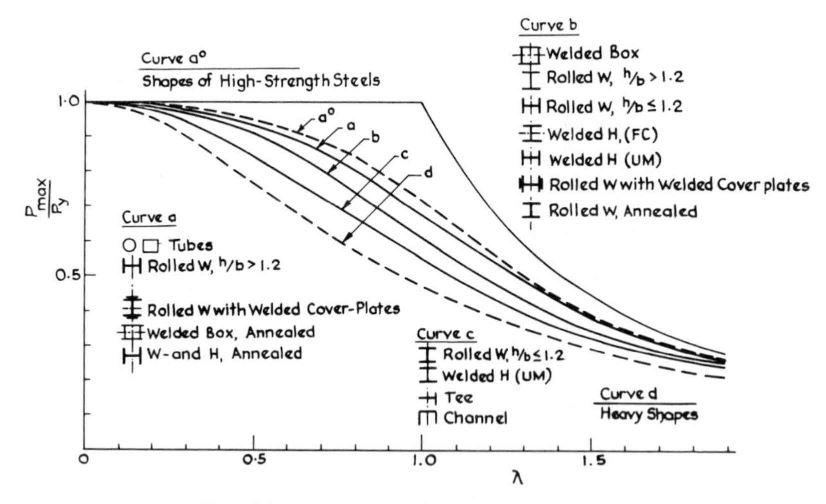

FIG. 34. European multiple column curves.

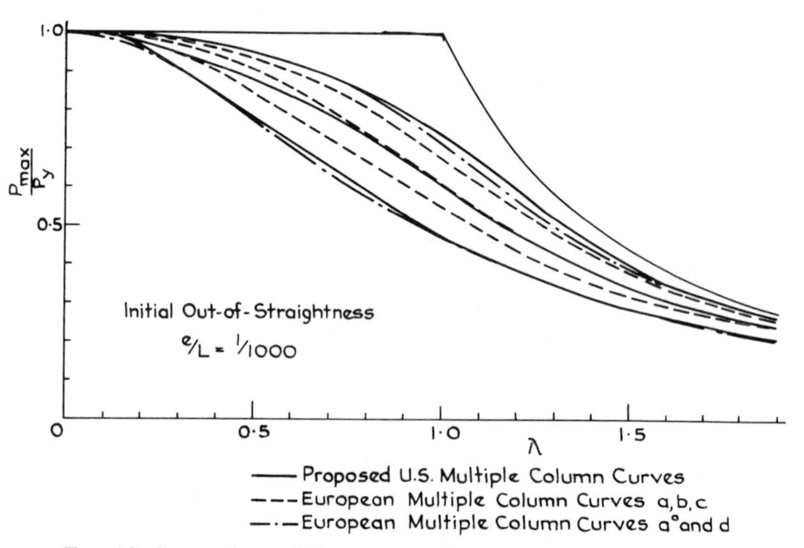

FIG. 35. Comparison of European and US multiple column curves.

Column curves used for design are derived directly from maximum column strength curves, and the difference between the two indicates the factor of safety used, as shown in Fig. 31 where the CRC curve is the basis of the AISC design curve.

The most recent studies on design criteria are concerned with the development of load-factor and limit-states design which give particular attention to the probabilistic basis of the design parameters, see Ref. 34. This topic is discussed in Chapter 2.

REFERENCES

1. ALPSTEN, G. A. *Thermal Residual Stresses in Hot-Rolled Steel Members*, Fritz Laboratory Report No. 337.3, Lehigh University, December 1968
2. ALPSTEN, G. A. and TALL, L. Prediction of Behavior of Steel Columns Under Load, *Proc. Final Report*, Symposium on Concepts of Safety of Structures and Methods of Design, IABSE, London, September 1969
3. ALPSTEN, G. A. and TALL, L. Residual Stresses in Heavy Welded Shapes, *Welding J.*, **49**, April 1970
4. ALPSTEN, G. A. Residual Stresses, Yield Stress, and Column Strength of Hot-Rolled and Roller-Straightened Steel Shapes, *Proc., 1972 International Colloquium on Column Strength, Paris*, IABSE, **23**, 1975
5. AMERICAN INSTITUTE OF STEEL CONSTRUCTION *Specifications for the Design, Fabrication, and Erection of Structural Steel for Buildings*, New York, AISC, 1961, revised 1963
6. AMERICAN INSTITUTE OF STEEL CONSTRUCTION *Steel Construction Manual*, 8th Ed., AISC, Chicago, 1980
7. BEEDLE, L. S. and TALL, L. Basic Column Strength, *Proc., Am. Soc. Civil Engrs.*, **86**, (ST-7), July 1960
8. BEEDLE, L. S., GALAMBOS, T. V. and TALL, L. *Column Strength of Constructional Steels*, Steel Design and Engineering Seminar, US Steel Corp., Pittsburgh, May 1961
9. BEER, G. and TALL, L. On the Strength of Welded Box Columns with Imperfections, *Trans., Fourth Australasian Conference on the Mechanics of Structures and Materials, Brisbane*, 1973
10. BEER, H. and SCHULZ, G. Bases Theoriques des Courbes Europèene de Flambement, *Construction Métallique*, No. 3, September 1970, Paris
11. BJORHOVDE, R. *The Philosophy of Column Design*, Fritz Laboratory Report No. 337.19, Lehigh University, December 1968
12. BJORHOVDE, R. and TALL, L. *Survey of Utilization and Manufacture of Heavy Columns*, Fritz Laboratory Report No. 337.7, Lehigh University, October 1970
13. BJORHOVDE, R. and TALL, L. *Maximum Column Strength and the Multiple Column Curve Concept*, Fritz Laboratory Report No. 337.29, Lehigh University, October 1971
14. BJORHOVDE, R. *Deterministic and Probabilistic Approaches to the Strength of Steel Columns*, PhD Dissertation, Lehigh University, May 1972, University Microfilms, Inc., Ann Arbor, Michigan and London.
15. BJORHOVDE, R., BROZZETTI, J., ALPSTEN, G. A. and TALL, L. Residual Stresses in Thick Welded Plates, *Welding J.*, **51**, August 1972

16. BJORHOVDE, R. and TALL, L. Development of Multiple Column Curves, *Proc., 1972 International Colloquium on Column Strength, Paris,* IABSE, **23**, 1975
17. BJORHOVDE, R. and TALL, L. The Probabilistic Characteristics of Maximum Column Strength, *Proc., 1972 International Colloquium on Column Strength, Paris,* IABSE, **23**, 1975
18. BLEICH, F. *Buckling Strength of Metal Structures*, McGraw-Hill, New York, 1952
19. BROZZETTI, J., ALPSTEN, G. A. and TALL, L. Welding Parameters, Thick Plates, and Column Strength, *Welding J.*, **50**, August 1971
20. BROZZETTI, J. and TALL, L. Effect of Welding Parameters on Built-Up Column Strength, *Proc., 1972 International Colloquium on Column Strength, Paris,* IABSE, **23**, 1975
21. COLUMN RESEARCH COUNCIL *The Basic Column Formula*, CRC Technical Memorandum No. 1, May 1952
22. CAMPUS, F. and MASSONNET, C. *Recherches sur le Flambement de Colonnes en Acier A37, un Profil en Double Tee, Sollicitées Obliquement*, IRSIA, No. 17, Belgium, April 1956
23. COLUMN RESEARCH COUNCIL *Guide to Design Criteria for Metal Compression Members,* CRC, Urbana, Illinois, and Crosby Lockwood, London, 1960
24. DUBERG, J. E. and WILDER, T. W. *Inelastic Column Behavior*, NACA Technical Note 2267, Washington, D.C., January 1951
25. ESTUAR, F. R. and TALL, L. Experimental Investigation of Welded Built-Up Columns, *Welding J.*, **42**, April 1963
26. ESTUAR, F. R. *Welding Residual Stresses and the Strength of Heavy Column Shapes*, PhD Dissertation, Lehigh University, August 1965. University Microfilms, Inc., Ann Arbor, Michigan, and London
27. EUROPEAN CONVENTION FOR CONSTRUCTIONAL STEELWORK. *European Recommendations for Steel Construction,* Milan, ECCS, March 1978
28. FEDER, D. and LEE, G. C. *Residual Stress and the Strength of Members of High Strength Steel,* Fritz Laboratory Report 269.2, Lehigh University, March 1959
29. FUJITA, Y. *Built-Up Column Strength,* PhD Dissertation, Lehigh University, August 1956. University Microfilms, Inc., Ann Arbor, Michigan, and London
30. FUJITA, Y. Ultimate Strength of Columns with Residual Stresses, *J. Soc. Naval Architects Japan*, January 1960
31. HOWARD, J. E. Some Results of the Tests on Steel Columns, in Progress at the Watertown Arsenal, *Proc. Am. Soc. for Testing and Materials,* **18**, 1908
32. HUBER, A. W. and BEEDLE, L. S. Residual Stress and the Compressive Strength of Steel, *Welding J.*, **33**, December 1954
33. HUBER, A. W. *The Influence of Residual Stress on the Instability of Columns,* PhD Dissertation, Lehigh University, May 1956. University Microfilms, Inc., Ann Arbor, Michigan, and London
34. HALL, D. H. Proposed Steel Column Strength Criteria, *Proc. Am. Soc. Civil Engrs*, **107** (ST-4), April 1981
35. INTERNATIONAL INSTITUTE OF WELDING, COMMISSION X. *Un Inventaire sur le Sujet: Tensions Residuelles et Instabilité, preparée par H. Louis, M. Marincek, et L. Tall,* Oslo, July 1962
36. JOHNSTON, B. G. Buckling Behavior Above Tangent Modulus Load, *Proc., Am. Soc. Civil Engrs,* **87**, (EM-6), December 1961

37. JOHNSTON, B. G., Ed. *CRC Guide to Design Criteria for Metal Compression Members*, 2nd Ed., Wiley, New York, 1966
38. JOHNSTON, B. G., Ed. *SSRC: Guide to Stability Design Criteria for Metal Structures*, 3rd Ed., Wiley, New York, 1976
39. KISHIMA, Y., ALPSTEN, G. A. and TALL, L. *The Strength of ASTM A572(50) Steel Welded Flame-Cut Columns*, Fritz Laboratory Report No. 321.4, Lehigh University, July 1970
40. LAZARO, A. and TALL, L. *The Philosophy of Column Design—A Preliminary Report*, Fritz Laboratory Report No. 337.2, Lehigh University, July 1968
41. MCFALLS, R. and TALL, L. A Study of Welded Columns Manufactures from Flame-Cut Plates, *Welding J.*, **48**, April 1969
42. NAGARAJARAO, N. R. and TALL, L. Columns Reinforced Under Load, *Welding J.*, **42**, April 1963
43. NITTA, A. *Ultimate Strength of High Strength Steel Circular Columns*, PhD Dissertation, Lehigh University, June 1960. University Microfilms, Inc., Ann Arbor, Michigan and London
44. O'CONNOR, C. Residual Stresses and Their Influence on Structural Design, *J. Inst. Engrs, Australia*, **27**, December 1955
45. ODAR, E., NISHINO, F. and TALL, L. *Residual Stresses in Rolled Heat-Treated T-1 Shapes*, WRC Bull. No 121, April 1967, New York
46. OSGOOD, W. R. The Effect of Residual Stress on Column Strength, *Proc. First National Congress Applied Mechanics*, June 1951
47. SALMON, E. H. *Columns, A Treatise on the Strength and Design of Compression Members*, Oxford Technical Publications, London, 1921
48. SHANLEY, F. R. Inelastic Column Theory, *J. Aeron. Sci.*, **14**, May 1947
49. TALL, L., HUBER, A. W. and BEEDLE, L. S. *Residual Stress and the Instability of Axially Loaded Columns*, Fritz Laboratory Report No. 220A.35, Lehigh University, February 1960. Published as Commission X Document, Colloquium, International Institute of Welding, Liége, Belgium, June 1960
50. TALL, L. *Stub Column Test Procedure*, Fritz Laboratory Report 220A.36, Lehigh University, February 1961. Revised as International Institute of Welding Document X-282-61, prepared by Working Group (H. Louis, M. Marincek, and L. Tall), Annual Conference, Oslo, July 1962
51. TALL, L. *The Strength of Welded Built-Up Columns*, PhD Dissertation, Lehigh University, May 1961. University Microfilms, Inc., Ann Arbor, Michigan, and London
52. TALL, L. and ESTUAR, F. R. Discussion to Ref. 1.36, *Proc. Am. Soc. Civil Engrs*, **88** (EM-5), October 1962
53. TALL, L. Residual Stresses in Welded Plates—A Theoretical Study, *Welding J.*, **43**, January 1964
54. TALL, L. Recent Developments in the Study of Column Behavior, *J. Inst. Engrs, Australia*, **36**, December 1964
55. TALL, L. and FEDER, D. Längsschweissspannungen in Platten und Ihr Einfluss auf die Grenzlast von geschweissten Stahlstützen, *Schweissen und Schneiden*, March 1965
56. TALL, L. *Welded Built-Up Columns*, Fritz Laboratory Report No. 249.29, Lehigh University, April 1966
57. TALL, L., Ed. *Structural Steel Design*, Wiley (Ronald Press), 2nd Ed., 1974, New York

58. TALL, L. The Calculation of Residual Stresses—In Perspective, *Trans. International Conf. on Residual Stresses in Welded Construction*, London, The Welding Institute, November 1977
59. TEBEDGE, N., ALPSTEN, G. A. and TALL, L. Measurement of Residual Stresses—A Comparative Study of Methods, *Proc. JBCSA Conference on the Recording and Interpretation of Engineering Measurements*, London, 1972
60. TEBEDGE, N. and TALL, L. Contraintes Residuelles dans les Profils en Acier—Synthése des Valeurs Mesurées, *Construction Métallique*, No. 2, Paris, June 1974
61. VAN DEN BROECK, J. A. English Translation of Euler's *On the Strength of Columns*, *Am. J. Phys.*, **15**, July 1947
62. YANG, C. H., BEEDLE, L. S. and JOHNSTON, B. G. Residual Stress and the Yield Strength of Steel Beams, *Welding J.*, **31**, April 1952
63. YU, C. K. and TALL, L. A Pilot Study on the Strength of 5 Ni–Cr–Mo–V Steel Columns, *Experimental Mechanics*, **8**, January 1968
64. YU, C. K. *Inelastic Columns with Residual Stresses*, PhD Dissertation, Lehigh University, June 1968. University Microfilms, Inc., Ann Arbor, Michigan, and London
65. YU, C. K. and TALL, L. *Welded and Rolled A514 Steel Columns—A Summary Report*, Fritz Laboratory Report No. 290.16, Lehigh University, June 1970
66. YU, C. K. and TALL, L. Significance of Application of Stub Column Test Results, *Proc. Am. Soc. Civil Engrs*, **97** (ST-7), July 1971
67. ZANDONINI, R. and TALL, L. *Strength of Welded Box Columns*, Fritz Laboratory Report No. 249.33, Lehigh University, October 1980

Chapter 2

CURRENT TRENDS IN THE TREATMENT OF SAFETY

I. H. G. DUNCAN
Buro Happold (South West) Consulting Engineers, Bristol, UK

W. I. LIDDELL
Buro Happold Consulting Engineers, Bath, UK

and

C. J. K. WILLIAMS
School of Architecture and Building Engineering, University of Bath, UK

SUMMARY

Probability theory has been adapted for practical application as a basis for comparing risk, adjusted by the process of calibration against practice and experience, and developed as a method of calculating partial safety factors.

This chapter reviews the application of probability theory in the current generation of structural codes. The British and American Safety formats are discussed and, despite their apparent differences, are shown to produce similar effects in practice.

It is noted that departures by national code committees from the guidelines laid down by the International Standards Organisation, ISO, do not help in achieving a unified approach to safety.

It is concluded that whilst current trends may lead to a more rational basis for safety calculations, and to more consistent levels of safety, it remains that safety is a wider subject than can be controlled by code rules.

2.1 INTRODUCTION

2.1.1 Safety

Throughout the discussion of safety one of the central difficulties has been the definition of terms in a way that is logically precise. This problem has been partially resolved by defining the conditions which would cause failure, and the peak operating conditions, on a probabilistic basis.

The measure of safety is the probability of operating conditions not reaching the failure conditions.

The *true* or actual *factor of safety* can be defined as:

$$\frac{\text{Actual load which would cause structure to collapse}}{\text{Actual maximum load ever to be carried by the structure}}$$

Provided that this ratio is greater than one, the structure is safe and will not collapse. However, the actual load which would cause collapse is not known unless the structure is loaded to failure. In practice, the collapse load is calculated or estimated from information of the geometry and materials of the structure. The maximum load is estimated from information gained from previous loading history of similar structures. The factor of safety is therefore:

$$\frac{\text{Estimated minimum load which would cause collapse}}{\text{Estimated maximum load ever to be carried by the structure}}$$

This could be described as the *estimated factor of safety* to distinguish it from the true factor of safety. This distinction is never made although the general public is most likely to understand the term 'factor of safety' as meaning the true factor of safety. Engineers have in the past used the term to apply to a wide variety of ratios of resistance to load effects. It is only in recent times that 'factor of safety' has come to mean the estimated factor of safety, and even then it is not universally the same. In order to have a discussion about the required values of factors of safety it is first necessary to define their function, and second to define the way in which estimates of maximum load and minimum strength are made.

2.1.2 Safety Format

In design calculations the geometry or form of the structure is usually selected to meet the requirements of span, height, or other functions. The maximum load ever to be carried by the structure is estimated from the intended loads or from previous records of loading phenomena. The forces in the components can be calculated from the loading and geometry using structural mechanics. A suitable size and material for the component can then be selected whose strength appropriately exceeds the calculated internal forces by an amount which is controlled by the factor of safety. The way in which the safety factors are included in the design calculation or code rules is known as the *safety format*.

In the past, this final step was always achieved by defining a working stress for a particular material which included the factor of safety. This procedure contained a remarkable amount of muddled thinking since the factor of safety was defined in a number of ways related to different properties and often contained factors introduced to compensate for known or partially known effects, e.g. dynamic loads, fatigue, or a difference between test conditions and the real situation. These compensation factors are often as large as the factors of safety introduced to account for uncertainty of loads and strengths and serve to disguise the real level of safety.

The amount of compensation may be linked to some other phenomenon such as crack size in brittle materials, which requires a different approach to prediction and control related to the maximum size of crack which can be expected to pass the inspection checks.

By the end of the 19th century statutory loads and working stresses were established as code rules. This safety format continued more or less unchanged until the 1960s when the concept of factored loads and factored strengths was introduced. This was accompanied by the definition of maximum load and minimum strength on a probabilistic basis and the use of partial factors for deriving the overall load and resistance factors. For a history of the development of this concept the reader is referred to Sir Alfred Pugsley's book *The Safety of Structures*.

Structural codes are now being rewritten in terms of factored loads and factored strengths in most countries. The aim of the new codes is to achieve a consistent level of safety defined by the probability of strength minus load effects being not less than zero. This new format is criticised by some engineers for being more complicated than many situations demand.

Collapse under static loads is not the only way in which a structure may fail to satisfy the requirements; excessive deflection or deformation, excessive vibrations or fracture from fatigue effects are other modes of failure which should be taken into account. Within the new code formats these various modes of failure are called *Limit States*, and engineers are required to check for these using different safety factors.

2.1.3 Safety Formats in Current Practice

There are two main approaches to the treatment of structural safety in codes which have been developing during the last decade or so; on the one hand the proposals contained in the CEB/FIB International Recommendations and on the other, the proposals of Sub-Committee E of ACI Committee 348 (Structural Safety) which were developed from earlier work by Cornell. The former, which are now embodied in the ISO–2394 proposals, is usually referred to as a semi-probabilistic approach, whereas the latter is related to one of a number of so called second-moment methods of reliability analysis.

The fundamental difference between these two approaches is that, in the semi-probabilistic approach, the safety of a structure is ensured by defining design loads and strengths or *resistances* which individually have such a remote chance of occurring that the probability of the load exceeding the load-carrying capacity in any part of a structure is considered to be sufficiently low. However, no explicit reliability calculations are undertaken and the levels of risk in different structures are unknown. With the second-moment methods, an attempt is made to specify the reliability of a structure or component, either in terms of a *reliability index* or in terms of a *target notional reliability*—an estimate of the acceptable probability of occurrence of each limit state during the life of the structure. Appropriate partial safety factors may be calculated for particular design situations.

These two approaches form the basis of the new European and the proposed American Code formats respectively, and below is summarised the way in which these theoretical bases are applied in Code format in the current British generation of Codes, and in the proposals for the American (AISC) Code on Steelwork (See Ravindra and Galambos (1978)).

(a) Limit State Design and ISO 2394

In 1973 the member bodies of the International Standards Organisation published ISO 2394 (*General Principles for the Verification of the Safety of Structures*). This document defines the terminology and safety format which is recommended for all future structural codes. In Britain the concrete and masonry codes are already written in this format, and work is proceeding on the codes for other materials.

In ISO–2394 the design method aims at guaranteeing adequate safety against the structure or structural element being rendered unfit for use without excessive material usage. It defines adequate safety as being provided when the probability of the structure attaining any particular state associated with unfitness for use is sufficiently small. The state associated with unfitness is defined as a *limit state*. These are placed in two categories:

1. The *ultimate limit states* which are those corresponding to the maximum load-carrying capacity and therefore related to strength.
2. The *serviceability limit states* which are related to the criteria governing normal use or durability.

There is a view that fatigue life is a third category of limit state.

Actions to be considered in determining the loading effects, S, on a structure are *direct actions*, e.g. selfweight and imposed loads, and *indirect actions* which may be imposed deformations caused for example by thermal or moisture movements, shrinkage, settlement, etc. Material strengths and the values of actions are defined in terms of their *characteristic values* R_k and Q_k respectively.

For the materials, the characteristic strengths are, by definition, those which have a probability, accepted *a priori*, of not being attained. To determine the characteristic strength, a statistical distribution of an appropriate type is assumed; generally, a normal distribution will be taken.

The characteristic strengths, R_k, are then defined by: $R_k = R_m - ks$, where R_m is the arithmetic mean of the different test results, s is the standard deviation, and k is a coefficient depending on the probability, accepted *a priori*, of obtaining test results less than R_k.

For loading that may be considered random, a characteristic value Q_k may be defined, if appropriate, by the relation $Q_k = Q_m(1 + k\delta)$, where Q_m is the value of the most unfavourable loading, with a 50% probability of its being exceeded, up to abnormally high values, once in the expected life of the structure, δ is the relative mean quadratic deviation of the distribution of the maximum loading, and k is a coefficient depending on the probability, accepted *a priori*, of maximum loadings being greater than Q_k.

However, when it is a question that the reduction of a load may endanger the stability of the structure, the characteristic value Q_k will be defined by the relation $Q_k = Q_m(1 - k\delta)$.

(In defining Q_m in terms of the building design life, a temporary building would be designed for lower loads than a more permanent building. This means that a person occupying a temporary building, say for one year, is exposed to more danger than if he were in a permanent building.)

When it is not possible to use a statistical distribution the characteristic loads must be chosen as a function of the use for which the construction is intended. These chosen loads are called *nominal loads* and are given in standards, codes of practice or other regulations. These 'nominal loads' should be introduced into the calculation as 'characteristic loads' Q_k.

(*i*) *Partial safety factors:* The condition for a satisfactory and safe design is that the design strength R^* must be greater than the design loading effects S^*. The design strength is defined by

$$R^* = \frac{R_k}{\gamma_m}$$

where γ_m, the *materials factor*, is in principle the function of two coefficients:

γ_{m_1} is intended to cover the possible reductions in the strength of the materials in the structure as a whole as compared with the characteristic value deduced from the control test specimen; and

γ_{m_2} is intended to cover possible weakness of the structure arising from any cause other than the reduction in the strength of the materials allowed by γ_{m_1}, including manufacturing tolerances.

The design loading effects S^* are determined from the characteristic actions, Q_k, by the relationship $S^* = \gamma_s . Q_k$ where γ_s, the load factor, is a function of three coefficients;

γ_{s_1} takes account of the possibility of unfavourable deviation of the loads from the characteristic loads, thus allowing for abnormal or unforeseen actions;

γ_{s_2} takes account of the reduced probability that various loadings acting together will all be simultaneously at their characteristic value; and

γ_{s_3} is intended to allow for possible adverse modifications of the loading effects due to incorrect design assumptions, e.g. introduction of simplified support conditions, hinges, neglect of thermal and other effects which are difficult to assess, constructional discrepancies such as dimensions of cross-section, deviation of columns from vertical, and accidental eccentricities.

γ_{s_1} and γ_{s_2} properly belong to the characteristic actions and can be used to define a design action, while γ_{s_3}, which allows for uncertainty in assumptions used in stress analysis, strictly belongs to the loading effects. Additional partial factors γ_{c_1} and γ_{c_2} which can be further applied to the loading effects, are defined as:

γ_{c_1} is intended to take account of the nature of the structure and its behaviour. For example structures or parts of structures in which partial or complete collapse can occur without warning, where redistribution of internal forces is not possible, or where failure of a single element can lead to overall collapse, and

γ_{c_2} is intended to take account of the seriousness of attaining a limit state from other points of view, for example economic consequences, danger to community, etc.

However, γ_c is usually set to one.

(*ii*) *Values of load and material factors:* Numerical values for the partial safety factors γ_m and γ_s for the ultimate limit state have now been fairly generally established in British Codes. Table 1 shows the partial load factors proposed in the new British Draft Steel Code. The overall load factors γ_s are the same as currently embodied in the British Concrete Code CP 110. For the serviceability limit state all values of partial safety factor are generally set to one, with reduction for certain load combinations.

Numerical values of material factors γ_m are selected according to their properties. For example, in the British Concrete Code (CP 110) γ_m is set to 1·5 for concrete and 1·15 for steel. In other codes, calibration has been carried out by setting the level of partial factors to produce structural sizes which would be similar to those designed to previous codes. For a detailed discussion of the significance of calibration see CIRIA Report No 63 and the Proceedings of the Seminar on Structural Codes—*Rationalisation of safety and serviceability factors.*

(*a*) *Load and Resistance Factor Design*

The countries which are notably absent from the ISO member bodies approving ISO–2394 are USA and Canada.

In Canada the 1975 National Building Code defines the loads and

TABLE 1

VALUES OF PARTIAL LOAD FACTOR, DRAFT BRITISH STEEL CODE

Type of load or combination of loads		*Load variation factor*	*Structural performance factor*	*Overall load factor*
Dead Load	Maximum	1+0·17=1·17	1·2	1·4
	Minimum	1−0·17=0·83	1·2	1·0
	Minimum for pattern loading	1·00	1·2	1·2
Imposed Load (in the absence of wind load)		1·33	1·2	1·6
Wind Load (with dead load only)		1·17	1·2	1·4
Wind and imposed load (acting together)		1·00	1·2	1·2
Forces due to temperature effects		1·00	1·2	1·2
Loads from overhead travelling cranes		1·33	1·25	1·67
Crane Loads (in the absence of wind load)		1·0	1·2	1·2
Wind and Crane Loads (acting together)		1·0	1·2	1·2

load factors to be used in design. Rules for calculating the resistance of structural elements are found in the various material codes. The procedure is presented in a paper by D. E. Allen (1975), with comments on calibration of the concrete and steel codes using the reliability Index β. In the United States, studies into the application of probability theory to structural safety have been proceeding for many years.

A limit state code for concrete ACI 318/77, was introduced in 1977. The currently proposed safety format for the American steel code is called Load and Resistance Factor Design (LRFD) This is described in an ASCE paper by Ravindra and Galambos (1978). The resulting format is similar to the limit state concrete code but with subtle differences to the definitions of loads and resistances which are difficult to grasp. The LRFD Design Criterion is

$$\phi R_{\mathrm{n}} \geqslant \sum_{k=1}^{j} \gamma_k Q_{\mathrm{rm}}$$

where ϕ is the resistance factor and R_n is the *nominal resistance* as calculated by code rules. The term *specified* is also used where the value is based on tests. In the first case, the material properties, and in the second case the resistance, are based on a specified probability level, e.g. 5 per cent probability of being exceeded, and so correspond to the characteristic strength of ISO–2394. In calculating the nominal resistance, variations in fabrication dimension and uncertainties in the assumptions used in calculating the component resistance are assessed as well as the variations in material properties. The resistance factor ϕ takes these into account.

The loading side of the design criterion is the sum of products $\gamma_k Q_m$ in which Q_m is the mean load effect and γ is a 'load factor'. The γ factors contain an allowance for the load being exceeded and for the uncertainties in the calculation of the load effects. The mean load effect, which includes dead load effects Q_{Dm} and live load effects Q_{Lm}, is calculated using mean loads with a specified recurrence period for the live loads.

The use of mean loads and their corresponding coefficient of variation is useful in assessing the relative probability of loads being exceeded. The rationale for using mean loads rather than characteristic loads in design calculations is harder to grasp. Climatic loads are almost universally analysed statistically using Fisher–Tippett Type 1 and 2 distributions to give values which are exceeded once in a specified return period. These are generally treated as characteristic loads with a specified probability of exceedance. Although a number of statistical analyses of floor loadings have been carried out to the extent that probability densities have been derived for some use categories, change of use of a building can dramatically change the load basis. It therefore seems likely that Building Control Authorities will demand statutory loadings related to use for a long time to come, if not indefinitely and these can more readily be defined and understood as characteristic loads. For a fuller discussion of load derivation the reader is referred to Ghiocel and Lungu (1975).

The measure of safety used in LRFD is the *safety index*. This is derived by considering structural safety as the probability of $(R-Q)$ being less than zero where R and Q are random variables representing resistance and load effect. Thus the probability of failure, $P_f = P(R - Q) \leqslant 0, = P(\log_e R/Q \leqslant 0)$. The safety index β is the inverse of the *coefficient of variation* of $\log_e R/Q$. It can be written as:

$$\beta = \frac{\ln(R_m/Q_m)}{\sqrt{V_R^2 + V_Q^2}}$$

where V_R and V_Q are the coefficients of variation of R_m and Q_m (see Section 2.3.3). The safety index is a relative measure of the structural safety. In order to develop a consistent set of design criteria β must be specified. In the absence of sufficient data to establish β from first principles the value can be selected to give the same degree of reliability as found in the existing design methods for a number of standard situations. This is the procedure adopted by the ASCE Committee on LRFD and they have recommended a value of $\beta = 3$ for normal structures. This value can be varied to account for the importance of the structure and the nature of the failure.

2.2 THE APPLICATION OF PROBABILITY THEORY TO STRUCTURAL SAFETY

2.2.1 Introduction

The field of mathematics which tries to ascribe numerical values to the likelihood of an event occurring is probability theory and the branch of probability theory which is applied to failures is sometimes called reliability theory.

The basic principle in applying probability theory to structural safety is very simple. Whether or not a structure will fail (crack, deflect excessively, collapse, etc.) depends on the actual value taken by a number of random variables such as loads, material strengths, dimensions, and a factor to account for the accuracy of structural analysis. The probability of failure is the probability that these random variables will have values which lead to failure. If a probability can be ascribed to each combination of variables which would lead to failure, then the probability of failure is just the sum of these probabilities.

Probability theory is a large branch of mathematics and there is not sufficient space in this chapter to discuss all the methods available. However, Section 2.2.2 gives some examples of the application of probability theory to structural safety and is intended to give an engineer an indication of the kind of information that can be obtained from probability theory. Section 2.2.3 introduces an approximate theory which is sometimes used to assess partial safety factors in practice.

Not all engineers are familiar with probability theory and therefore an Appendix is included as an introduction to those areas of probability theory most directly relevant to structural safety. Many of the best-known formulae used in probability theory only apply if the random variables under consideration are *independent* and it is important to know whether this assumption has to be made before using a particular formula.

2.2.2 Some Examples of the Application of Probability Theory to Structural Safety

The following examples are intended to show the sort of information that can be obtained from probability theory and also the limitations on what can be done. In particular it will be seen that it is usually very difficult to obtain the data necessary for a full analysis and, even if they can be obtained, a large amount of calculation is required in any but the simplest of cases.

(a) The Strength of a Chain

The strength of a chain is equal to that of its weakest link. Suppose that we have data from tests to failure of a large number of individual links and want to calculate the probability that a chain containing n links will fail at a particular load, x.

Let us first consider the case when we make the rather questionable assumption that the strengths of the individual links in the chain are independent of each other. This implies that the factors which caused the variations in the test results are assumed to act randomly from link to link and not produce batches of links with average strengths greater or less than the overall mean.

Let the density and distribution functions of the test results on individual links be $f_1(x)$ and $F_1(x)$ respectively where x is the failure load. The subscript 1 refers to the fact that individual links were tested, not sections of chain. The probability that one link will fail at a load, X, less than x is given by eqn. (A12).

$$P(X < x) = F_1(x)$$

Therefore the probability that the link *will not* fail is:

$$P(\text{one link does } \textit{not} \text{ fail at load } x) = 1 - F_1(x)$$

Now consider a chain consisting of two links. The chain will not fail at load x if both links do not fail. If the assumption of link strengths being independent is made then eqn. (A8) can be used to find the probability that both links do not fail:

$$P(\text{two-link chain does } \textit{not} \text{ fail at load } x) = [1 - F_1(x)][1 - F_1(x)].$$

Using the same argument repetitively:

$$P(n\text{-link chain does } \textit{not} \text{ fail at load } x) = [1 - F_1(x)]^n \quad (1)$$

or if $F_n(x)$ is the probability that an n-link chain *does* fail at load x:

$$1 - F_n(x) = [1 - F_1(x)]^n \quad (2)$$

Figure 1 shows $F_1(x)$, $F_{10}(x)$, $F_{100}(x)$ and $F_{1000}(x)$ for a particular choice of $F_1(x)$. This particular function, the second double exponential distribution, has the two special properties that the shape of all the

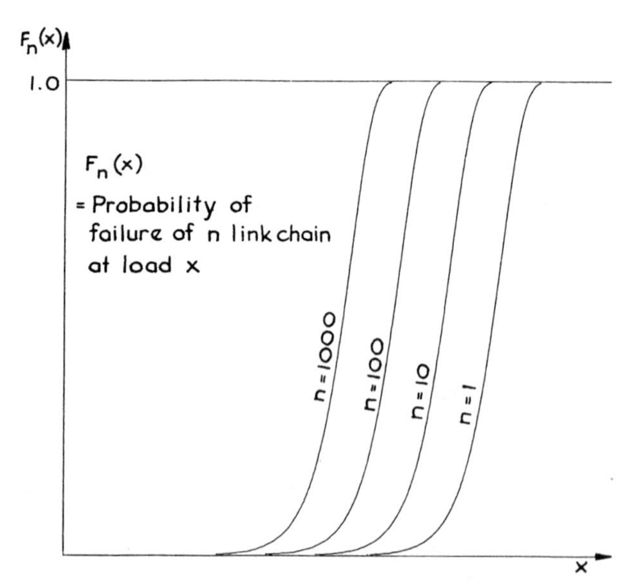

FIG. 1. Second double exponential distribution.

curves is the same and that they are shifted leftwards by an amount proportional to log n. In fact the first property implies the other, see Gumbel (1958). The same properties also apply to $f_n(x)$ since $f_n(x) = \mathrm{d}/\mathrm{d}x[F_n(x)]$—see Fig. 2. If any other function is chosen for $F_1(x)$, the shape of the curves will vary for different values of n. Whether or

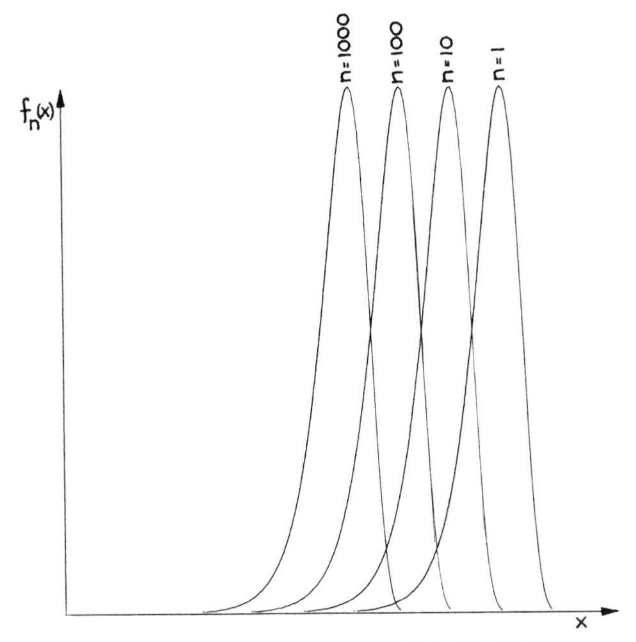

FIG. 2. Second double exponential distribution.

not the second double exponential distribution represents real chain links with any accuracy can only be ascertained from test results.

Looking at the curves one can see that $F_n(x)$ is controlled by the left-hand extreme of $F_1(x)$ and that this becomes more and more pronounced as n increases. This is as one would expect using common sense: one is mainly interested in the distribution of weak links.

Unfortunately, the weak links only represent a small proportion of test results and therefore *it is much more difficult to obtain data about the shape of* $F_1(x)$ *in this region than to obtain overall information such as mean and standard deviation*. When fitting a function to test data, the low results must be well fitted but it does not really matter about the higher results.

Let us now consider the case when the strengths of the individual links are not independent of each other. If we consider the factors which affect link strength such as material impurities and additives, machine adjustment, operative attention, etc., it can be expected that a number of links will be affected at a time.

If the link strengths are not independent we can no longer use eqn.

(A8) to calculate the probability that a chain will not fail. Instead we have to use an extension of eqn. (A7):

$$P\begin{pmatrix}n\text{-link chain does}\\ \text{not fail at load } x\end{pmatrix} = P\begin{pmatrix}\text{1st link does not}\\ \text{fail at load } x\end{pmatrix} \times$$

$$\times\ P\begin{pmatrix}\text{2nd link does not fail given}\\ \text{that 1st link does not fail}\end{pmatrix}$$

$$\times \ldots \times\ P\begin{pmatrix}i\text{th link does not fail given that}\\ \text{1st } (i-1) \text{ links do not fail}\end{pmatrix}$$

$$\times \ldots \times\ P\begin{pmatrix}n\text{th link does not fail given that}\\ \text{1st } (n-1) \text{ links do not fail}\end{pmatrix} \tag{3}$$

In order to calculate the individual probabilities on the right-hand side, one requires data on the interdependence of link strengths. If the interdependence is only over a short range, or in other words the expected strength of a link is only influenced by the strength of a small number of links immediately preceding it, then the probabilities on the right-hand side of eqn. (3) will quickly tend to a constant value. Thus if interdependence is over a range of m links:

$$P\begin{pmatrix}n\text{-link chain does not}\\ \text{fail at load } x\end{pmatrix} = P\begin{pmatrix}\text{1st link does not fail}\\ \text{at load } x\end{pmatrix} \times$$

$$\times\ P\begin{pmatrix}\text{2nd link does not fail}\\ \text{given that 1st link}\\ \text{does not fail}\end{pmatrix}$$

$$\times\ P\begin{pmatrix}(m-1\text{th}) \text{ link does not fail}\\ \text{given that the first}\\ (m-2) \text{ links do not fail}\end{pmatrix}$$

$$\times\ P\begin{pmatrix}i\text{th link does not fail given}\\ \text{that the previous } (m-1)\\ \text{links do not fail}\end{pmatrix}^{n-m+1} \tag{4}$$

A cable or a bar can be thought of as being made up of a large number of short elements of cable or bar fitted end to end. This is exactly analogous to the chain problem. Small lengths of bar are normally tested and if the assumption of independence of element strengths is made then eqn. (2) could be used to find the probability that a long bar will fail.

This will predict a mean strength for long bars which is much lower than the mean strength of test specimens. However, it is more normal to assume that the mean strength of long bars is the same as the mean strength of test specimens. This is equivalent to assuming that there is a very strong interdependence between the strengths of the elements of any one bar.

(b) Load Sharing Systems

There are many types of structure where the total strength is dependent upon the strengths of a number of parts which work together in parallel. For example, the reinforcing bars in a concrete beam or slab work together to carry bending moment.

Let us consider a simple case where the strength of the structure is the straightforward sum of the strengths of n similar parts. This if Y is the strength of the structure and X_1 to X_n the strength of the individual parts:

$$Y = X_1 + X_2 + \dots X_n$$

This is a particular case of eqn. (A31) with $a_0 = 0$ and a_1 to $a_n = 1$. Again we will first assume that the strengths X_1 to X_n are independent and since the individual parts are similar we will assume that they all have the same distribution of strengths with mean μ and standard deviation σ.

By eqns. (A32) and (A33), Y will have mean $n\mu$ and standard deviation $\sqrt{n}\sigma$. In addition if the distribution of X_i is normal, then the distribution of Y will also be normal. If n is large then by the Central Limit Theorem the distribution of Y will be approximately normal even if that of X_i is not. Figure 3 shows an example where X_i has a normal distribution and $n = 2$.

Thus as one would expect, the mean strength of a number of such structures will be the sum of the mean strengths of the n parts contained in each structure. However, the standard deviation of the strength of the structures will only be $\sqrt{n}$ times the standard deviation of the strength of the parts.

In discussing safety factors we are interested in possible variations in strength as a proportion of a representative or 'characteristic' strength. The magnitude of possible variations in strength is proportional to the standard deviation ($\sqrt{n}\,\sigma$) while the characteristic strength will be defined by some formula involving the mean strength and probably the standard deviation and will therefore be a function of n and $\sqrt{n}$.

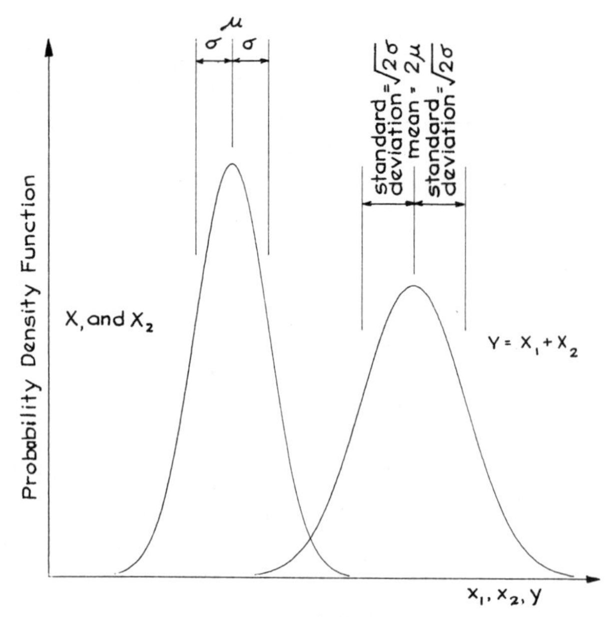

FIG. 3. Normal distribution.

However, the term containing n will tend to dominate and we can therefore say that:

$$\frac{\text{Variations in strength of structure}}{\text{Characteristic strength of structure}}$$

are approximately proportional to $1/\sqrt{n}$ where n is the number of parts sharing the load.

Now consider the case where a very large sample of parts have strengths with mean, μ, and standard deviation, σ, but where a smaller sample of n parts tends to have less variation in strength and therefore the strengths cannot be considered independent. This could happen when the variations in strength have a long range compared to n—see Fig. 4.

A structure whose strength is the sum of the strengths of n parts will now have a strength which is approximately equal to n times the strength of any one of its parts. The mean and standard deviation of the strengths of a number of such structures will therefore be $n\mu$ and $n\sigma$. The ratio

$$\frac{\text{Variations in strength of structure}}{\text{Characteristic strength of structure}}$$

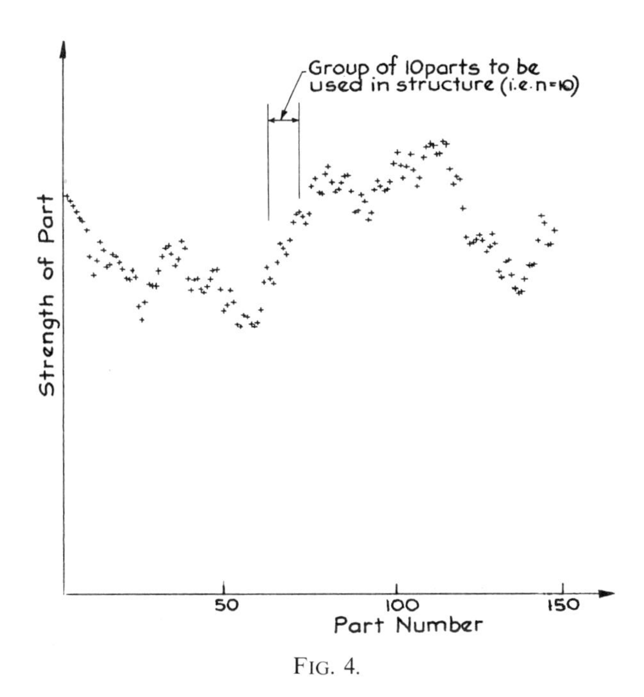

FIG. 4.

will therefore be largely independent of n. We can therefore say that load-sharing systems are only effective in reducing uncertainty as to structure strength if the strengths of the parts of the structure are largely independent. If they are independent, the possible variations in strength of the structure will be roughly proportional to the characteristic strength of the structure divided by the square root of the number of parts sharing the load.

(c) Variation of Loading with Time and the Concept of Return Period

Most types of load are not constant with time. This especially applies to loads due to climatic factors such as wind and snow. It is therefore necessary to be able to answer a question such as 'What is the probability that a wind speed, x, will be equalled or exceeded in a period of N years?'. We will not concern ourselves here with how the wind speed is measured or the length of time over which gust speeds are averaged. Assuming sufficient records are available we could construct a histogram plotting the number of years with maximum wind speed in a number of ranges—see Fig. 5. A probability density function, $f(x)$, and distribution

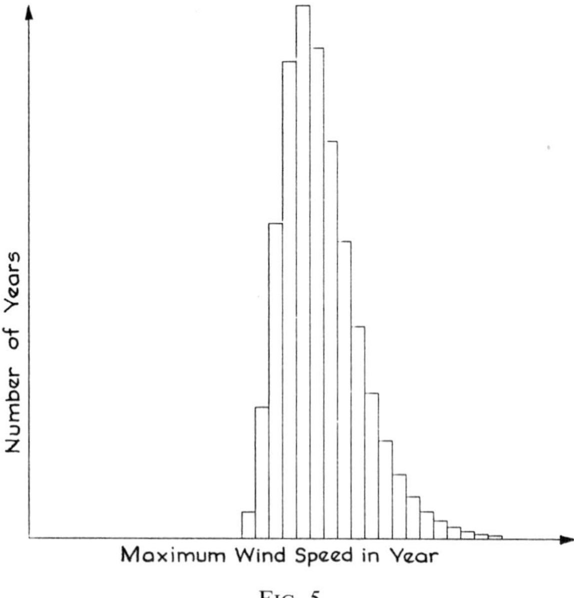

FIG. 5.

function, $F(x)$, can be derived from the histogram. If X is the maximum wind speed in any one year, then:

$$P(X < x) = F(x)$$

The *return period* of a certain wind speed, x, is written $T(x)$ and is defined to be the average number of years between years whose maximum wind speed equals or exceeds x. Note that this is not the same as the average time between winds equalling or exceeding x since if a number of winds equal or exceed x in any one year, they are only counted as one. Because of this the date at which one year finishes and the next begins should preferably not be in a period of high winds or one severe storm could be counted twice instead of once. The wind year should therefore not start on 1 January if this is normally a period of high winds.

The probability that the maximum wind speed in any one year is less than x is equal to $F(x)$. Therefore the probability that the maximum wind speed is greater than or equal to x is $1-F(x)$. Thus in a period of n years, where n is large, we would expect the wind speed x to have been equalled or exceeded in $n \times [1 - F(x)]$ years. The average number of years between years in which the wind speed x is equalled or exceeded is then:

$$\frac{n}{n \times [1 - F(x)]}$$

Thus the return period, $T(x)$, is given by:

$$T(x) = \frac{1}{1 - F(x)} \tag{5}$$

Thus if a certain wind speed has a 50-year return period the probability that it will be equalled or exceeded in any one year is:

$$1 - F(x) = \frac{1}{T(x)} = \frac{1}{50}$$

So far we have not had to make the assumption that the maximum wind speeds in successive years are independent of each other. However, to calculate the probability of a wind speed being equalled or exceeded in a period of n years we either need to make this assumption or have some data concerning the interdependence of wind speeds in successive years. The assumption of independence is normally made in which case the probability of the wind speed not equalling or exceeding x in a period of n years is equal to:

$$F_n(x) = [F_1(x)]^n \text{ where } F_1(x) = F(x) \tag{6}$$

Equation (6) is of very similar form to eqn. (2) applied to the strength of a chain.

In general $F_n(x)$ will have a different shape to $F_1(x)$ and therefore $f_n(x)$ has a different shape to $f_1(x)$. However, there is a special case, the first double exponential distribution, where the shape does not change, the graphs merely move to the right as n increases by an amount proportional to log n. The first double exponential distribution applies to largest values in exactly the same way that the second double exponential distribution applies to smallest values. They are simply mirror images of each other. The first double exponential distribution has a number of alternative names including extreme type 1, Fisher–Tippet Type 1, and Gumbel and is often used for the prediction of maximum wind speeds and snowfalls. Figures 6 and 7 show $F_n(x)$ and $f_n(x)$, for the first double exponential distribution with $n = 1$, 10, 100, and 1000.

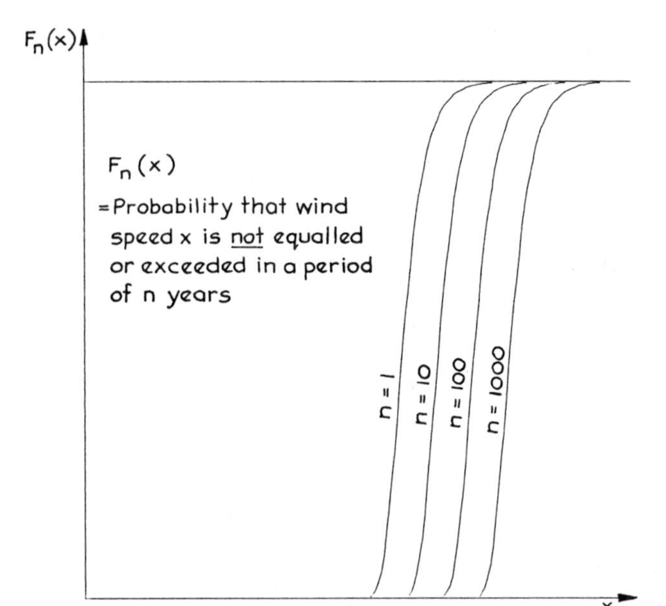

FIG. 6. First double exponential distribution.

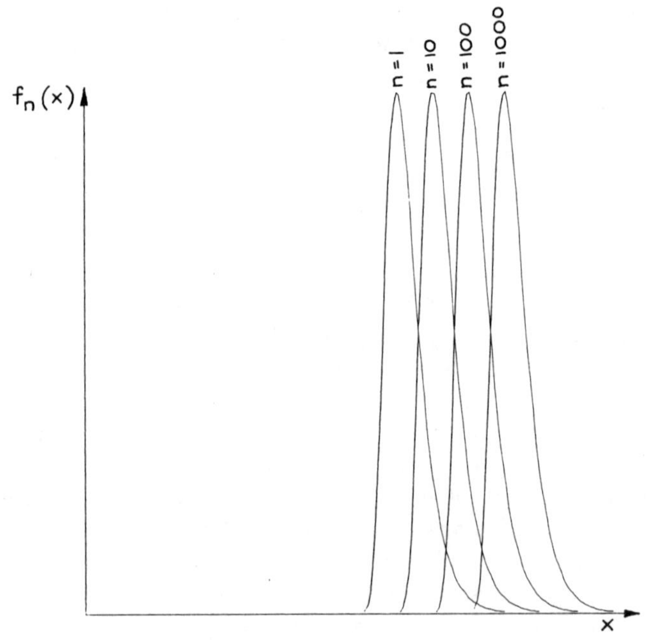

FIG. 7. First double exponential distribution.

$F_n(x)$ is the probability that the wind speed x is not equalled or exceeded in a period of n years and therefore the probability that the wind speed x is equalled or exceeded at least once in a period of n years is equal to:

$$\begin{aligned} 1 - F_n(x) &= 1 - [F_1(x)]^n = 1 - [F(x)]^n \\ &= 1 - \left[1 - \frac{1}{T(x)}\right]^n \\ &= 1 - \left[\left(1 - \frac{1}{T(x)}\right)^{T(x)}\right]^{n/T(x)} \qquad (7) \\ &= 1 - e^{-n/T(x)} \text{ if } T(x) \text{ is greater than about 20} \\ &= 0{\cdot}63 \text{ if } n = T(x) \end{aligned}$$

Thus the probability that a wind speed whose return period is n years will be equalled or exceeded at least once in a period of n years is equal to 0·63 if n is not too small.

(d) Floor Loadings

The chain and maximum wind speed are analogous in that in each case we are interested in the extreme value of a variable in a 'one-dimensional space'. In the case of the chain we required the strength of the weakest link in a line of links and in the case of wind speed we required the maximum wind speed in a line of years.

The load on a column or beam supporting an area of floor is similarly analogous to a load-sharing system in that the load on an element is the sum of the loads over a certain area just as the strength of a load-sharing system is the sum of the strengths of its parts. In sub-section (b) it was shown that the uncertainty in strength divided by total strength of a structure with n load-sharing parts is proportional to $1/\sqrt{n}$ if the strengths of the parts are independent of each other. If the strengths of the parts are not independent of each other then there is less reduction in uncertainty and if the strengths of the parts are all equal there is no reduction in uncertainty.

It has long been known that a similar relationship holds for floor loadings especially live load, and most standards and codes specify a base load which is increased for small areas (often by means of a point load) and reduced for large areas. The loading on small areas is often increased for another reason. A small area of a bridge (for example) can support the entire axle load from a lorry, but it is physically impossible for every such small area to be simultaneously supporting an axle.

The major problem in assessing floor loads is in predicting the exact use to which an area will be put. If this is known other similar areas can be analysed statistically and the probability of a given load being exceeded can be calculated.

(*e*) *Probability of Collapse*

The probability of collapse or some other failure of a structure will depend on the values taken by a number of random variables including material strengths, dimensions, and loads. In order to establish the probability of failure it is necessary to know whether the structure will fail when these random variables have particular values. This can be expressed by introducing a function:

$$Y = t(X_1 \dots X_n) \tag{8}$$

where $X_1 \dots X_n$ are the random variables expressing strengths, loads, etc., and Y is a new random variable. The function $t(X_1 \dots X_n)$ is normally chosen such that $Y < 0$ means failure of the structure.

For example, in the case of a slender pin-ended column (where effects of yielding, initial bow, etc., can be neglected) the function:

$$Y = \frac{\pi^2 EI}{L^2} - P = t(E, I, L, P)$$

will give $Y < 0$ if the random variables E, I, L, and P (= load on column) are such that the column will fail. The function is not unique and the function:

$$Y = \pi^2 EI - PL^2 = t(E, I, L, P)$$

would do equally well.

Very often the relationship between the values of the random variables and whether the structure will fail is not known with precision. In this case extra random variables can be introduced to try and account for the lack of precision. For example in the case of the slender pin-ended column the function:

$$Y = \frac{Q\pi^2 EI}{L^2} - P = t(Q, E, I, L, P)$$

could be used where the random variable Q is intended to express the possible errors of using the Euler formula.

The probability of failure is equal to the probability that $Y < 0$. Hence from eqn. (A27):

$$P(\text{failure}) = P(Y < 0) = \int_{t(x_1 \ldots x_n) < 0} \ldots \int f(x_1 \ldots x_n)\, dx_1 \ldots dx_n \tag{9}$$

where $f(x_1 \ldots x_n)$ is the probability density function of $X_1 \ldots X_n$ and the integration is carried out over the region where $t(x_1 \ldots x_n) < 0$.

This equation looks complicated, but the basic idea is very simple. Certain combinations of the random variables $X_1 \ldots X_n$ will lead to failure, others will not, and this is expressed by whether $t(x_1 \ldots x_n) < 0$. Certain combinations of values of $X_1 \ldots X_n$ are more likely than others and this is expressed by $f(x_1 \ldots x_n)$ as can be seen from eqn. (A15).

The integration in eqn. (9) is simply adding together the probabilities of getting values of $X_1 \ldots X_n$ which correspond to failure. Equation (9) gives the 'exact' probability of failure and makes no assumption regarding independence of the random variables $X_1 \ldots X_n$.

(f) Effect of Duration of Loading

To illustrate the effect of duration of loading we will consider the case of a structure with constant strength, S, subject to a load, W, which is independent of S but varies with time. Let:

$$P(S < s) = G(s)$$
$$P(W_1 < w) = H_1(w)$$

where $G(s)$ and $H_1(w)$ are the distribution functions for S and W_1. W_1 is the random variable expressing the maximum load in any one year. The corresponding probability density functions are $g(s)$ and $h_1(w)$. Assuming that the maximum values of load from year to year are independent, the probability that the maximum load, W_n, in a period of n years is less than w is:

$$P(W_n < w) = H_n(w) = [H_1(w)]^n$$

Therefore the probability that the structure does not collapse in a period of n years is equal to:

$$P(\text{no collapse in } n \text{ years}) = P(W_n < S) = \iint_{w < s} g(s) h_n(w)\, dw\, ds$$

where $h_n(w)$ is the density function corresponding to $H_n(w)$ and the integration is taken over the area where $w<s$. If S and W are limited to positive values the area of integration is that shaded on Fig. 8. Therefore:

$$P(\text{no collapse in } n \text{ years}) = \int_0^\infty \left[\int_0^s g(s) h_n(w)\, \mathrm{d}w \right] \mathrm{d}s$$

$$= \int_0^\infty g(s) H_n(s)\, \mathrm{d}s$$

$$= \int_0^\infty g(s) [H_1(s)]^n\, \mathrm{d}s$$

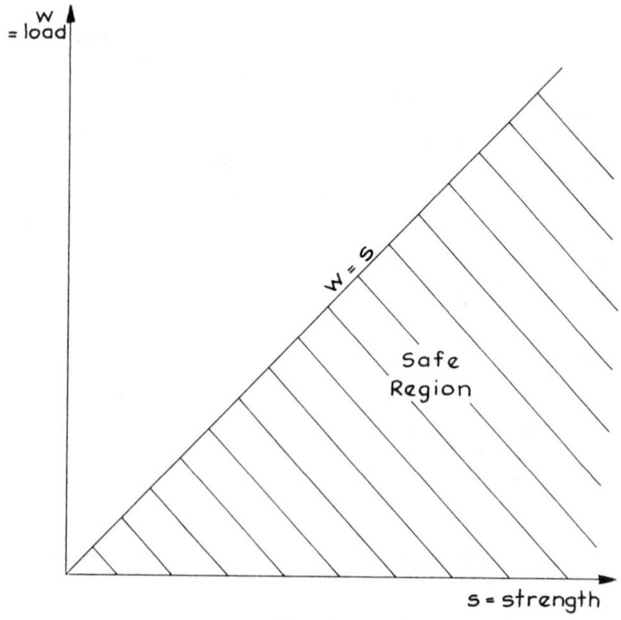

FIG. 8.

From eqn. (A7):

P(no collapse in $n+1$ years) $= P$(no collapse in n years $\cap$ no collapse in $(n+1)$th year)

$$= P(\text{no collapse in } n \text{ years}) P \left[\begin{array}{l} \text{no collapse in } (n+1)\text{th year} \\ \text{given that there was no} \\ \text{collapse in the previous } n \\ \text{years} \end{array} \right]$$

Therefore:

$$P\left[\begin{array}{l}\text{no collapse in } (n+1)\text{th year given} \\ \text{that there was no collapse in the} \\ \text{previous } n \text{ years}\end{array}\right] = \frac{P(\text{no collapse in } n+1 \text{ years})}{P(\text{no collapse in } n \text{ years})}$$

$$= \frac{\int_0^\infty g(s)[H_1(s)]^{n+1}\,\mathrm{d}s}{\int_0^\infty g(s)[H_1(s)]^n \mathrm{d}s} \qquad (10)$$

This expression will always be less than 1·0, but will tend to increase as n increases, depending on the form of the functions g and H_1. Engineers often have to assess the safety of existing structures and the above discussion gives a justification for including the fact that a structure is still standing in the assessment. This should only be used if the structure is not getting weaker with time.

2.2.3 Mean Value First Order Second Moment Method

In sub-section (e) above the random variable Y (eqn. (8)) was introduced such that the probability of failure is the probability that $Y < 0$. This probability is equal to the shaded area on Fig. 9. This represents the 'exact' probability of failure. However, the probability density function $f(x_1 \ldots x_n)$ of the random variables $X_1 \ldots X_n$ representing material strengths, loads, etc., is rarely known with precision and even if it is the evaluation of the integral (eqn. (9)) is usually very difficult.

The ratio of the mean value of $Y(=\mu_y)$ to the standard deviation of $Y(=\sigma_y)$ gives a measure of the reliability of a structure and can be estimated more easily. The ratio μ_y/σ_y is called the reliability index or safety index and is given the symbol β. It is the reciprocal of the coefficient of variation of Y.

The value of the probability of failure depends only on the value of β and the shape of the probability density function for Y. An increase in β corresponds to the curve in Fig. 9 moving to the right or getting taller and thinner. Therefore, for a particular type of probability density function, an increase in β corresponds to a decrease in the probability of failure. Table 1 gives the relationship between β and the probability of failure for the case of the normal distribution.

If the type of probability density function is not known, the values of β can still be used to compare the safety of two situations provided that it is thought that the type of density function is the same in each case. Then if the two situations have the same value of β, they both have the same

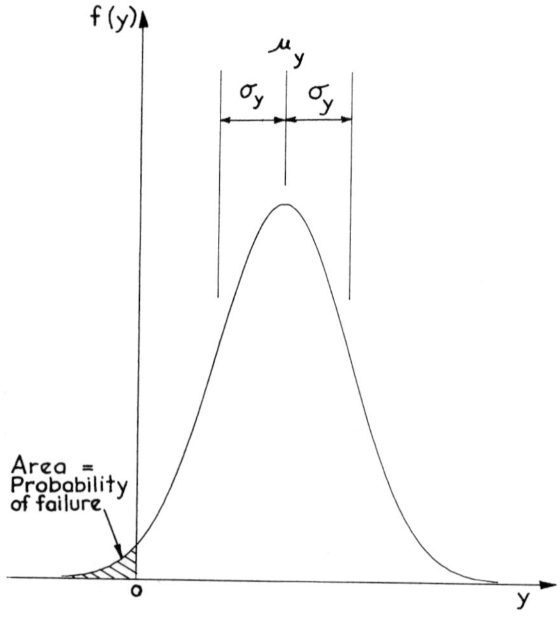

FIG. 9.

TABLE 1
RELATIONSHIP BETWEEN P (FAILURE) AND β FOR THE NORMAL DISTRIBUTION

β	*Probability of failure*
0	0·5
0·5	0·31
1·0	0·16
2·0	0·023
3·0	0·0013

probability of failure even though the actual value of the probability of failure is unknown. This process of comparing the safety of different situations or the same situation using different safety factor formats is called *calibration.*

The value of β can be estimated in terms of the means and standard deviations of the random variables $X_1 \ldots X_n$ using the *mean value first order second moment method*. The mean and standard deviation (i.e. the square root of the variance) of Y are given by eqns. (A29) and (A30). However, in the special case where Y is a linear function of $X_1 \ldots X_n$ such as that given by eqn. (A31), eqns. (A32) and (A33) can be used. If $X_1 \ldots X_n$ are independent, the covariances σ_{ij} are zero and eqn. (A33) can be replaced by (A34).

Thus if the failure condition can be expressed in the form:

$$\text{Structure fails if } Y = t(X_1 \ldots X_n) = a_0 + \sum_{i=1}^{n} a_i X_i < 0 \tag{11}$$

and if $X_i \ldots X_n$ are independent then:

$$\beta = \frac{\mu_y}{\sigma_y} = \frac{a_0 + \sum_{i=1}^{n} a_i \mu_i}{\left[\sum_{i=1}^{n} a_i^2 \sigma_i^2 \right]^{1/2}} \tag{12}$$

where μ_i and σ_i are the mean standard deviation of X_i.

Most failure conditions cannot be expressed in a purely linear form. However, they can be approximated by taking only the linear terms of the Taylor series:

$$t(x_1 \ldots x_n) = t(b_1 \ldots b_n) + \sum_{i=1}^{n} (x_i - b_i) \frac{\partial t}{\partial x_i} + \frac{1}{2!} \sum_{i=1}^{n} \sum_{j=1}^{n} (x_i - b_i)(x_j - b_j) \frac{\partial^2 t}{\partial x_i \partial x_j} + \ldots. \tag{13}$$

where the partial derivatives are evaluated at $x_i = b_i$.

If eqns. (11) and (13) are compared, it can be seen that:

$$a_0 = t - \sum_{i=1}^{n} b_i \frac{\partial t}{\partial x_i} \text{ and } a_i = \frac{\partial t}{\partial x_i} \text{ evaluated at } x_i = b_i$$

Equation (12) can be rewritten:

$$\beta = \frac{\mu_y}{\sigma_y} = \frac{t + \sum_{i=1}^{n} (\mu_i - b_i) \frac{\partial t}{\partial x_i}}{\left[\sum_{i=1}^{n} \left(\sigma_i \frac{\partial t}{\partial x_i} \right)^2 \right]^{1/2}} \tag{14}$$

The value of β given by eqn. (14) will be exact if the function $t(X_1 \ldots X_n)$ is linear but will otherwise depend on the values of b_i chosen to evaluate t and $\partial t/\partial x_i$. In the mean value first order second moment method b_i is chosen as μ_i, the mean of X_i. This has the effect of removing the summation term from the numerator in eqn. (14).

The equation $t(x_1 \ldots x_n)$ is a hypersurface (a surface in n-dimensional space) separating the regions corresponding to values of $X_1 \ldots X_n$ which lead to failure from those which do not. The probability of failure is the probability of getting values of $X_1 \ldots X_n$ in the failure region. The approximation introduced by the mean value first order second moment method is to replace the true hypersurface by a hyperplane. If b_i is chosen as values different to μ_i, a different hyperplane results and CIRIA Report 63 (1976) discusses the accuracy of various hyperplanes.

The LRFD (Load and Resistance Factor Design) method is based on the mean value first order second moment method and expresses the failure condition in the form:

$$\text{Structure fails if } \log_e\left[\frac{R}{Q}\right] < 0 \tag{15}$$

where R is the resistance of the element or structure and Q is the load effect. R and Q are random variables which are functions of more basic random variables such as material strengths, dimensions, and loads.

If we write $R = X_1$ and $Q = X_2$ the terms in eqn. (14) become:

$b_1 = \mu_1 = R_m$, the mean value of R
$b_2 = \mu_2 = Q_m$, the mean value of Q
$\sigma_1 = \sigma_R$, the standard deviation of R
$\sigma_2 = \sigma_Q$, the standard deviation of Q

$$t = \log_e\left[\frac{R_m}{Q_m}\right]$$

$$\frac{\partial t}{\partial x_1} = \frac{1}{R_m}$$

$$\frac{\partial t}{\partial x_2} = -\frac{1}{Q_m}$$

$$\sigma_1 \frac{\partial t}{\partial x_1} = \frac{\sigma_R}{R_m} = V_R, \text{ the coefficient of variation of } R$$

$$\sigma_2 \frac{\partial t}{\partial x_2} = -\frac{\sigma_Q}{Q_m} = -V_Q, (-1) \times \text{the coefficient of variation of } Q.$$

If these quantities are substituted into eqn. (14), the LRFD expression for

β results:

$$\beta = \frac{\log_e \left[\frac{R_m}{Q_m} \right]}{\sqrt{V_R^2 + V_Q^2}}. \tag{16}$$

2.3 OTHER FACTORS INFLUENCING SAFETY

It is evident from examination of case histories that a majority of structural failures occur because of causes other than inadequacies in the margins of safety, or safety factors, incorporated in the Codes of Practice. It has been estimated, for instance, that the total risk may be of the order of ten times the probability of failure calculated by reliability theory. Control of structural safety is therefore not entirely within the jurisdiction of the Codes of Practice and those who employ their recommendations, but is largely influenced by the quality of the implementation and control procedures associated with the project.

It is a common view that the purpose of an investigation into an accident is to determine its 'cause'. But those who have taken part in more than a few such investigations of structural accidents are well aware that the apportionment of blame, whether applied to designer, material, or to user is in a human society much less effective as a deterrent to similar accidents than the appreciation of the broader factors producing the 'climate' in which the accident was set. This concept of climate was introduced by Sir Alfred Pugsley in *The Safety of Structures* and developed in a paper (1973) on proneness to structural accidents. The climate referred to includes factors such as the designers' knowledge of the materials and structural behaviour, the organisation and communication within both the design team and the construction team, financial and political pressures. Pugsley suggested that by listing the parameters defining the Engineering Climate of the Project and by assessing each as say, good, average, or bad it would be possible to make a systematic analysis of proneness to structural failure.

This work was taken up by Blockley (1977) who proposed eight categories of basic causes of structural failure as follows.

1. Structures, the behaviour of which is reasonably well understood by the designers, but which fail because random extremely high value of load or extremely low value of strength occurs.

2. Structures which fail due to being overloaded or to being understrength but where the behaviour of the structure is poorly understood by the designers and the system errors in the calculation procedural model are as large as the random errors in the parameters describing the model.
3. Structural failures where some independent random hazard is the cause, e.g. earthquake.
4. Failures which occur because the designers do not allow for some basic mode of behaviour inadequately understood by existing technology.
5. Failures which occur because the designers fail to allow for some basic mode of behaviour well understood by existing technology.
6. Failures which occur through an error during construction.
7. Failures which occur in a deteriorating climate surrounding the whole project, resulting in pressures on the personnel involved; pressures may be of a financial, political, or industrial nature.
8. Failures which occur because of a misuse of a structure or because the owners of the structure have not realised the critical nature of certain factors.

He assessed 23 past structural failures from reports of detailed inquiries by giving a grading to the confidence in and relative importance of a number of statements describing the state of each of the above parameters. In all cases there was more than one primary cause, the best being 2, the average 4 or 5, and the worst 13. The most frequently observed cause was errors in the calculation model followed by a likelihood of construction errors, inadequate contractors' site staff, and inadequate site supervision.

Whether or not these methods can be used for prediction is debatable. One disadvantage is that the list of categories and grades of assessment would vary according to the views of the assessor. It is clear from the categories listed above that only those causes in categories (1) and (3) are amenable to probabilistic treatment, and hence to formal safety factor treatment. It should be noted that the partial factor γ_{s3} is not intended to allow for inadequate understanding of the structural behaviour or for errors, but only for simplifications in the analytical model.

The benefit of such studies is that they highlight the importance of the 'Engineering Climate' particularly control, communication, and checking within the design and construction teams, even under financial or time pressures.

2.4 CONCLUSIONS

Structural codes are currently undergoing a change in format from working stresses to limit state and partial factors. It is possible that by increasing the complexity of codes an extra category could be added to the causes of failure, that of misinterpretation of the code principles or rules.

There clearly are advantages in the logic of the limit state code format. The application of the new codes should allow simple structures to be designed simply, while with extra calculation effort special repetitive structures can be designed to the greatest economy with consistent levels of safety.

One of the first tasks towards avoiding misinterpretation is to use consistent nomenclature and consistent methods of defining loads and strengths. ISO-2394 went a long way towards this, and although it may require some amendments and revisions we believe that it is a sound document. In our survey for this chapter we found that the nomenclature and symbols were often disregarded. LRFD is a particular case of disguising the ISO limit state format and adding the perfectly reasonable concept of the safety index. By not presenting their proposals within the context of the ISO document the authors of the LRFD papers add another layer to the confusion inherent in the change.

It is clear that even simplified probability methods will not come into general use for design calculations. Nevertheless the semi-probabilistic format allows loads and material strengths to be defined by probability theory. The application of the Reliability Index should help code committees select appropriate values of load and material factors.

REFERENCES

ALLEN, D. E. (1975). Limits States design, a probabilistic study, *Can. J. Civ. Engg.*, **2**, 36.

AMERICAN CONCRETE INSTITUTE (1977). Building Code Requirements for Reinforced Concrete (ACI 318/77).

BLOCKLEY, D. I. (1977). Analysis of Structural failures, *Proc. Instn. Civ. Engrs.* Part I.

BRITISH STANDARDS INSTITUTION (1972). The Structural Use of Concrete CP110: Part 1.

BRITISH STANDARDS INSTITUTION (1978). The Structural Use of Masonry BS 5628, Part 1.

COMITE EUROPEAN DU BETON/FEDERATION INTERNATIONALE DE LA PRECONTRAINTE (1970). International recommendations for the design and construction of Concrete Structures. Cement & Concrete Association, London.

CONSTRUCTION INDUSTRY RESEARCH and INFORMATION ASSOCIATION (1976). Rationalisation of Safety and serviceability factors in Structural Codes. Report No 63.

CONSTRUCTION INDUSTRY RESEARCH and INFORMATION ASSOCIATION (1976). Structural Codes—The rationalisation of Safety and serviceability factors. Proceedings of the Seminar.

CORNELL, C. A. (1969). Structural Safety Specifications based on second moment reliability Analysis. IABSE Symposium on Concepts of Safety of Structures and Methods of Design, London.

GHIOCEL, D. and LUNGU, D. (1975). *Wind, snow and temperature effects in structures based on probability*, Abacus Press, Tunbridge Wells.

GUMBEL, E. M. (1958). *Statistics of extremes*, Columbia University Press, New York.

INTERNATIONAL STANDARDS ORGANISATION (1973). General principles for the verification of the safety of structures, ISO 2394.

PUGSLEY, SIR ALFRED (1965). *The Safety of Structures*, Edward Arnold, London.

PUGSLEY, SIR A. (1973). The prediction of proneness to Structural Accidents, *Structural Engineer*, London.

RAVINDRA, M. K. and GALAMBOS, T. V. (1978). Load and Resistance Factor Design for Steel, *Journal of the Structural Division, ASCE.*

APPENDIX—PROBABILITY THEORY

A1 Introduction

Probability theory is used to estimate the probability or likelihood of an event occurring. Any event can be thought of as the outcome of an 'experiment'. If we consider that a certain event has probability, p, it means that we would expect it to occur roughly pN times if we were to perform an experiment N times where N is a reasonably large number. The event may or may not occur in each experiment and therefore the total number of events must lie between 0 and N. Thus the probability, p, must be between 0 and 1. Very often an experiment is to be performed only once in which case we have to consider the hypothetical case of performing the experiment a large number of times.

It could be argued that as more knowledge is obtained about a proposed experiment, the probability ascribed to a particular outcome should tend either to 0 or 1. Certainly most engineering failures could have been predicted; at least hindsight usually shows that there were signs predicting the failure but that they went un-noticed.

Whether or not probability has any true meaning is a question to be answered in the realms of physics and metaphysics. This question does not concern us here. Probability theory is a useful tool for making quantitative predictions about the future.

A2 Addition of Probabilities

Very often we are interested in the probability of two or more events occurring simultaneously, in which case we need rules for combining probabilities. The rules are stated here in terms of two events, but can be extended to any number.

Consider an experiment in which there are four possible outcomes:

1. Event A only occurs
2. Event B only occurs
3. Events A and B both occur
4. Neither event A nor event B occurs

Then:

$$P(A \cap \hat{B}) + P(\hat{A} \cap B) + P(A \cap B) + P(\hat{A} \cap \hat{B}) = 1 \quad \text{(A1)}$$

Where P(event) = the probability of the event occurring, $\hat{B}$ means event B *not* occurring and the symbol $\cap$ means 'and'.

We may be interested in the probability that event A occurs with or without event B. Event A occurs in both outcomes (1) and (3) and hence:

$$P(A) = P(A \cap \hat{B}) + P(A \cap B) \quad \text{(A2)}$$

Similarly:

$$P(B) = P(\hat{A} \cap B) + P(A \cap B) \quad \text{(A3)}$$

On the other hand we may require the probability that one or both of events A and B occur. This happens in outcomes (1), (2), and (3) and therefore:

$$P(A \cup B) = P(A \cap \hat{B}) + P(\hat{A} \cap B) + P(A \cap B) \quad \text{(A4)}$$

Where $A \cup B$ means 'A and/or B'.
Thus combining eqns. (A2), (A3) and (A4):

$$P(A \cup B) = P(A) + P(B) - P(A \cap B) \quad \text{(A5)}$$

Finally, since any event must occur or not occur:

$$P(A) + P(\hat{A}) = 1 \quad \text{(A6)}$$

A3 Multiplication of Probabilities

Let us first consider the case where the two events A and B which may result from an experiment are not independent. For example, consider the experiment 'take a random selection from the adult population'. Event A could be the result 'man' (in which case event $\hat{A}$ would be 'woman') and event B could be 'height greater than or equal to 1·8 m' (in which case event $\hat{B}$ would be 'height less than 1·8 m').

Events A and B are not independent in that men tend to be taller than women and thus the probability that event B will occur will depend on whether event A does or does not occur. To deal with this we introduce the conditional probability of event B occurring given that event A has occurred. This is written $P(B|A)$. Then:

$$\left.\begin{aligned} P(A \cap B) &= P(A)P(B|A) \\ &= P(B)P(A|B) \end{aligned}\right\} \quad \text{(A7)}$$

On the other hand, if the two events are independent $P(B|A)=P(B)$ and $P(A|B)=P(A)$ thus:

$$P(A \cap B)=P(A)P(B) \quad \text{(A8)}$$

It is very important to note that eqn. (A8) only applies when events A and B are independent.

A4 Random Variables

A random variable is a variable whose value depends on 'chance'. For example the strength, X, of a material will vary from sample to sample in some random manner. However, even though we cannot predict the exact strength of a sample before it is tested, we should be able to assess the probability that X will lie within a given range provided that we have sufficient data from previous tests.

Random variables are usually denoted by capital letters to distinguish them from ordinary variables. A random variable can be continuous or be discrete and limited to certain values, usually integers. We will only consider continuous random variables.

A5 Presentation of Data Distributions

Data from material tests, loading records, etc., have to be presented in a way which expresses the variability of the results in order to estimate the probability of material strength or load reaching certain values in the future. The total range of results is split up into equal intervals and the number of results in each interval is plotted on a graph known as a histogram (see Fig. A1).

The number of results in each interval is somewhat arbitrary since it depends on the total number of results and the range of each interval.

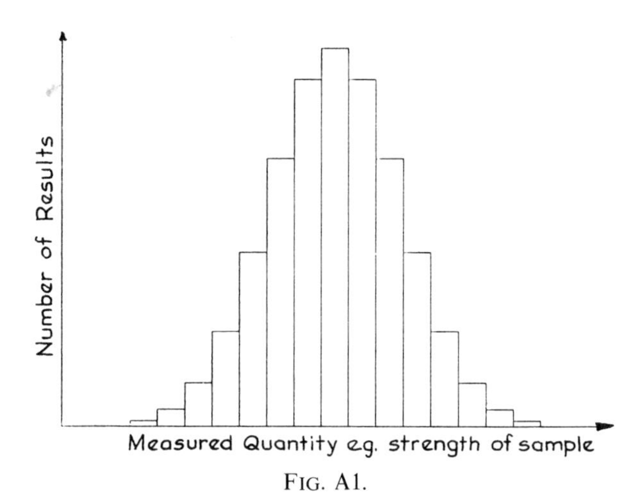

FIG. A1.

This problem is removed by dividing the number of results in each interval by the total number of results multiplied by the interval range. The total area of the blocks of the histogram will then be 1·0.

If a sufficiently large number of results is available and the histogram is plotted with a sufficiently small range for each interval, then the outline of the histogram should tend to a smooth curve such as that shown in Fig. A2. If the random variable is called X, the vertical ordinate is termed $f(x)$, the *probability density function.* Note that the capital letter X is used for the random variable which may take an infinite number of different values (if X is continuous random variable) and the small letter x is an ordinary variable which is used to denote any particular value which X may take.

The curve $f(x)$ usually consists of one 'hump', but not always so and may or may not be symmetrical. The area under the curve is always equal to 1·0, i.e.

$$\int_{-\infty}^{\infty} f(x)\mathrm{d}x = 1{\cdot}0$$

If X is limited to positive values, the lower limit of integration would be 0 instead of $-\infty$. The mean value of X is written μ and is given by:

$$\mu = \int_{-\infty}^{\infty} xf(x)\mathrm{d}x \tag{A9}$$

A little thought shows that this is consistent with the mean of the

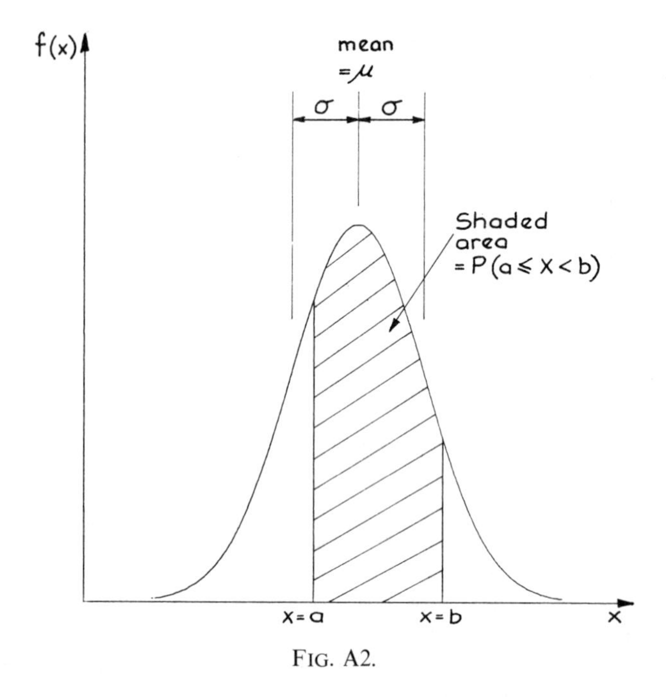

FIG. A2.

original set of (say) N test results X to X. (There may be a slight difference since plotting the results on a histogram does not make full use of the accuracy of each result.)

$$\text{Sample mean} = \frac{1}{N}\sum_{i=1}^{N} x_i$$

If the test results are used to predict the results of further similar experiments then the mean of the results to date is termed the expectation of the future results.

Geometrically, the mean μ represents the horizontal distance of the centroid of the area under the curve from the origin.

The standard deviation, σ, and variance, σ^2, are given by:

$$\sigma^2 = \int_{-\infty}^{\infty} (x-\mu)^2 f(x)\mathrm{d}x \tag{A10}$$

The formula for σ^2 is the same as that for the second moment of area of the area under the curve about a vertical axis through its centroid. The area under the curve equals 1·0 and therefore σ is equal to the 'radius of gyration' of the area under the curve about the same axis.

The mean, μ, 'positions' the curve along the x axis and the standard deviation, σ, defines the 'spread' of the curve. If σ is small the curve is tall and narrow, if σ is large the curve is low and wide, but the area under the curve is always 1·0. The mean and standard deviation do not fully define the curve; curves of very different shapes can share the same mean and standard deviation. The best known probability density function is the normal or Gauss distribution,

$$f(x)=\frac{1}{\sigma\sqrt{2\pi}}\,\mathrm{e}^{-1/2[(x-\mu)/\sigma]^2}$$

The normal distribution fits many types of test results reasonably well, but there are many others which it does not fit and in these cases use of the normal distribution can lead to most misleading predictions.

Once the probability density function has been obtained from a series of test results or possibly from some theoretical considerations, it can be used to predict the results of similar tests in the future. Consider the question 'What is the probability that the result of a future test, X, will lie in the range $a \leqslant X < b$?' In other words, what is the value of $P(a \leqslant X < b)$? If the only indication as to the result of the test is the past results represented by $f(x)$, all we can say is that:

$$P(a \leqslant X < b)=\int_a^b f(x)\mathrm{d}x \qquad \text{(A11)}$$

which equals the shaded area on Fig. A2.

If we adopt the convention:

$$P(X < x)=F(x)=\int_{-\infty}^{x} f(u)\mathrm{d}u \qquad \text{(A12)}$$

then:

$$P(a \leqslant X < b)=F(b)-F(a) \qquad \text{(A13)}$$

If eqn. (A12) is differentiated with respect to x:

$$\frac{\mathrm{d}}{\mathrm{d}x}[F(x)]=f(x) \qquad \text{(A14)}$$

$F(x)$ is called the *distribution function or cumulative distribution function* of the random variable X. $f(x)$ is always positive or zero and therefore $F(x)$ cannot decrease as x increases. The total area under the curve $f(x)$ is equal to 1·0 and therefore $F(x)$ must lie in the range 0 to 1—see Fig. A3.

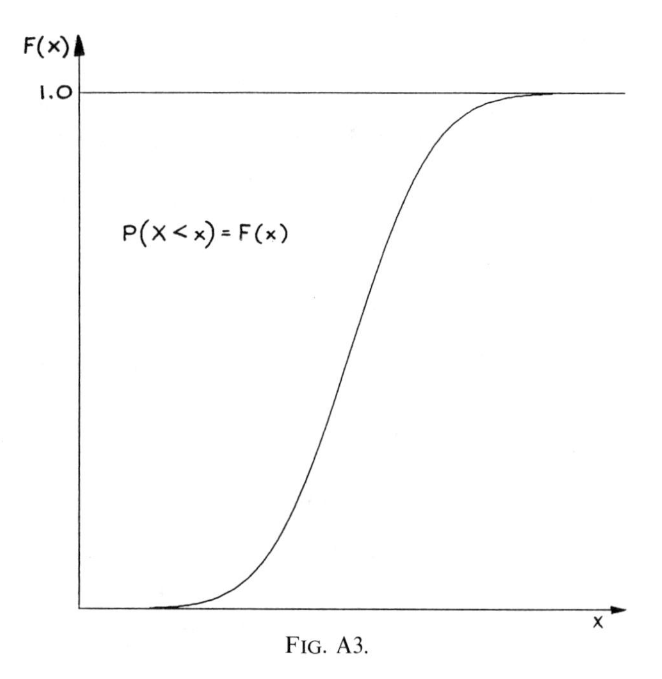

FIG. A3.

A6 Distributions of Several Random Variables

We are very often concerned with the interaction of a number of random variables such as material strengths, dimensions, loads, etc. Equations (A11) and (A12) can be extended to the case of a number of random variables, X_1 to X_n:

$$P(a_1 \leqslant X_1 < b_1 \cap a_2 \leqslant X_2 < b_2 \ldots \cap a_n \leqslant X_n < b_n)$$

$$= \int_{a_n}^{b_n} \cdots \int_{a_2}^{b_2} \int_{a_1}^{b_1} f(x_1, x_2 \ldots x_n)\, dx_1\, dx_2 \ldots dx_n \qquad \text{(A15)}$$

$$P(X_1 < x_1 \cap X_2 < x_2 \ldots \cap X_n < x_n) = F(x_1, x_2 \ldots x_n)$$

$$= \int_{-\infty}^{x_n} \cdots \int_{-\infty}^{x_2} \int_{-\infty}^{x_1} f(u_1, u_2 \ldots u_n)\, du_1\, du_2 \ldots du_n \qquad \text{(A16)}$$

$f(x_1, x_2 \ldots x_n)$ and $F(x_1, x_2 \ldots x_n)$ are again called the probability density function and distribution function. The probability density function can be obtained from test data (if sufficient data are available) by dividing the 'space' $x_1, x_2 \ldots x_n$ into small 'volumes' and counting the number of results in each volume. The distribution function can then be found by integration (at least in theory).

In order for particular values of $X_1, X_2 \ldots X_n$ to form a point in $x_1, x_2 \ldots x_n$ space it is necessary to take the n values together. This obviously makes sense if $X_1, X_2 \ldots X_n$ are dimensions and material properties in which case the properties of one particular sample would be taken together. Similarly weather records would be taken together if they referred to the same place at the same time.

If the random variables $X_1, X_2 \ldots X_n$ can be split into two groups $X_1, X_2 \ldots X_m$ and $X_{m+1}, X_{m+2} \ldots X_n$ such that the variables in each group are independent of the variables in the other group, then the probability density function and distribution function can be rewritten:

$$f(x_1, x_2 \ldots x_n) = g(x_1, x_2 \ldots x_m)\, h\,(x_{m+1}, x_{m+2} \ldots x_n) \qquad \text{(A17)}$$

$$F(x_1, x_2 \ldots x_n) = G(x_1, x_2 \ldots x_m) H(x_{m+1}, x_{m+2} \ldots x_n) \qquad \text{(A18)}$$

This result can be extended to any number of groups of random variables such that the variables in each group are independent of the variables in the other groups.

If eqn. (A16) is differentiated then:

$$\frac{\partial^n[F\,(x_1, x_2 \ldots x_n)]}{\partial x_1\, \partial x_2 \ldots \partial x_n} = f\,(x_1, x_2 \ldots x_n) \qquad \text{(A19)}$$

If $n=2$, the probability density function, $f(x_1, x_2)$, can be plotted as a surface. The surface will usually consist of one 'hill' and may or may not have any axes of symmetry—see Fig. A4. The total volume under the surface is equal to 1·0 and $P(a_1 \leqslant X_1 < b_1 \cap a_2 \leqslant X_2 < b_2)$ is given by the volume shown on Fig. A5.

Sometimes we require the probability that a particular random variable, X_i, will lie in a certain range regardless of the values taken by the other random variables:

$$P(a_i \leqslant X_i < b_i) = P(-\infty < X_1 < \infty \ldots \cap a_i \leqslant X_i < b_i \ldots \cap -\infty < X_n < \infty)$$

$$= \int_{-\infty}^{\infty} \cdots \int_{a_i}^{b_i} \cdots \int_{-\infty}^{\infty} f\,(x_1 \ldots x_i \ldots x_n)\, \mathrm{d}x_1 \ldots \mathrm{d}x_i \ldots \mathrm{d}x_n$$

$$= \int_{a_i}^{b_i} f_i(x_i)\, \mathrm{d}x_i \qquad \text{(A20)}$$

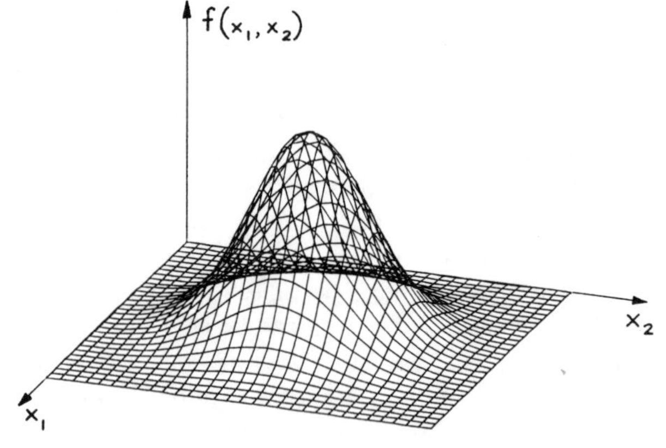

FIG. A4.

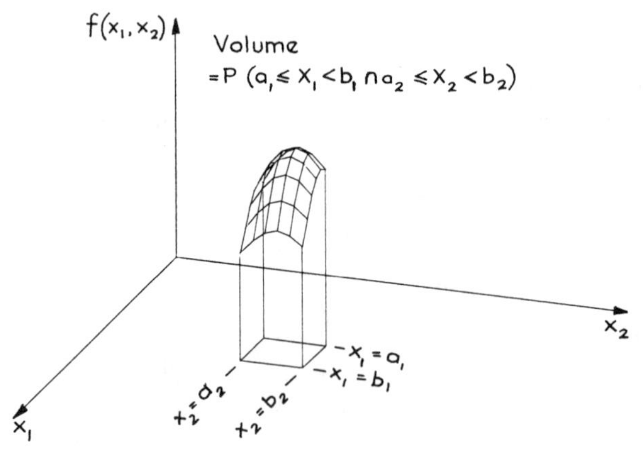

FIG. A5.

Where:

$$f_i(x_i) = \int_{-\infty}^{\infty} \cdots \int_{-\infty}^{\infty} \int_{-\infty}^{\infty} \cdots \int_{-\infty}^{\infty} f(x_1 \ldots x_{i-1}, x_i, x_{i+1} \ldots x_n) \mathrm{d}x_1 \ldots \mathrm{d}x_{i-1}\, \mathrm{d}x_{i+1} \ldots \mathrm{d}x_n \tag{A21}$$

and hence:

$$P(X_i < x_i) = F_i(x_i) = \int_{-\infty}^{x} f_i(u)\, \mathrm{d}u. \tag{A22}$$

$f_i(x_i)$ and $F_i(x_i)$ are the density function and distribution function of the marginal distribution of X_i.

The mean value of X_i or the expectation of future values of X_i can be written μ_i and is given by (c.f. eqn. (A9)):

$$\mu_i = \int_{-\infty}^{\infty} x_i f_i(x_i)\,\mathrm{d}x_i \qquad \text{(A23)}$$

The variance of X_i, σ_i^2, is given by (c.f. eqn. (A10)):

$$\sigma_i^2 = \int_{-\infty}^{\infty} (x_i - \mu_i)^2 f_i(x_i)\,\mathrm{d}x_i \qquad \text{(A24)}$$

Lastly the covariance, σ_{ij}, of the two random variables X_i and X_j is given by:

$$\sigma_{ij} = \int_{-\infty}^{\infty} \cdots \int_{-\infty}^{\infty} (x_i - \mu_i)(x_j - \mu_j) f(x_1 \ldots x_n)\,\mathrm{d}x_1 \ldots \mathrm{d}x_n \qquad \text{(A25)}$$

It can be seen that $\sigma_{ij} = \sigma_{ji}$ and if $j = i$ then $\sigma_{ii} = \sigma_i^2$. If the two random variables X_i and X_j are independent $\sigma_{ij} = 0$, however if $\sigma_{ij} = 0$, X_i and X_j are not necessarily independent.

A.7 Functions of Several Random Variables

A random variable, Y, can be defined as a function of the random variables $X_1 \ldots X_n$:

$$Y = t(X_1 \ldots X_n) \qquad \text{(A26)}$$

The distribution function for Y will be given by:

$$P(Y < y) = F(y) = \underset{t(x_1 \ldots x_n) < y}{\int \cdots \int} f(x_1 \ldots x_n)\,\mathrm{d}x_1 \ldots \mathrm{d}x_n \qquad \text{(A27)}$$

The integration is taken over the region where $t(x_1 \ldots x_n)$ is less than y. The probability density function for Y can then be found by differentiation:

$$f(y) = \frac{\mathrm{d}}{\mathrm{d}y}[F(y)] \qquad \text{(A28)}$$

The function $t(X_1 \ldots X_n)$ may be an ordinary function or may contain conditional statements such as:

$$\frac{X_1}{X_2} - 7.X_3. \text{ (the smallest of } X_2 \text{ to } X_8)$$

The mean (or expectation), μ_y, of Y is:

$$\mu_y = \int_{-\infty}^{\infty} \cdots \int_{-\infty}^{\infty} t(x_1 \ldots x_n) f(x_1 \ldots x_n) \mathrm{d}x_1 \ldots \mathrm{d}x_n \qquad \text{(A29)}$$

and the variance, σ_y^2, of Y is:

$$\sigma_y^2 = \int_{-\infty}^{\infty} \cdots \int_{-\infty}^{\infty} [t(x_1 \ldots x_n) - \mu_y]^2 f(x_1 \ldots x_n) \mathrm{d}x_1 \ldots \mathrm{d}x_n \qquad \text{(A30)}$$

For the special case

$$Y = t(X_1 \ldots X_n) = a_0 + a_1 X_1 \ldots + a_n X_n = a_0 + \sum_{i=1}^{n} a_i X_i \qquad \text{(A31)}$$

where $a_0, a_1 \ldots a_n$ are constants, two results follow:

$$\mu_y = a_0 + a_1 \mu_1 \ldots + a_n \mu_n = a_0 + \sum_{i=1}^{n} a_i \mu_i \qquad \text{(A32)}$$

where μ_i is the mean value of X_i and secondly

$$\sigma_y^2 = \sum_{i=1}^{n} \sum_{j=1}^{n} a_i a_j \sigma_{ij} \qquad \text{(A33)}$$

where σ_{ij} is the covariance of X_i and X_j if $i \neq j$ or the variance of X_i if $j = i$.

If the variables X_1 to X_n are independent $\sigma_{ij} = 0$ if $i \neq j$ and (A33) reduces to

$$\sigma_y^2 = \sum_{i=1}^{n} a_i^2 \sigma_i^2 \qquad \text{(A34)}$$

where σ_i^2 is the variance of X_i.

Chapter 3

BOX AND CYLINDRICAL COLUMNS UNDER BIAXIAL BENDING

W. F. Chen
School of Civil Engineering,
Purdue University, USA

SUMMARY

This chapter summarises the recent achievements in the analysis and design of welded built-up box and fabricated cylindrical steel tubular columns subjected to biaxial bending moment combined with axial compression. These columns contain imperfections that are far more complicated than in the hot-rolled members. The current design codes, such as AISC, however, do not consider these large welded built-up box and fabricated cylindrical tubular members as commonly used in offshore structures and tall building frames.

Four major subjects on the behaviour and strength of large welded built-up box and fabricated cylindrical tubular columns are described. These include:

1. *Welded built-up box columns,*
2. *Fabricated cylindrical columns under axial load,*
3. *Fabricated cylindrical columns under external pressure, and*
4. *Cyclic inelastic buckling of tubes.*

3.1 INTRODUCTION

The reader is assumed to be familiar with the more elementary aspects of the theories of columns and beam-columns and some of the underlying philosophy, principles, or approaches commonly used in the analysis and design of these members. A recent two-volume treatise on *Theory of Beam-Columns* by Chen and Atsuta (1976, 1977) may prove helpful in this respect as an introduction to the field of columns and beam-columns. The two recent review papers on 'Limit States Design of Steel Beam-Columns' by Chen and Cheong-Siat-Moy (1980) and 'Recent Advances on Analysis and Design of Steel Beam-Columns in USA' by Chen (1981) summarise the most advanced state of the art in this area, among others (Chen, 1977, Galambos, 1981), and some of the statements they contain will be repeated here for the sake of continuity of presentation.

The first part of the chapter discusses the column problem to highlight the particular difficulties associated with large welded built-up box and fabricated cylindrical columns. A brief description follows of a rigorous analysis capable of dealing with these types of columns as commonly used in building and offshore framed structures. Studies of the four topics described in the summary using this analysis are then presented. Design methods, based on proposals put forward by these studies, are then given. Comparisons are also made with available tests on these columns. It is concluded that these proposed methods are suitable for adoption in practice.

3.2 THE COLUMN PROBLEM

The use of welded built-up box and fabricated cylindrical tubular columns is a growing phenomenon in structural engineering. Currently, such columns with relatively large sizes are commonly used in offshore structures and building frames, and have increasing applications in various other civil engineering structures. However, these columns contain imperfections that are far more complicated than in the hot-rolled members such as channel, angle, and wide flange, because they involve a more complicated manufacturing and fabricating process than that of the hot-rolled counterparts. For example, fabricated cylindrical tubular columns commonly used in offshore structures are usually fabricated from flat steel plate that is formed into circular elements, such that the width of the plate becomes the height of the element. The ends of the plate are then longitudinally joined by welding. The completed element is known as a 'can' and is generally about 3 m (10 ft) long. By joining a number of cans end to end with circumferential or girth welds, a column of any length may be formed. The longitudinal welds in the cans are staggered, and it is common for the longitudinal weld in one can to be about 180° out of phase with the longitudinal weld in the next can (Fig. 1).

It can be readily appreciated that the extensive forming and welding

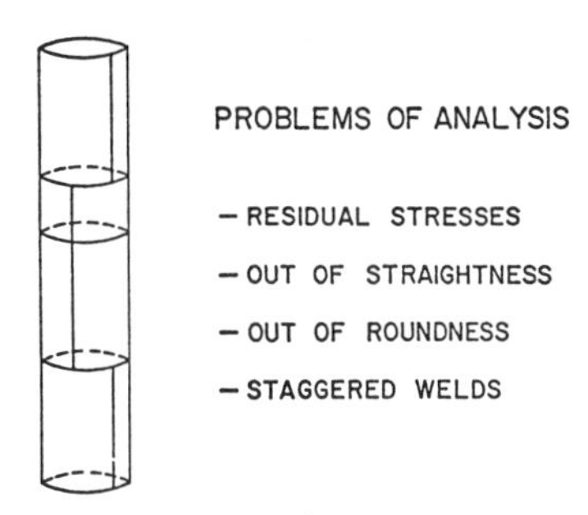

FIG. 1. Fabricated tubular column.

processes used during fabrication of a tubular column by this method introduce major residual stresses into the completed column in at least two directions: circumferential and longitudinal. In addition, it introduces out-of-roundness of the cross-section. Furthermore, the transverse welding of the 'cans' to form a long column results in a significant out-of-straightness of the column.

A realistic design of axially loaded steel tubular columns must, therefore, consider the fact that an actual tubular column is geometrically and materially imperfect, and is also frequently subjected to bending moments resulting from unavoidable end eccentricities and lateral forces. Thus, all fabricated tubular columns must be analysed as beam-columns.

The beam-column analysis can be divided into two steps: (1) The moment–curvature behaviour of a short column; and (2) the load–deflection behaviour of a long column (Chen and Atsuta, 1976). The moment–curvature relationship is of prime importance in the analysis of any long beam-column. The slope of this relation curve gives the required stiffness of a beam-column. For an elastic cross-section, this stiffness, EI, is a constant. This presents no difficulties for solutions. However, for an elastic–plastic cross-section, the moment-curvature relationship becomes nonlinear. The value of EI in this case can be considered as the current slope between moment, M, and curvature ϕ, which depends on the magnitude of moment, M.

The existence of residual stresses and out-of-roundness will affect the slope of a moment–curvature relationship. As expected, the longitudinal residual stress will cause the cross-section to reach the elastic–plastic regime much earlier than that without the effect of longitudinal residual stress.

Herein, for each of the four subjects considered, the moment–curvature behaviour of a short tubular column is first studied considering the effects of residual stresses and out-of-roundness caused by forming and welding. This is followed by the study of elastic–plastic behaviour of long columns, considering the effect of out-of-straightness. The effect of actual end restraint provided by joints or connections on the strength and behaviour of columns is not considered. Details of this important effect are given elsewhere (Chen, 1980). Using the computer model developed, column strength curves for various tubular columns under different loading conditions are obtained and compared with the current Column Research Council–American Institute of Steel Construction (CRC–AISC) design formulas.

3.3 ANALYSIS OF SHORT COLUMN

Tangent stiffness method is generally applied to obtain moment–curvature–thrust (or M–ϕ–P) relationship for a tubular steel column segment subjected to axial compression and biaxial bending moment. In this part of analysis, both the initial out-of-roundness of the tube and the longitudinal and circumferential residual stresses are considered using Tresca yield criterion. The effects of residual stresses and initial imperfection on the elastic–plastic behaviour and the strength of tubular steel column segments are presented in terms of the M–ϕ–P curves.

3.3.1 Tangent Stiffness Formulation

For a box and cylindrical tubular column the appropriate set of generalised stresses are bending moments M_x and M_y and axial force P (Fig. 2). The corresponding set of generalised strains are bending curvatures ϕ_x and ϕ_y and axial strain, ε_0,

$$\{f\} = \begin{Bmatrix} M_x \\ M_y \\ P \end{Bmatrix} \tag{1}$$

and

$$\{x\} = \begin{Bmatrix} \phi_x \\ \phi_y \\ \varepsilon_0 \end{Bmatrix} \tag{2}$$

The generalised stresses and strains are shown in Fig. 2 in positive directions. The orientation of axes x and y is defined by their relative position to the longitudinal weld. The objective of this analysis is to

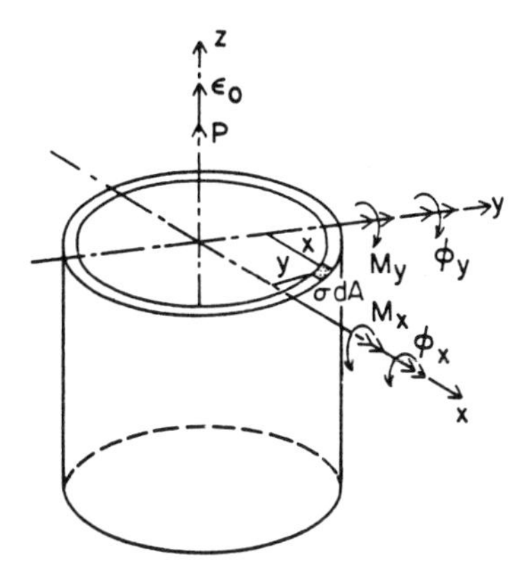

FIG. 2. Column segment under axial load and biaxial bending moment. (From Toma and Chen, 1979.)

calculate the deformation history $\{x\}$ of the cross-section corresponding to a given path of loading history $\{f\}$.

Since the plastic behaviour depends highly on the previous load history of the structure, it is possible only to establish analytically the relationship between the infinitesimal generalised stress increments $\{df\}$ or $\{\dot{f}\}$ and the corresponding infinitesimal generalised strain increments $\{dx\}$ or $\{\dot{x}\}$.

Assuming plane sections remain plane after bending and introducing the effective modulus concept, the following incremental relationship between $\{\dot{f}\}$ and $\{\dot{x}\}$ can be derived for a given column cross-section (see for example, Chen and Atsuta, 1977).

$$\begin{Bmatrix} \dot{M}_x \\ \dot{M}_y \\ \dot{P} \end{Bmatrix} = \begin{bmatrix} Q_{11} & Q_{12} & Q_{13} \\ Q_{21} & Q_{22} & Q_{23} \\ Q_{31} & Q_{32} & Q_{33} \end{bmatrix} \begin{Bmatrix} \dot{\phi}_x \\ \dot{\phi}_y \\ \dot{\varepsilon}_0 \end{Bmatrix} \tag{3}$$

or

$$\{\dot{f}\} = [Q]\ \{\dot{x}\} \tag{4}$$

In eqn. (3), Q_{ij} is defined as

$$\begin{aligned} &Q_{11} = \int E_{\text{eff}} y^2 \, dA;\ Q_{12} = Q_{21} = -\int E_{\text{eff}} xy \, dA;\ Q_{22} = \int E_{\text{eff}} x^2 \, dA; \\ &Q_{13} = Q_{31} = \int E_{\text{eff}}\, y dA;\ Q_{33} = \int E_{\text{eff}} dA;\ Q_{23} = Q_{32} = -\int E_{\text{eff}} x dA \end{aligned} \tag{5}$$

For the case when the entire cross-section is elastic, the double symmetry of a cross-section requires that

$$Q_{ij}=0; \ (i \neq j) \tag{6}$$

For a partially yielded section, eqn. (6) no longer holds. None of the elements of the tangent stiffness matrix will be zero, except when the section is completely yielded. Unlike in the elastic problems, the matrix $[Q]$ of the partially yielded section is a function of current state of stress and strain as well as the properties of the material and the cross-section. For an elastic–perfectly plastic material under uniaxial state of stress, the value of E is zero in the yielded zone. Therefore, only the area of the elastic core will contribute to the integration in eqn. (5) . This implies that further increment of external forces is resisted by the remaining elastic area of the section only. This is the case for most commonly used column cross-sections. However, for the case of fabricated cylindrical tubular cross-sections as used in offshore structures, the yielding of a material element is caused by both circumferential and longitudinal stresses. The value of E in eqn. (5) must therefore be replaced by an 'effective' modulus, E_{eff}, in the regions of biaxial yielding. Further consideration of the value of E_{eff} will be given in what follows. Once the tangent stiffness matrix $[Q]$ corresponding to a given state of stress can be evaluated, it is a simple matter to find the path of generalised strains $\{x\}$ for a given path of generalised stresses $\{f\}$ through a step-by-step incremental calculation.

3.3.2 Effective Young's Modulus

When both circumferential and longitudinal stresses are either in tension or compression, then the Tresca yield condition indicates that the limiting stress of an element is the uniaxial yield stress, σ_y, beyond which the element can assume no more load, but merely deforms plastically (see the insert of Fig. 3). However, in the tension–compression regimes of the Tresca yield diagram, the element stress condition is such that the elemental stress state can change while the element still remains on the yielded curve, but merely shifts its position on the sloping lines of the Tresca yield diagram. For this case, the limiting longitudinal stress reaches, eventually, the uniaxial yield stress, σ_y, of the material. In other words, an element which has yielded in either of the tension–compression zones of the Tresca yield diagram does have some resistance for an increased section curvature.

Herein, the concept of effective modulus E_{eff} is used. In the elastic

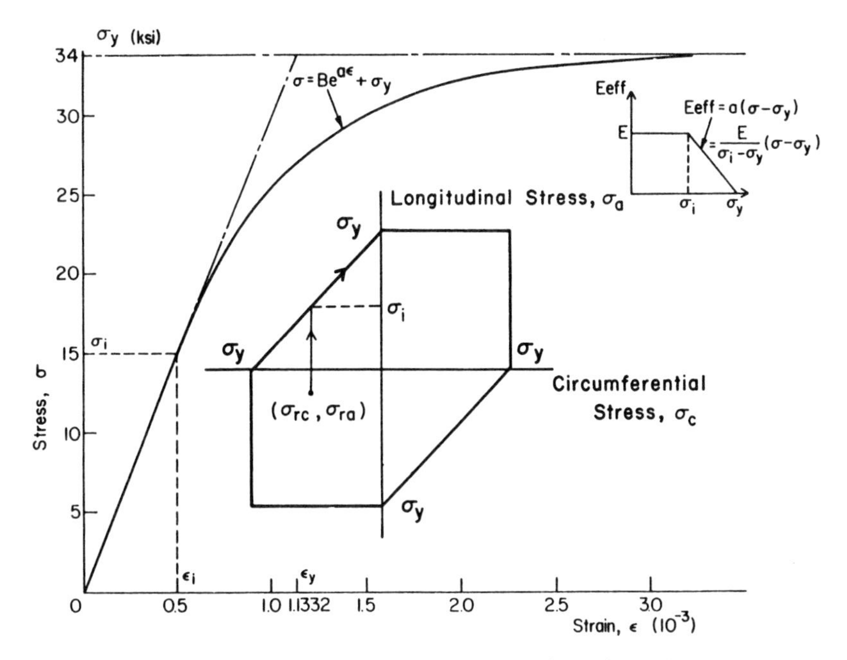

FIG. 3. Stress–strain relationship in the elastic and plastic range.

range, the element is assumed to have a constant elastic modulus, E, until the element yields at a longitudinal stress, σ_i. We also know that at the uniaxial yield stress, σ_y, the effective E value for a perfect plastic material is zero. These two conditions were linked with a straight line variation as shown in the inset of Fig. 3, to provide a reasonable estimate of decreasing elemental resistance to an increased applied longitudinal loading in the tension–compression regimes of the Tresca yield diagram. Thus, the effective Young's modulus may be assumed to vary linearly from E at the first yield stress, σ_i, to zero at the uniaxial yield value, σ_y. The linear variation of Young's modulus in the plastic range results in the expression

$$E_{\text{eff}} = \frac{E}{\sigma_i - \sigma_y} (\sigma - \sigma_y) \tag{7}$$

For the tubular columns, it is assumed that the tube can expand or contract freely so that in the elastic range the circumferential stresses, σ_c, will not be affected by the axial stress changes, $\dot{\sigma}_a$. In the biaxial stress space (σ_a, σ_c), this means that the stress point moves in the direction parallel to the σ_a axis (see inset Fig. 3). As the axial stresses, σ_a, are increased as the result of the axial load combined with biaxial bending, the stress point located initially at the residual stresses σ_{rc} and σ_{ra} moves vertically upward until it reaches the Tresca yield curve, then moving in the direction toward its corresponding final point σ_y.

After the initial yielding of an element, the linear variation of the Young's modulus is assumed as described previously. From the definition of effective Young's modulus, and using eqn. (7), we obtain

$$E_{\text{eff}} = \frac{d\sigma}{d\varepsilon} = \frac{E}{\sigma_i - \sigma_y}(\sigma - \sigma_y) \tag{8}$$

Integrating eqn. (8) and using the initial yield values

$$\sigma = \sigma_i = E\varepsilon_i \text{ at } \varepsilon = \varepsilon_i \tag{9}$$

the equivalent uniaxial stress–strain relationship is obtained (Chen *et al.*, 1979)

$$\sigma = \sigma_y[1 - (1 - \bar{\sigma}_i)\,e^{(\bar{\sigma}_i - \bar{\varepsilon})/(1 - \bar{\sigma}_i)}] \geqslant \sigma_i \tag{10}$$

in which

$$\bar{\varepsilon} = \frac{\varepsilon}{\varepsilon_y};\ \bar{\sigma}_i = \frac{\sigma_i}{\sigma_y} \tag{11}$$

As an example, the elastic and plastic stress–strain relationships are plotted in Fig. 3 for the following set of typical values

$$\begin{aligned} \sigma_y &= 34\text{ ksi (235 MPa)};\ \varepsilon_y = 1{\cdot}1332 \times 10^{-3} \\ \sigma_i &= 15\text{ ksi (104 MPa)};\ \varepsilon_i = 0{\cdot}5 \times 10^{-3} \end{aligned} \tag{12}$$

It is of interest to note that for an elastic–perfectly plastic material, the biaxial stress interaction between the longitudinal and circumferential stresses results in a work-hardening type of material behaviour with the initial yield value of $\sigma_i = 15$ ksi (104 MPa) and ultimate strength of $\sigma_y = 34$ ksi (235 MPa). The slope of the stress–strain curve gives the effective tangent modulus, E_{eff}, of the tube.

In the present analysis, each element in the tubular cross-section is treated independently and each contributes to the total sectional stiffness as individual; the interaction among the elements is not considered. To account for this interaction of elements, a more rigorous approach such as shell analysis or finite element method is needed in addition to the incremental theory of plasticity (Chen, 1982). This will make the problem exceedingly complicated.

3.4 ANALYSIS OF LONG COLUMN

3.4.1 General

For the analysis of long column and beam-column for which elastic analysis may be applied, the governing differential equations may be solved rigorously by the use of formal mathematics. In the plastic or nonlinear range, the differential equations are often intractable and recourse must be made to numerical methods to obtain solutions. Newmark's method appears to be the most convenient numerical method to obtain the maximum strength of a column or beam-column by first tracing the load deflection (or load rotation) curve of a beam-column and then determining the peak point from this curve. Details of Newmark's method are given elsewhere (e.g. Chapter 13, Chen and Atsuta, 1976).

Only a brief description of the method of analysis is given in what follows. The effect of out-of-straightness on buckling strength of column is considered in this part of analysis.

3.4.2 Newmark's Method

Newmark's numerical integration method is a useful means to compute the deflected shape of a loaded column from a given curvature distribution. The moment–curvature–thrust relationship for the cross-section must be known before applying this method.

The method of solution is iterative, in that initial deflected shape is assumed, and bending moments are computed for each division point or station. This part of the procedure ensures that equilibrium conditions are satisfied.

The curvatures at each station are then computed from a known M–P–ϕ relation. When the curvatures are known, the new deflected shape is calculated and compared with that initially assumed. If close agreement is obtained, the initially assumed deflection comprises a valid solution. Otherwise, the initially assumed deflection is modified and the procedure is repeated.

This procedure must be repeated for every increment of the axial load until the resultant deflection diverges, at which point the axial load exceeds the maximum strength of a column (buckling load). Thus, the maximum strength of a column or beam-column is obtained as the peak point of the axial load vs. the end rotation (or deflection or shortening) curve of a beam-column.

3.5 WELDED BUILT-UP BOX COLUMNS

3.5.1 Residual Stress Distribution in Welded Box Section

As shown in Fig. 4, the residual stress from the ends of each plate to a distance of 0·2 times the plate width from each end is σ_y (tension), and

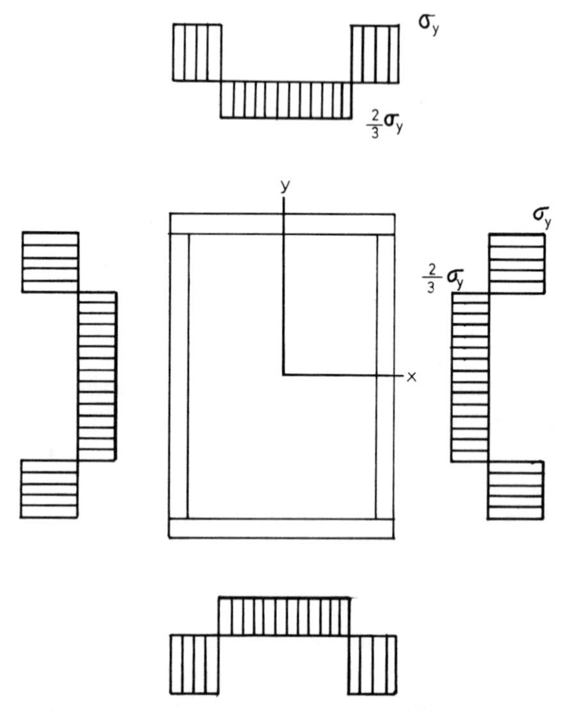

FIG. 4. Residual stress distribution in welded built-up box section.

the remainder of each plate has the value $-(2/3)\ \sigma_y$ (compression) (see Tall, 1961). This stress distribution, of course, satisfies the equilibrium condition that the resultant force due to the summation of the residual stresses alone over the entire cross-section must be zero.

3.5.2 Moment–Curvature–Thrust Relationships For Box Section

The moment–curvature curves plotted in Figs. 5 and 6 are for the resultant moment $m_t = M_t/M_{px}$ and resultant curvature $\bar{\phi}_t = \phi_t/\phi_{yield}$ for a given set of values of $\theta = \tan^{-1}(M_y/M_x) = 15°$ and 30° and $p = P/P_y = 0$, 0·2, 0·4, 0·6, and 0·8 where M_{px} is plastic moment capacity of the section about x-axis, ϕ_{yield} is the initial yield curvature, and P_y is the axial yield load of the section. In these Figures, the column section is first loaded axially to some value; and then the axial force P is held constant while

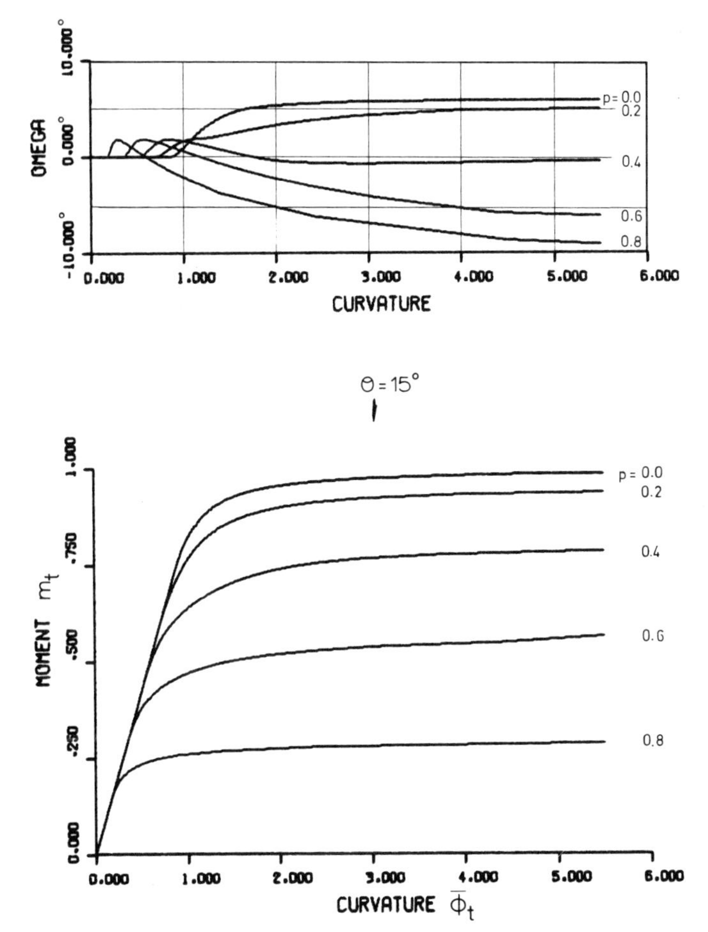

FIG. 5. Moment–curvature–thrust relations for welded built-up section with $\theta = 15°$.

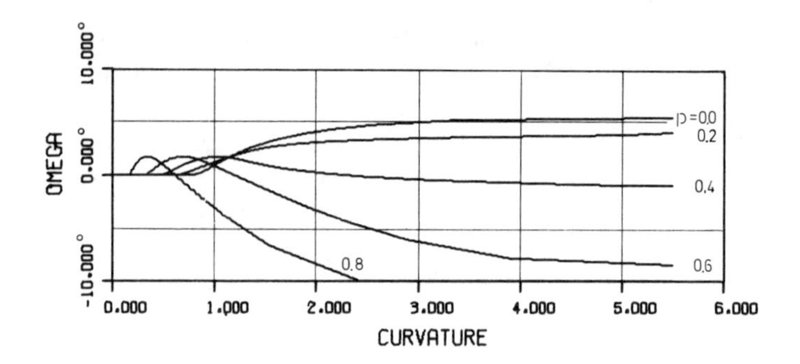

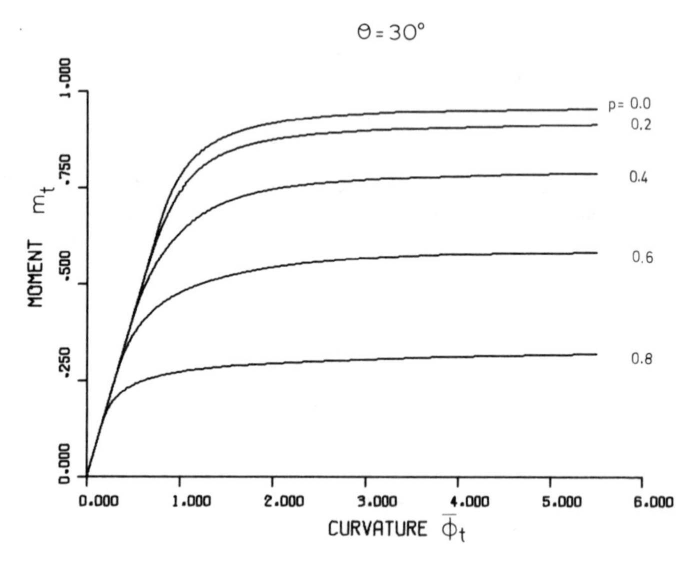

FIG. 6. Moment–curvature–thrust relations for welded built-up section with $\theta = 30°$.

the bending moments M_x and M_y or $M_t = \sqrt{M_x^2 + M_y^2}$ are increased proportionally in magnitude from zero. The corresponding bending curvatures ϕ_x and ϕ_y (or $\phi_t = \sqrt{\phi_x^2 + \phi_y^2}$) and axial strain ε_0 at the centre of the cross-section can be obtained by the tangent stiffness method.

The maximum difference between the angles θ and $\psi = \tan^{-1}(\phi_y/\phi_x)$; i.e. between the directions of the resultant moment M_t and resultant curvature ϕ_t vectors, $\omega = \psi - \theta$, is also shown in Figs. 5 and 6. It can be seen that the moment and curvature vectors nearly coincide in direction throughout the entire range of loading. The maximum difference between the two vectors is of the order of ten degrees.

3.5.3 Maximum Strength Interaction Curves for Box Column

The maximum load-carrying capacity of welded built-up box section columns has been obtained by utilising Newmark's numerical integration method and the M–ϕ–P curves obtained. The results obtained for long columns under symmetric as well as unsymmetric loading conditions have been compared with the results of tests on small-scale actual columns (Marshall and Ellis, 1970, Pillai and Ellis, 1974), providing the needed confirmation of the validity of the method of analysis. Figure 7

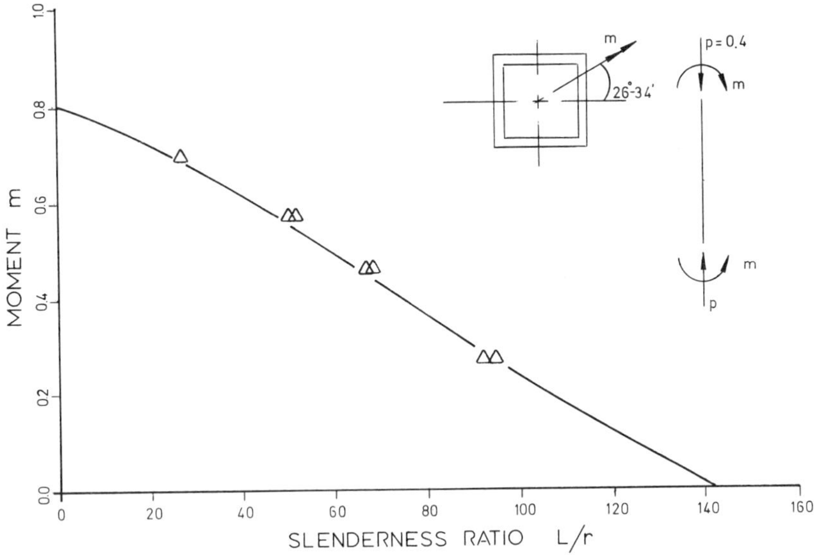

FIG. 7. Comparison of predicted strength of symmetrically loaded box column with tests.

shows the typical comparison of the theoretical predictions with experimental results corresponding to the symmetric loading case. It can be readily seen that the correlation is very good. Similar conclusion can be also made for the case of unsymmetric loading condition.

3.5.4 Equivalent Moment Factor

The concept of extending the use of the interaction curves obtained for the case of columns symmetrically loaded to the case of unsymmetrically loaded columns has been in use for many years (Chen and Atsuta, 1976). A coefficient C_m has been introduced such that:

$$(M_x)_{eq} = C_{mx} M_{ax} \tag{13}$$

$$(M_y)_{eq} = C_{my} M_{ay} \tag{14}$$

where C_m is defined as:

$$C_{mx} = 0{\cdot}6 - 0{\cdot}4 \frac{M_{bx}}{M_{ax}} \geq 0{\cdot}4 \tag{15}$$

$$C_{my} = 0{\cdot}6 - 0{\cdot}4 \frac{M_{by}}{M_{ay}} \geq 0{\cdot}4 \tag{16}$$

M_{ax}, M_{ay}, M_{bx}, M_{by} are the end moments of the column, with M_{ax} and M_{ay} being the larger ones.

Figure 8 shows the theoretical P–M interaction curves for symmetrical ($\kappa = -1$) biaxial bending about an axis 30° to one of the principal axes.

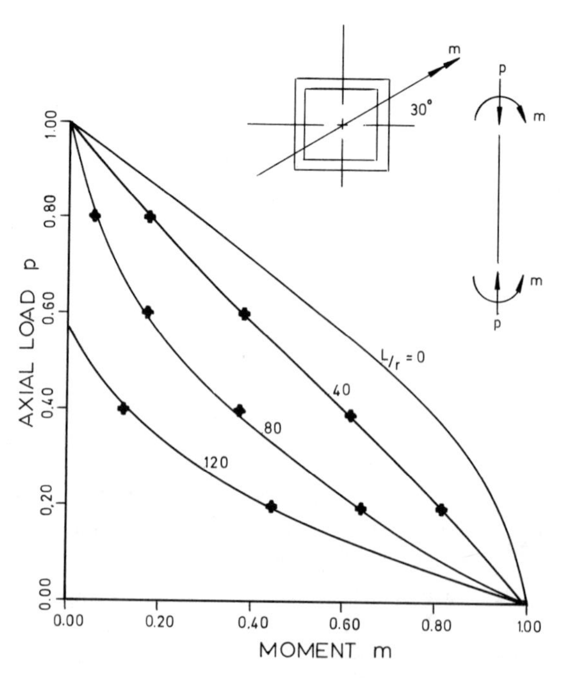

FIG. 8. Comparison of theoretical strength between symmetric and unsymmetric loading cases using equivalent moment factor C_m.

The crosses represent the equivalent moments of the theoretical predictions computed by eqns. (13) and (14) for the case of an end moment ratio $\kappa = M_a/M_b = 0$. As we can see, the agreement between the theoretical values for the symmetrical case and the equivalent moment values for the non-symmetrical case is very good.

3.5.5 Approximate Formula for the Strength of Short Box Columns

Figure 9 shows the family of strength curves for a short square box column corresponding to a constant value of axial force $p = 0{\cdot}2$. These

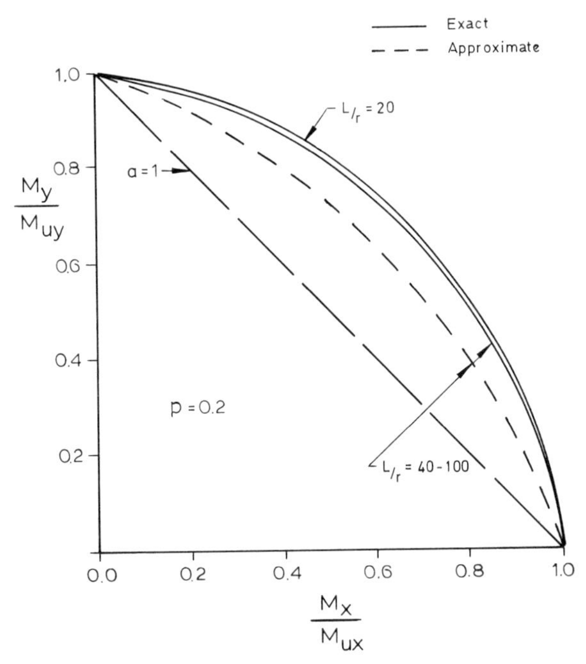

FIG. 9. Normalised interaction curves for box column with $p = 0{\cdot}2$.

curves can be approximated by the general non-dimensional interaction equation:

$$\left(\frac{M_x}{M_{ux}}\right)^{\alpha} + \left(\frac{M_y}{M_{uy}}\right)^{\alpha} = 1{\cdot}0 \tag{17}$$

where M_{ux} and M_{uy} are the ultimate moment carrying capacities about the x and y axes when M_y and M_x are zero respectively. That is, for a given value of P, the values of M_{ux} and M_{uy} can be taken from the $M_y=0$ and $M_x=0$ axes. These values also correspond to M_{pc}, the reduced plastic moment capacity of the member due to axial load.

The exponent α is a factor whose value is dependent on the axial load and the slenderness ratio of the column. The values of α for short column ($L/r \approx 0$) vary from 1·7 when $p=0$ to 5·5 when $p=0{\cdot}8$. Tebedge and Chen (1974) have suggested the following relationship for α for short H-columns

$$\alpha = 1{\cdot}60 - p/(2 \ln p) \tag{18}$$

This value of α gives a very conservative estimate of the strength of a short column when used for the hollow box section, especially for higher values of p.

Another approximation of α which compares more favourably with the exact value of α and is less conservative than eqn. (18) is the following (Chen and McGraw, 1977)

$$\alpha = 1{\cdot}7 - p/\ln p \tag{19}$$

It is of interest to note that the current SSRC guide (Johnston, 1976) makes use of eqn (17) with $\alpha = 1{\cdot}0$ for all values of p. This formula is then clearly an over-conservative estimate of the strength of a biaxially loaded short column (see Fig. 9).

Using the approximate values of α developed in eqn. (19), the approximate family of interaction curves for all values of p can be determined by use of eqn. (17). A comparison between the approximate methods and the exact results has been made, and it is found that for low values of $p=P/P_y$, the agreement is almost exact, at higher values of p it is slightly conservative (see Fig. 10).

3.5.6 Approximate Formula for Stability of Long Box Columns

The non-dimensional interaction equation (eqn. (17)) proposed earlier may also be used for long columns (Ross and Chen, 1976). However, we are now faced with the problem that α is dependent upon both the axial load p and the slenderness ratio L/r. Figure 10 shows the relationship between α and p, for a given value of L/r. It can be seen that for a low value of L/r (say $L/r=20$) the effect of p is to raise the value of α necessary to satisfy eqn. (17). As the slenderness ratio is increased, the effect of p is lessened; at high values of L/r, α decreases as p increases. An attempt has been made to show this relationship by use of the following

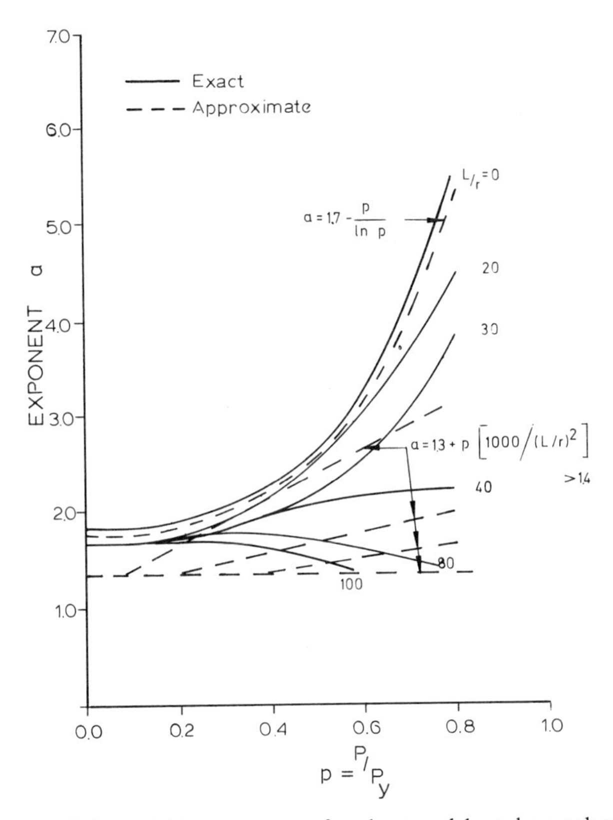

FIG. 10. Values of the exponent α for short and long box columns.

equation (Chen and McGraw, 1977):

$$\alpha = 1{\cdot}3 + \frac{1000\,p}{(L/r)^2} \geq 1{\cdot}4 \qquad \text{For } L/r > 10 \tag{20}$$

Although this is by no means an exact representation, the values of α produced by this equation are reasonable and conservative; though less conservative than the use of $\alpha = 1$ by the SSRC equation. Figures 9 and 11 compare the interaction curves for various values of p and L/r using the exact and approximate values of α. Excellent agreement is shown for low slenderness ratios. As L/r increases, the approximate equation yields conservative curves.

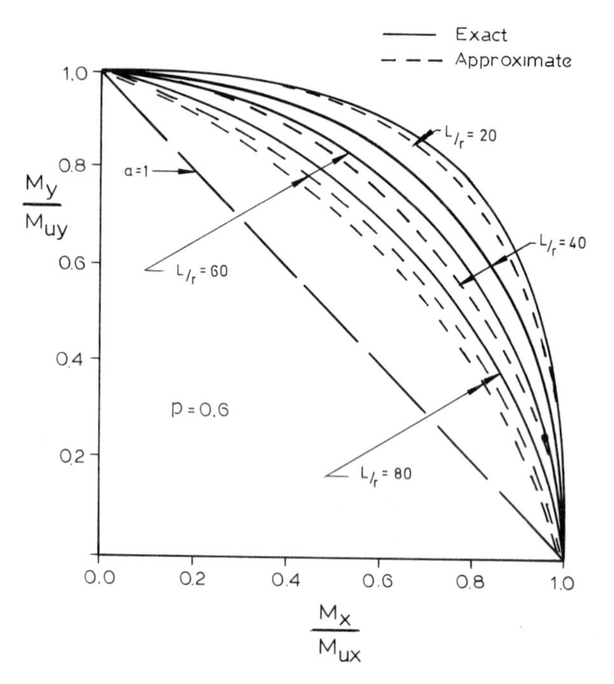

FIG. 11. Comparison of interaction curves for long box columns with $p=0{\cdot}6$.

3.6 FABRICATED CYLINDRICAL COLUMNS UNDER AXIAL LOAD

3.6.1 Residual Stress Distributions in Fabricated Tube

There are two major types of residual stresses existing in fabricated tubular columns—longitudinal and circumferential residual stresses. During the forming process by which a flat plate is converted to a cylinder, significant circumferential residual stresses are introduced which vary through the wall thickness of the cylinder. The longitudinal welding process introduces into the 'can' an additional longitudinal residual stress pattern. The measured residual stresses were reported by Chen and Ross (1976, 1977). The experimental results are summarised below.

The longitudinal residual stress distribution is shown in Fig. 12. This distribution represents an average value through the wall thickness. The curved solid line through the experimental results is a proposed fit and

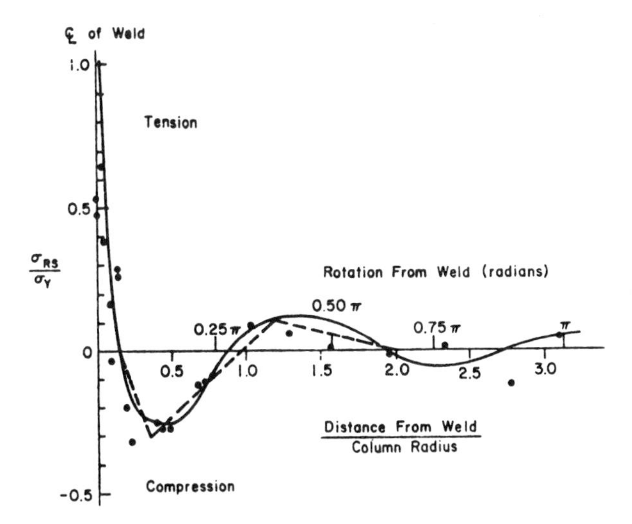

FIG. 12. Longitudinal residual stress distribution obtained from slicing method (Lehigh Test).

the dotted lines are a straight-line approximation suggested as a simplified alternative.

Herein, this straight-line distribution is adopted for the calculation of M–ϕ–P curve and buckling load.

In Fig. 13, the circumferential residual stresses through the wall thickness are given. If a section through the column wall is taken as shown in Fig. 13(a), then a typical result is shown in Fig. 13(b). Since the results between the various test sites around the circumference do not differ significantly, the suggested circumferential residual stress distribution through the wall thickness is as shown in Fig. 13(c).

3.6.2 Out-of-Roundness of Fabricated Tube

The effect of out-of-roundness in the fabricated tubes on the behaviour and strength of tubular columns is included in the moment–curvature studies. Such imperfections may be caused by either initial fabrication of each 'can' and also by subsequent welding of cans to form the cylinders. Herein, a geometrical imperfection is introduced in the form of

$$w = w_i \cos 2\theta \tag{21}$$

in which w_i = the initial maximum out-of-roundness; and θ = the angle around the circumference. If the maximum and minimum diameters of

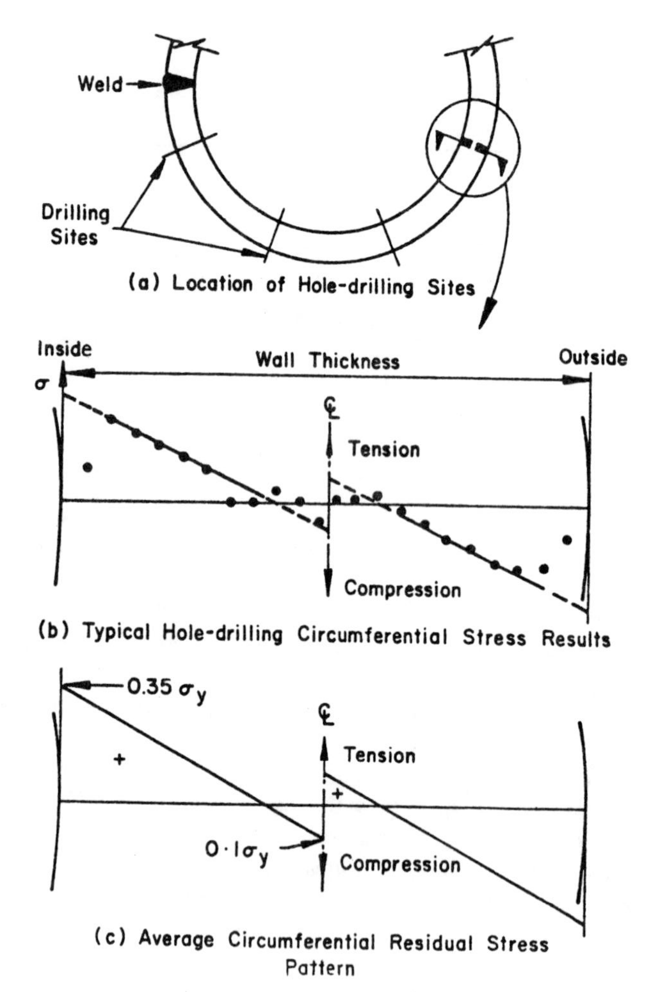

FIG. 13. Circumferential residual stress pattern (Lehigh Test).

the tube are denoted by D_{max} and D_{min} respectively, the initial deflection w_i can be expressed in the form

$$\frac{D_{max} - D_{min}}{D} = \frac{4w_i}{D} = 0{\cdot}01 \tag{22}$$

where we have assumed the maximum out-of-roundness to be 1%. It is to be noted that measurements of the out-of-roundness of the 10 fabricated tubular specimens reported by Chen and Ross (1977) indicate that, in general, there is less than 1% difference between two perpendicular diameters at all positions along the column length. This 1% out-of-roundness adopted in the present analysis is the maximum tolerance specified by American Petroleum Institute (API) specifications (1972) for allowable fabrication imperfections for out-of-roundness.

3.6.3 Moment–Curvature–Thrust Relationships for Tube Section

As described previously, tubular members as used in offshore structures are cold-formed from plate, welded along a single longitudinal seam, and used without any kind of stress relief. The resulting asymmetrical pattern of longitudinal residual stress as shown in Fig. 12 and its effect on the moment–curvature behaviour are examined here so that the critical direction of tube bending with respect to the location of longitudinal weld can be determined.

Moment–curvature relations for a tubular cross-section with different directions of applied bending moment with respect to the longitudinal weld are shown in Fig. 14 corresponding to a constant axial load

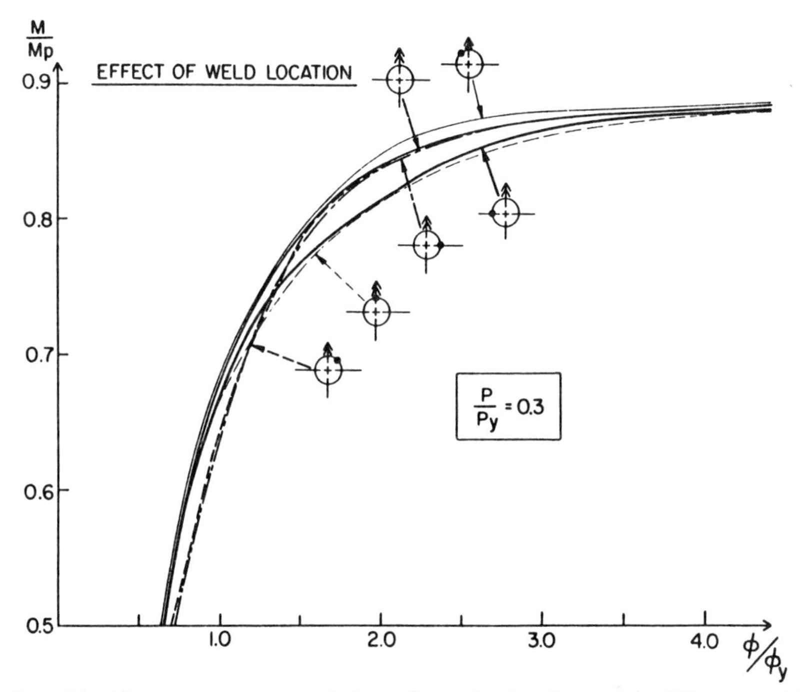

FIG. 14. Moment–curvature relations for tube bending with different weld location, $P/P_y = 0{\cdot}3$. (From Toma and Chen, 1979.)

$P/P_y=0{\cdot}3$. The bending moment vector M as shown in the insets of the Figure follows the right-hand screw rule, i.e., the moment acts in the diametrical plane which is perpendicular to the moment vector shown in the sense that follows the right-hand screw rule. The moment–curvature curves have been computed by considering only the longitudinal residual stress as shown in Fig. 12. These curves are non-dimensionalised by the fully plastic hinge moment, M_p, and initial yield curvature, ϕ_y, of a perfect tube with no residual stress. It is well known that the intensity of axial force P reduces the plastic hinge moment of a tube cross-section from M_p for which $P=0$ to M_{pc} for which $P\neq 0$. This fully plastic limit moment M_{pc} is a sectional property, independent of the residual stress. The limit moments, M_{pc}, as given by the maximum moments in various moment–curvature curves are compared with those previously found by the limit analysis method and a good agreement is found for all cases (see Chapter 5, Chen and Atsuta, 1976).

It is found that the intensity of axial load has a major influence on the determination of the critical plane of bending. For low axial force ($P/P_y \leq 0{\cdot}3$) with high moments, the critical bending plane is found to be the plane which is perpendicular to the diametrical plane containing the longitudinal weld. In other words, the critical direction of bending moment is the one whose vector passes through the longitudinal weld. This conclusion, however, is not correct when the intensity of moment is only of moderate value, but the differences among various orientations studied are found to be not large.

As the axial load increases beyond $P/P_y=0{\cdot}3$, however, the critical bending plane changes somewhat. Fortunately, the stiffness (i.e., slopes of the moment–curvature relations) and maximum moment capacity do not appear to differ much from the critical case described previously (Toma and Chen, 1979). For simplicity, we shall assume in the following parameter study that the critical bending moment is the one whose moment vector passes through the longitudinal weld.

3.6.4 Interaction Equations for Strength of Short Tube

Using the same type of equations as for H-columns, the corresponding equations for a fabricated tubular cross-section are obtained

$$\frac{M}{M_p}=1-1{\cdot}18\left(\frac{P}{P_y}\right)^2; \text{ for } 0\leq\frac{P}{P_y}\leq 0{\cdot}65 \tag{23}$$

$$\frac{M}{M_p}=1{\cdot}43\left(1-\frac{P}{P_y}\right); \text{ for } 0{\cdot}65\leq\frac{P}{P_y}\leq 1{\cdot}0 \tag{24}$$

These two equations give the maximum moment capacity of a tubular cross-section in the presence of axial force.

3.6.5 Column Strength Curves and Equations for Long Tube

Two theoretical column strength curves for a fabricated tubular column are computed considering the effects of longitudinal and circumferential residual stresses of the type shown in Fig. 12 and Fig. 13, 1% and 2% of out-of-roundness and 0·1% and 0·2% of out-of-straightness, respectively. The theoretical curves are also compared with CRC and AISC design curves in Fig. 15. The CRC and SSRC design curves are compared with

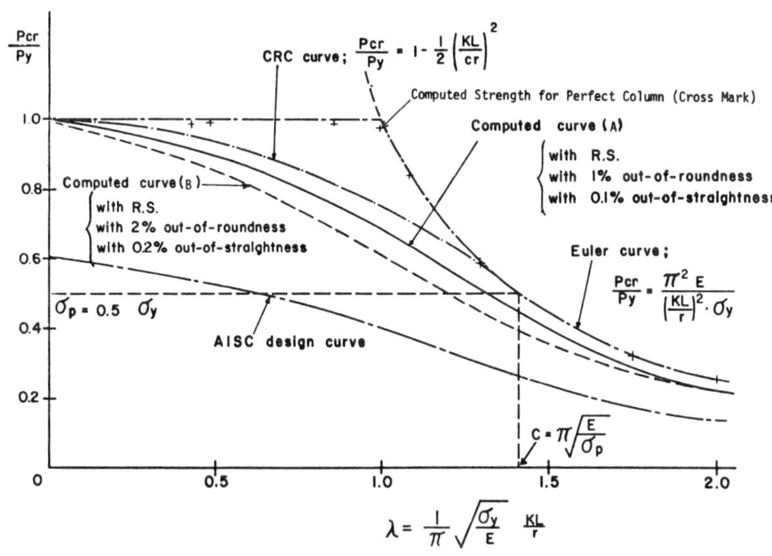

FIG. 15. Calculated column strength curve compared with CRC curve. (From Toma and Chen, 1979.)

Lehigh tests (Chen and Ross, 1977) in Fig. 16. In Fig. 15 the cross denotes the theoretical maximum strength of a perfect tubular column using the present computer model and its error to the exact solution is found to be less than 1·5%. The initial deflection in the perfect column analysis is assumed to have a sinusoidal shape with an amplitude of 0·1 in. (2·5 mm).

The computed column strength curve in Fig. 15 is seen slightly below the CRC strength curve (e.g., Chapter 14, Chen and Atsuta, 1976). This implies that the safety factor contained in the present AISC design curve

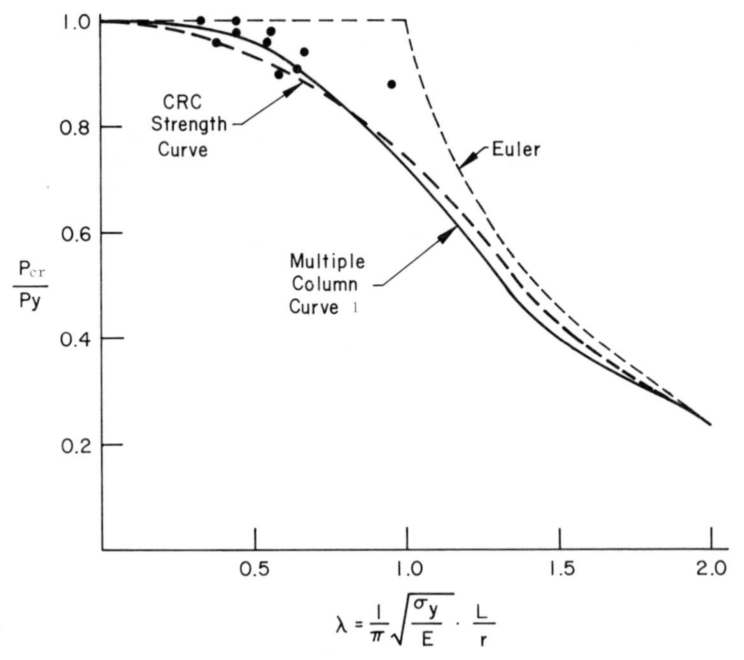

FIG. 16. Comparison of test results with column buckling curves—static yield stresses assumed (Lehigh test).

will be somewhat reduced for fabricated tubular columns. Comparing the computed curve (A) with curve (B) in Fig. 15, it is seen that the increased imperfections have a significant weakening effect on the load-carrying capacity of a fabricated tube.

The results of the 10 full-sized column tests are shown graphically in Fig. 16. For comparison, the accepted CRC ultimate-column-strength equation and the SSRC multiple column curve (1) are also shown, and good agreement is generally observed.

It is found that the computed curve (A) in Fig. 15 can be closely approximated by the following equations:

$$\frac{P}{P_y} = 1{\cdot}0 - 0{\cdot}091\lambda - 0{\cdot}22\lambda^2; \text{ for } 0 \leq \lambda \leq 1{\cdot}41 \qquad (25)$$

and

$$\frac{P}{P_y} = 0{\cdot}015 + \frac{0{\cdot}84}{\lambda^2}; \text{ for } 1{\cdot}41 \leq \lambda \leq 2{\cdot}0 \qquad (26)$$

where

$$\lambda = \frac{1}{\pi}\sqrt{\frac{\sigma_y}{E}}\frac{KL}{r} \tag{27}$$

3.7 FABRICATED CYLINDRICAL COLUMNS UNDER EXTERNAL PRESSURE

For deepwater oil platforms, an additional important consideration for tubular members is the interaction of axial and bending stresses with compressive hoop stresses caused by the external hydrostatic pressure. Owing to the reversible nature of storm forces, this axial stress can be tensile as well as compressive. As with other factors, the external pressure may reduce significantly the axial load-carrying capacity of actual fabricated tubes to less than that predicted by the conventional column results such as the AISC–CRC column curve.

Recently, the effect of external hydrostatic pressure on the strength and behaviour of fabricated steel tubular members has been studied extensively by Chen and Toma (1979). In this study, column strength curves for columns loaded axially at the ends and column interaction curves for columns loaded eccentrically at the ends are obtained considering the influence of hydrostatic pressure. It is found that external pressure reduces the load carrying capacity of fabricated tubular columns significantly when the out-of-straightness and pressure are large. Also, it is found that sectional capacity is reduced significantly by the presence of external pressure when axial load is small.

A brief summary of this development is given below. Design formulas are presented for axially loaded as well as eccentrically loaded columns with external pressure by a simple modification of the present AISC design formulas.

3.7.1 M–P–ϕ–Q Relations for Tube Section

The relation between moment M and curvature ϕ for a given thrust P and hydrostatic pressure Q is studied here. The individual and combined effects of longitudinal and circumferential residual stresses (Fig. 12 and Fig. 13), circumferential hoop compression caused by hydrostatic pressure, and out-of-roundness of the tube on the behaviour and strength of a fabricated 'can' (Fig. 1) can be assessed by studying the moment–curvature–thrust–hydrostatic pressure relationships (M–P–ϕ–Q curves).

This study has been reported by Chen and Toma (1979). In Fig. 17, typical M–P–ϕ–Q curves are shown. It can be seen that external pressure reduces moment carrying capacity of a fabricated tubular column section significantly when the axial force is small.

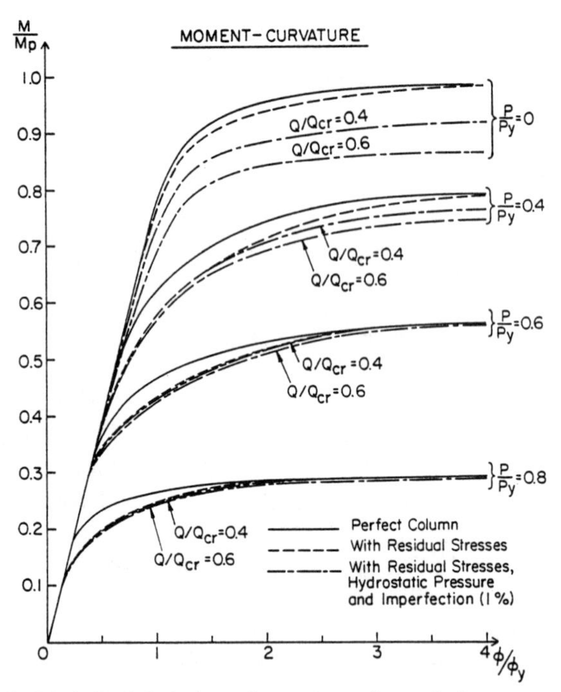

FIG. 17. M–ϕ–P–Q Relations for an imperfect tubular cross-section.

The fully plastic moment, M_{pQ}, reduced by the presence of hydrostatic pressure, Q, may be approximated empirically by the equation

$$\frac{M_{pQ}}{M_p}=1-\frac{i_R+3}{20}\left(\frac{Q}{Q_{cr}}\right) \tag{28}$$

in which i_R is the value of out-of-roundness of a cross section expressed in percent, and Q is external pressure. The plastic moment without the influence of hydrostatic pressure is denoted by M_p. It has the value (ASCE, 1971)

$$M_p=\frac{\sigma_y}{6}D^3\left[1-\left(1-\frac{2t}{D}\right)^3\right] \tag{29}$$

and the critical external pressure, Q_{cr}, depends on the thickness/diameter ratio. It is defined by the elastic buckling solution (Timoshenko, 1961)

$$Q_{cr} = \frac{2E(t/D)^3}{(1-v^2)} \tag{30}$$

where E = Young's modulus, v = Poisson's ratio, t = wall thickness, and D = tube outside diameter.

3.7.2 Columns Loaded Axially at Ends

(a) Column Strength Curves

Two theoretical column strength curves for a fabricated tubular column are computed considering the effects of longitudinal and circumferential residual stresses of the type shown in Figs. 12 and 13, hydrostatic pressure, 1% and 2% of out-of-roundness, and 0·1% and 0·2% of out-of-straightness. The theoretical curves are compared with CRC and AISC design curves in Fig. 18 and also with the SSRC multiple column curves in Fig. 19. In Fig. 18 the cross denotes the theoretical maximum strength of a perfect tubular column using the computer model, and its error to the

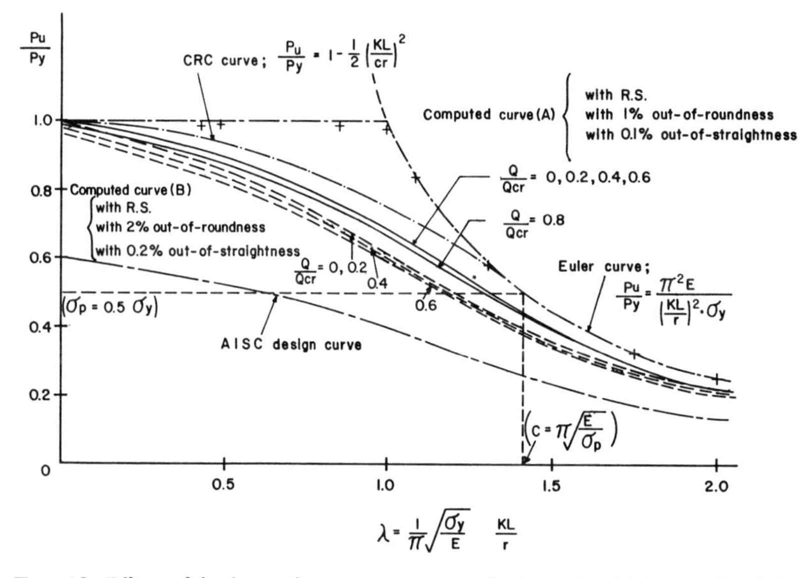

FIG. 18. Effect of hydrostatic pressure on axial strength of fabricated tubular columns.

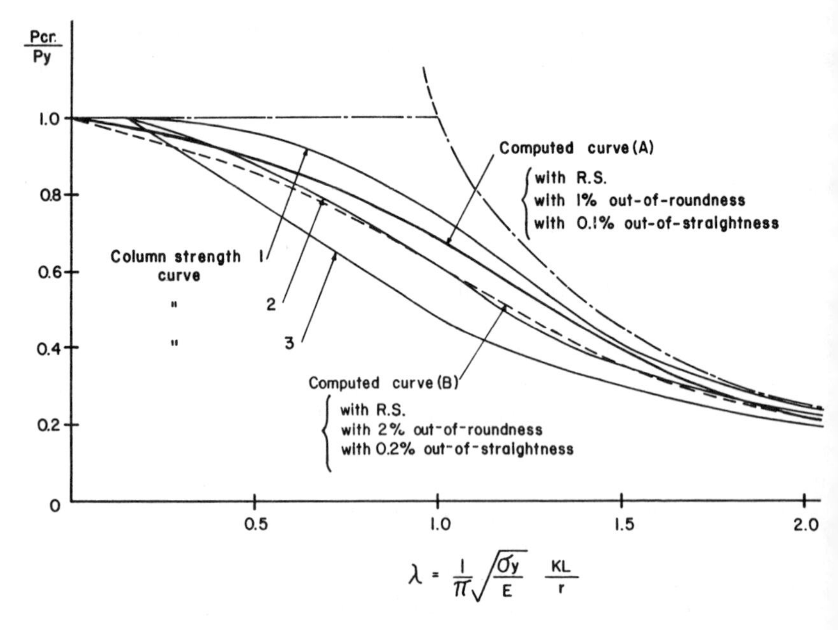

FIG. 19. Comparison of tubular column strength curves with SSRC multiple column curves.

exact solution is found to be less than 1·5%. The initial deflection in the perfect column analysis is assumed to have a sinusoidal shape with an amplitude of 0·1 inch.

Here, as in Fig. 15, all the computed column strength curves in Fig. 18 lie below the CRC strength curve. This implies that the safety factor contained in the present AISC design curve will be somewhat reduced for fabricated tubular columns under external pressure.

Comparing the computed curves (A) with curves (B) in Fig. 18, it is seen that the increased imperfections have a significant weakening effect on the load-carrying capacity of a fabricated tube under external pressure.

The effect of hydrostatic pressure on the column strength is also shown in Fig. 18 with the pressure varying from $Q/Q_{cr}=0$ to 0·8. In the case of out-of-straightness 0·1% and out-of-roundness 1% (or computed curves A), the effect of hydrostatic pressure on column strength is negligible for hydrostatic pressure up to the value $Q/Q_{cr}=0{\cdot}6$. However, when $Q/Q_{cr}=0{\cdot}8$, this hydrostatic pressure effect becomes somewhat significant. In the case of out-of-straightness 0·2 and out-of-roundness 2% (or computed curves B), this effect is seen to be slightly amplified by the increased imperfections.

Comparing the multiple column strength curves (Johnston, 1976) with computed curves in Fig. 19 shows that the computed curve (B) follows closely the multiple column curve (2), while the computed curve (A) lies between multiple column curves (1) and (2).

(b) Column Strength Equations for Columns Loaded Axially at Ends

Equations for the column strength curves shown in Figs. 18 and 19 are developed here.

It is found that the computed curve A, with $Q/Q_{cr}=0$, which represents a typical case of fabricated tubular columns as used in offshore structures, can be closely approximated by the following equations (Chen and Toma, 1979):

$$\frac{P_u}{P_y}=1\cdot0-0\cdot091\lambda-0\cdot22\lambda^2 \quad \text{for } 0\leq\lambda\leq1\cdot41$$

and

$$\frac{P_u}{P_y}=0\cdot015+\frac{0\cdot84}{\lambda^2} \quad \text{for } 1\cdot41\leq\lambda\leq2\cdot0 \tag{32}$$

in which P_u and P_y are the ultimate strength of an axially loaded long column and the axial capacity of a cross-section, respectively, and λ is the modified slenderness ratio,

$$\frac{1}{\pi}\sqrt{\frac{\sigma_y}{E}}\frac{KL}{r}.$$

For the computed curve B with $Q/Q_{cr}=0$, the corresponding equations have the form.

$$\frac{P_u}{P_y}=1\cdot0-0\cdot25\lambda-0\cdot13\lambda^2 \quad \text{for } 0\leq\lambda\leq1\cdot41 \tag{33}$$

$$\frac{P_u}{P_y}=0\cdot052+\frac{0\cdot67}{\lambda^2} \quad \text{for } 1\cdot41\leq\lambda\leq2\cdot0 \tag{34}$$

To include the effect of external hydrostatic pressure Q on the ultimate axial capacity P_{uQ} of an imperfect long column, the modified column strength curves with $Q\neq0$ as shown in Fig. 18 may be obtained simply from the previous cases with $Q=0$ (or P_u/P_y in eqns. (31) to (34)), by the relation

$$\frac{P_{uQ}}{P_y} = \left(\frac{P_u}{P_y}\right)\left(\frac{P_{uQ}}{P_u}\right) \tag{35}$$

where the strength reduction factor P_{uQ}/P_u due to the external hydrostatic pressure Q may be approximated empirically in the form

$$\frac{P_{uQ}}{P_u} = 1 - 2i_S^2\left(\frac{Q}{Q_{cr}}\right)^2 \tag{36}$$

where i_S is the value of out-of-straightness of a column expressed in percent. Hence, for a column with 0·1% of out-of-straightness, the value of i_S to be used in eqn. (36) is 0·1.

3.7.3 Columns Loaded Eccentrically at Ends

The analysis of a beam-column subjected to eccentric loads applied at the ends is essentially the same as that of the axially loaded case. In the following discussions, the imperfections and residual stresses adopted are 1% out-of-roundness, 0·1% out-of-straightness, and residual stress distributions of the type shown in Figs. 12 and 13.

(a) Different End Eccentricity Conditions

Interaction curves for the combinations of axial force and end moment that can be safely supported by the fabricated tubular column are shown in Fig. 20 for both perfect and imperfect columns. Hydrostatic pressure is not considered here. Figure 20(a) gives the curves for a loading condition in which two equal end eccentricities cause the column to be bent in a single curvature ($M_B/M_A = -1$). Figure 20(b) is the case where only one end eccentricity exists ($M_B/M_A = 0$).

The effect of imperfections and residual stresses on the maximum strength of a fabricated tubular column may be seen in these figures by comparing the solid curves with the dashed curves. It is seen that the difference between the perfect and imperfect curves is quite significant. The difference becomes small when the values of slenderness ratio are decreased.

(b) Effect of Hydrostatic Pressure on Eccentrically Loaded Columns

The effect of hydrostatic pressure on the interaction curves of eccentrically loaded columns has been studied for various end moment conditions and slenderness ratios. The results for $M_B/M_A = -1$ are presented

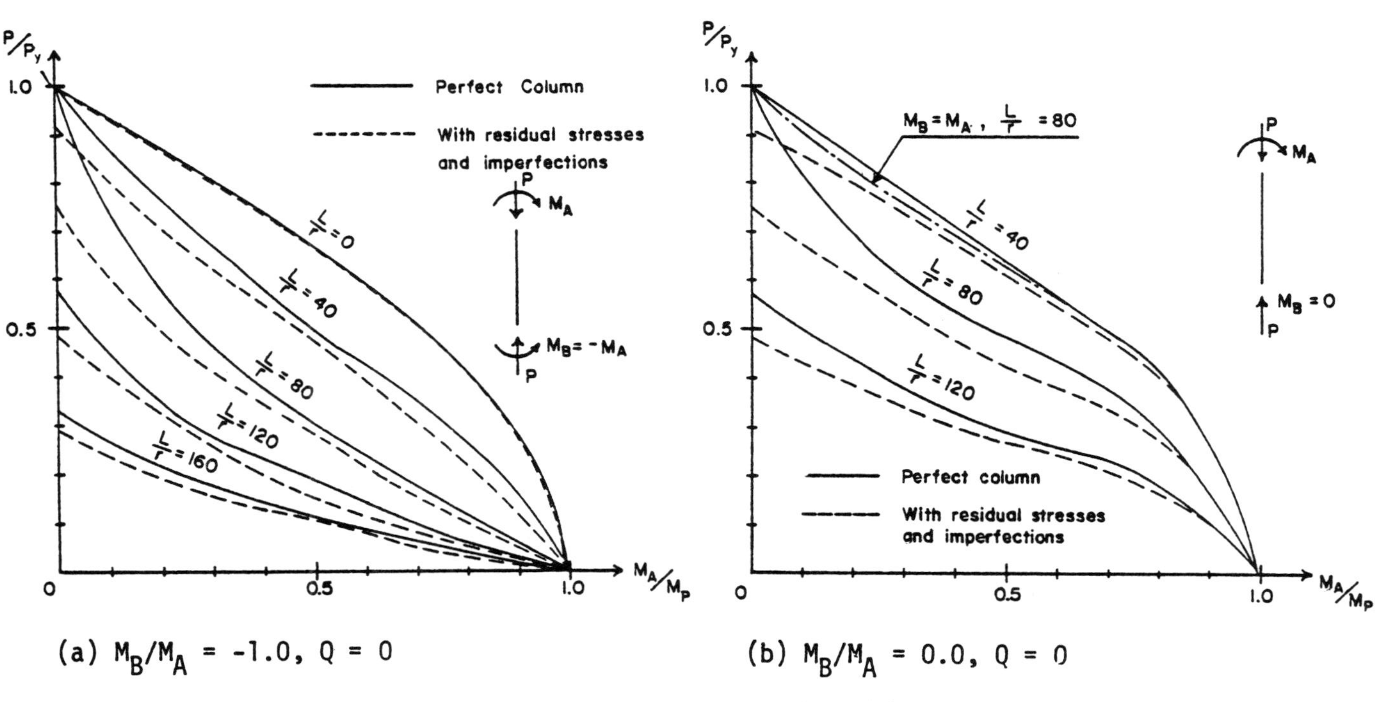

FIG. 20. Interaction curves for eccentrically loaded columns.

in Fig. 21 for $L/r = 80$ and 120. It is seen that the difference between the upper and the lower curves corresponding to the external hydrostatic pressure $Q/Q_{cr} = 0$ to 0·8 is quite significant when the values of axial force are small. Although the curves in Fig. 21(b) which have a larger slenderness ratio, appear to form a rather narrow band for all values of Q/Q_{cr}, the percentage of variance to the maximum strength without the effect of hydrostatic pressure has about the same value as those having similar slenderness ratios.

(*c*) *Interaction Equations for Columns Loaded Eccentrically at Ends*

To develop a simple interaction equation for combined bending and axial compression, the linear interaction of thrust P and the maximum bending moment M_{max} at failure may be applied as a criterion (Chen and Atsuta, 1976).

$$\frac{P}{P_u} + \frac{M_{max}}{M_u} = 1 \tag{37}$$

in which P_u is the ultimate thrust of the beam-column in absence of the end moments, that is, the buckling strength of an axially loaded column, while M_u is the ultimate end moment in absence of the thrust P, that is the plastic moment of the section M_p. In the elastic range, the presence of an axial load amplifies the primary bending moment by the ratio, $1/(1 - P/P_e)$, where P_e is Euler buckling load. If this 'elastic' factor is used to estimate the maximum moment at the centre of a plastic beam-column, M_{max} can be estimated by

$$M_{max} = \frac{M}{1 - P/P_e} \tag{38}$$

In order to cover the loading cases other than the symmetric end moments, $M_B/M_A = -1$, a reduction factor, C_m, is used to convert the unsymmetric case to the equivalent symmetric condition

$$M = C_m M_A \tag{39}$$

where M_A is the numerically larger one of end moments M_A and M_B. Thus, the interaction formula, eqn. (37), becomes the following familiar AISC design formula (Johnston, 1976):

$$\frac{P}{P_u} + \frac{C_m M_A}{M_p(1 - P/P_e)} = 1 \tag{40}$$

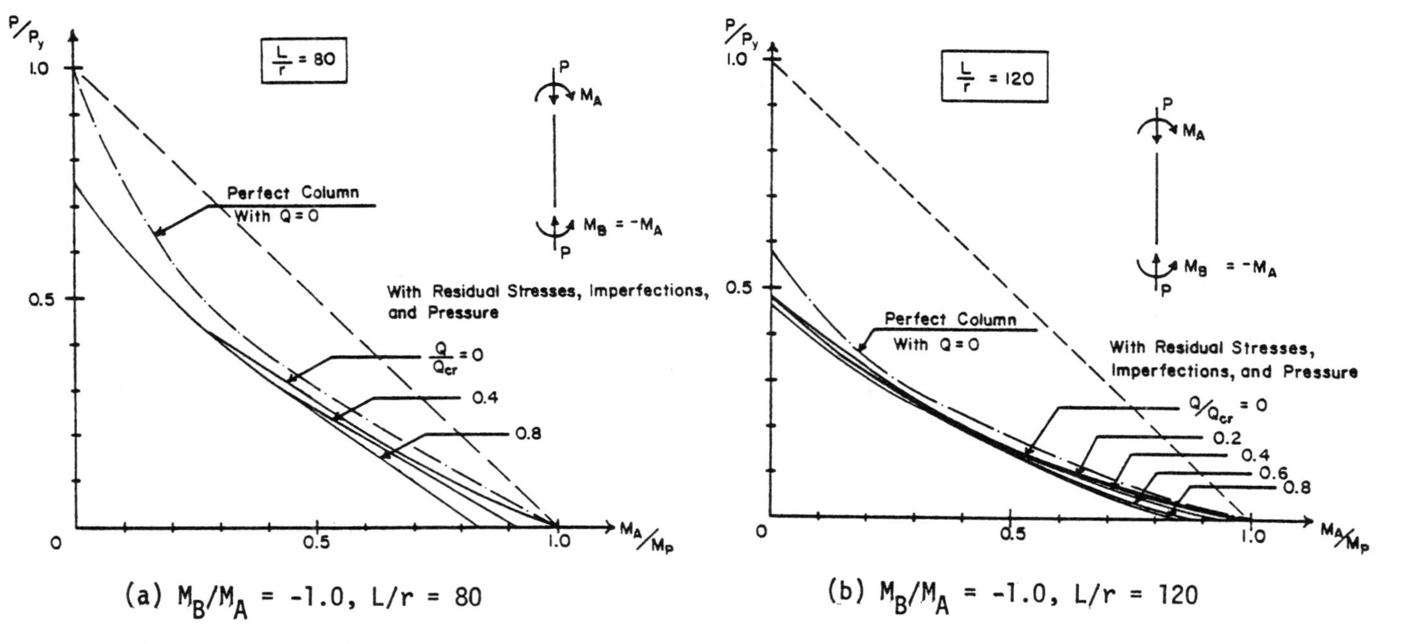

FIG. 21. Pressure effect on interaction curves for eccentrically loaded columns, equal moment at both ends.

in which the reduction factor C_m is given by AISC (Chen and Atsuta, 1976)

$$C_m = 0{\cdot}6 - 0{\cdot}4\,\frac{M_B}{M_A} \geq 0{\cdot}40 \tag{41}$$

and the Euler buckling load P_e is

$$P_e = \frac{\pi^2 EA}{(KL/r)^2} \tag{42}$$

The fully plastic bending moment capacity for M_p for a thin wall tube is a section-dependent quantity and has the value given by eqn. (29). The ultimate strength of an axially loaded long column P_u is defined by the CRC column strength curve in the present AISC specifications (1969).

Since the end points of the exact curves shown in Fig. 21 are P_{uQ} and M_{pQ} where the curve crosses the P/P_y and M/M_p axes respectively, the exact interaction curves of Fig. 21 are replotted on axes of P/P_{uQ} vs. M/M_{pQ} in Fig. 22. The general form of AISC equation (eqn. (40)) can now be modified by the interaction equation:

$$\frac{P}{P_{uQ}} + \frac{C_m M_A}{M_{pQ}(1 - P/P_e)} = 1 \tag{43}$$

where the ultimate axial strength P_{uQ} and the plastic moment capacity M_{pQ} considering the effect of hydrostatic pressure are used to replace those without the effect of hydrostatic pressure P_u and M_p. Expressions for estimating the values of P_{uQ} and M_{pQ} for fabricated tubular columns are given by eqn. (36) and eqn. (28) respectively and exact values are plotted as the end points in Fig. 21.

Using the CRC column strength equation and the proposed equations for tubular columns (eqn. (31) and eqn. (32)) with $Q=0$ for P_u in the AISC Formula (eqn. (40)), a comparison between the AISC formula and the exact numerical results for a fabricated imperfect column with residual stresses is also shown in Fig. 22. Clearly, the proposed column strength equation gives a closer estimation for the strength of eccentrically loaded tubular columns as used in deepwater platforms.

The maximum loads determined theoretically by the computer model are compared in Fig. 23 with the present AISC interaction formula (eqn. (40)) and in Fig. 24 with the proposed modified AISC formula (eqn. (43)). To calculate the ultimate strength P_u and M_p in the AISC formula, the proposed column strength equations, eqns. (31) and (32), and the plastic

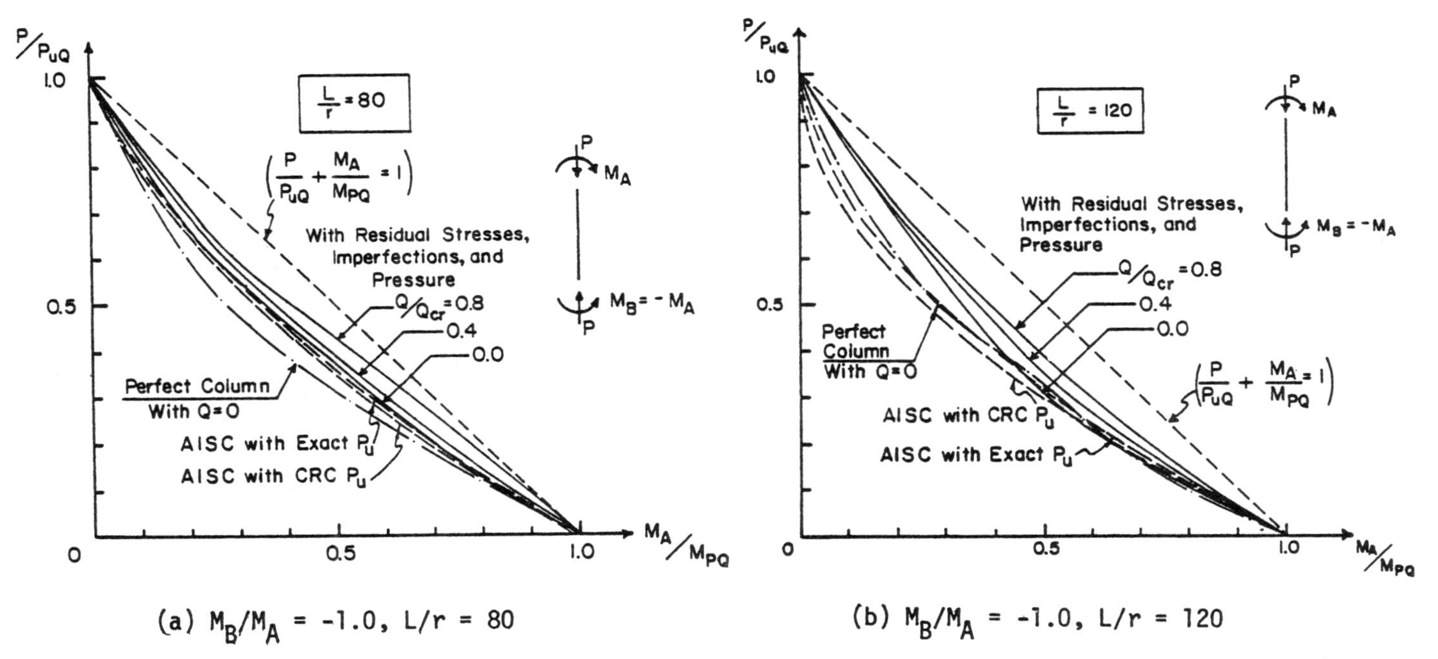

FIG. 22. Pressure effect and comparison of interaction curves between AISC Formula (1969) and exact solutions, equal moment at both ends.

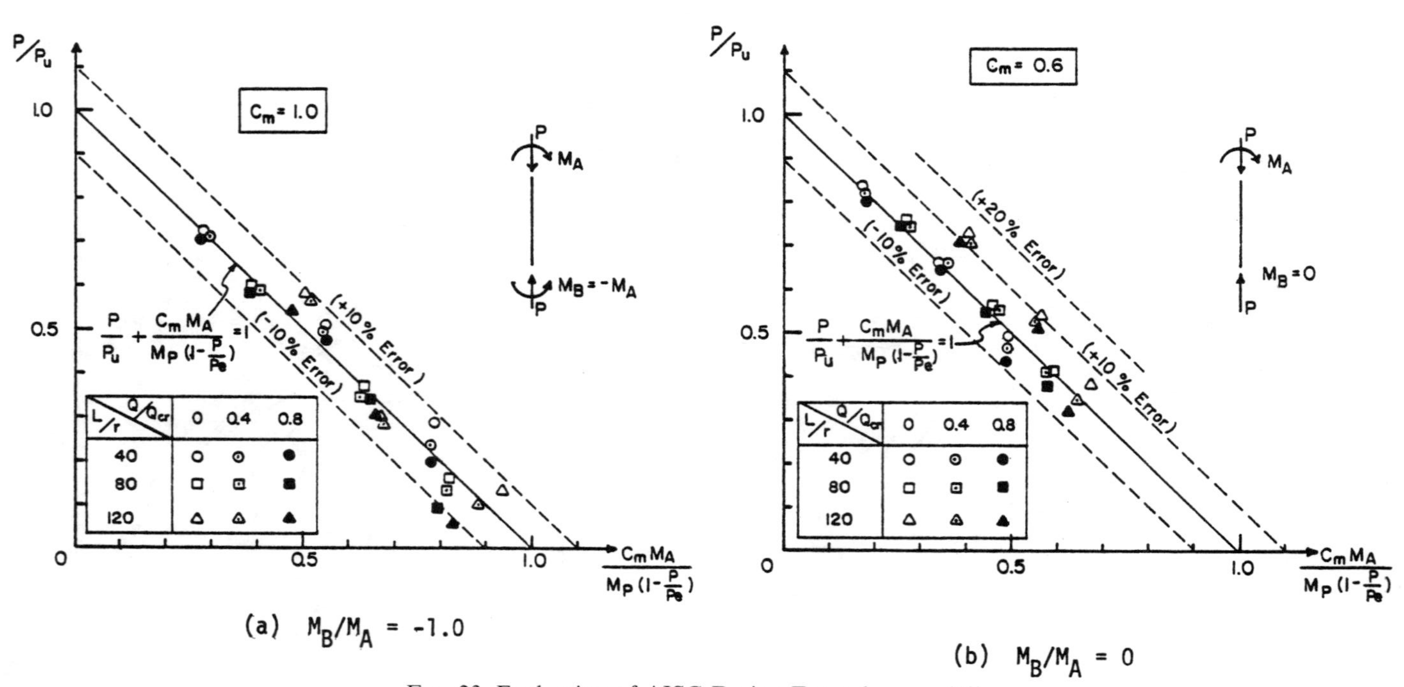

FIG. 23. Evaluation of AISC Design Formula, eqn. (40).

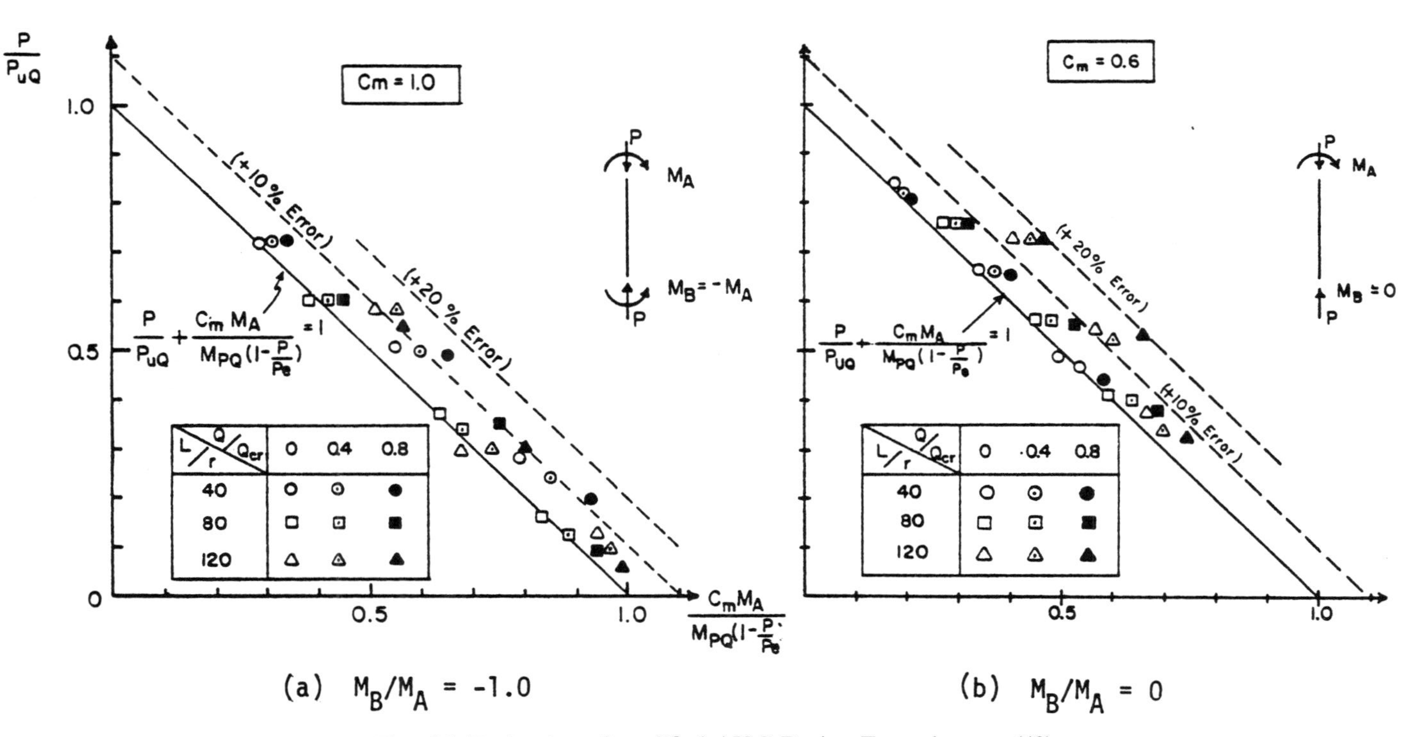

FIG. 24. Evaluation of modified AISC Design Formula, eqn. (43).

moment equation, eqn. (29) are used. In the modified AISC formula, eqns. (36) and (28) developed empirically for fabricated tubular columns considering the effect of external hydrostatic pressure are used for P_{uQ} and M_{pQ} respectively. In general, the present AISC interaction formula is seen to give an average value of the ultimate strength of all the cases investigated considering the effect of hydrostatic pressure. The computed exact values scatter within $\pm 10\%$ band of the present AISC formula, eqn. (40), which is seen to be remarkably accurate for the case of high intensity of axial load, conservative for moderate axial load, and unconservative for low axial load. The proposed modified AISC formula (eqn. (43)), is compared in Fig. 24 and is seen to give a good conservative estimate of all the cases studied. The error is generally less than 10% on the conservative side.

3.8 CYCLIC INELASTIC BUCKLING OF TUBES

Offshore structures which are used mostly as platforms and supports for the facilities of oil production, must be able to sustain the heavy weight of oil production facilities. However, in most cases, the weight itself does not control the design of offshore structures, rather, the horizontal forces induced by wave or earthquake are more critical. Since these design extreme environmental loads occur infrequently during the life of the structure, it is important for the designer to ensure that the structure is adequately designed against collapse under extreme loading conditions for the planned location. Thus, offshore structures are designed for their ultimate strengths.

To investigate the ultimate strength of a structure, one must know first the precise behaviour of its components. This includes the member behaviour under axial and/or lateral loads, and the connection behaviour under various load conditions. The axial load deformation behaviour of pipes has been the subject of intensive investigations in recent years. For the most part, the investigations for tubular members under axial load have been directed in the past mainly toward the study of behaviour of tubes up to ultimate strength (Toma and Chen, 1979, Wagner *et al.*, 1976, 1977, and Sherman, 1978). This type of study, however, is not entirely realistic for plastic collapse analysis. For collapse analysis under extreme load conditions, the entire behaviour including post-buckling or cyclic loading must be studied. Some investigations on this problem have been carried out in recent years (see for example,

Higginbothan, 1976, Jain *et al.*, 1978, Kahn and Hanson, 1976, Prathuangsit *et al.*, 1978, Popov *et al.*, 1978–80, Chen and Toma, 1980, Toma, 1980, and ASCE, 1978).

3.8.1 Moment–Curvature Relations

Figure 25 shows the M–ϕ relations for proportional loading cases obtained by using the tangent stiffness method described previously (Chen and Atsuta, 1976). The stress–strain relation of the material is assumed to be of the elastic–perfectly plastic type. It can be seen from the figure that the transition from elastic to plastic is gradual and smooth, and that the M–ϕ relations are symmetric and their traces are the same under second cycle of loading. However, with the effect of residual stress, the loops are shifted slightly.

3.8.2 Load-Deflection Behaviour

Figure 26 shows a typical cyclic behaviour of pin-ended column with $L/r = 80$. Newmark's integration method is used here to obtain the curves. The solid line and the dashed line represent the load–shortening and load–deflection relation, respectively. An initial out-of-straightness of 0·1% is used here. Four different reversed loading paths starting from the post-buckling branch are calculated. The reversed load–deflection curves (dashed lines) are almost parallel to each other in the elastic unloading–tensioning range, and after a plastic hinge is formed at centre cross-section under tension, the slope of the curves becomes flatter. Eventually, all the reversed load–deflection curves merge to the same curve as shown in Fig. 26.

The load–deflection curves show a similar behaviour only in the elastic unloading–tensioning range for different reversed loading paths. In the plastic tension range, the curves scatter widely but converge to the same yield strength. It is found that the larger the negative axial strain attained, the larger the positive strain for the reversed loading which can be applied. Note that the smaller the displacements of the curves, the closer the curves are to the elastic slopes.

Fixed-ended beam-columns are studied by using the hinge-by-hinge method. In this method, the behaviour of a column is divided into number of stages depending upon the plastic hinges formed. Each stage is analysed elastically by solving directly the governing differential equation. Figure 27 shows a typical behaviour of the fixed-ended column with $KL/r = 120$ subjected to different unloading paths corresponding to the lateral load $Q/Q_y = 0{\cdot}1$. It can be seen from the figure that the post-

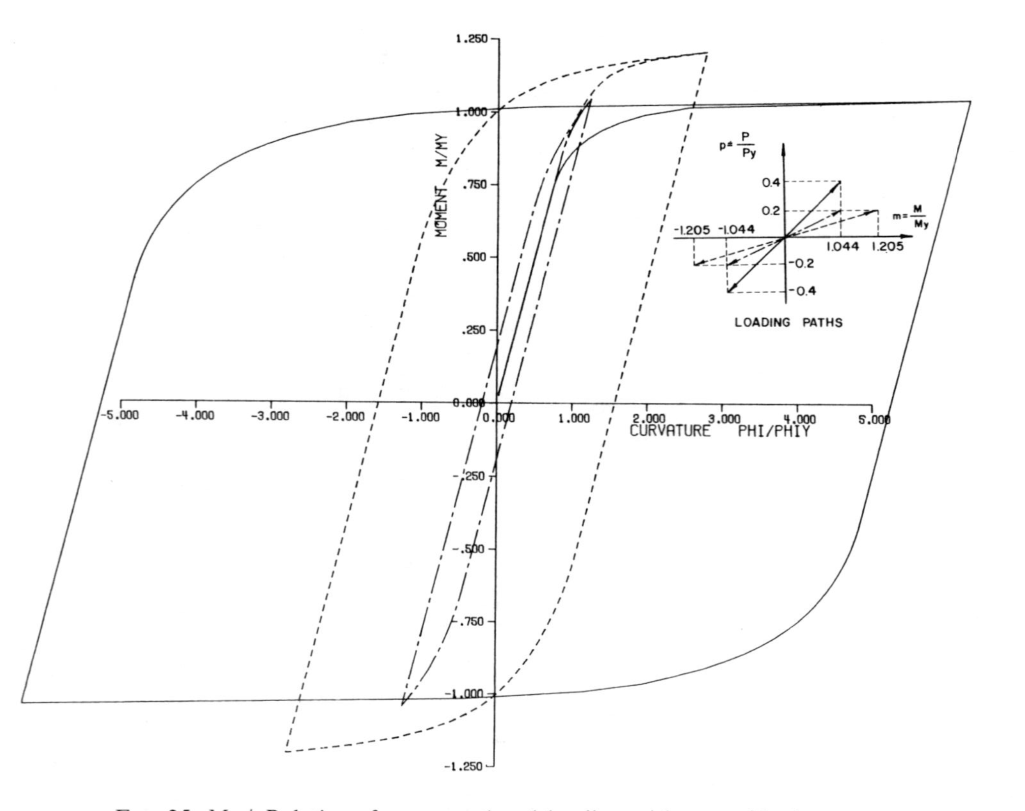

FIG. 25. M–ϕ Relations for proportional loading without residual stress.

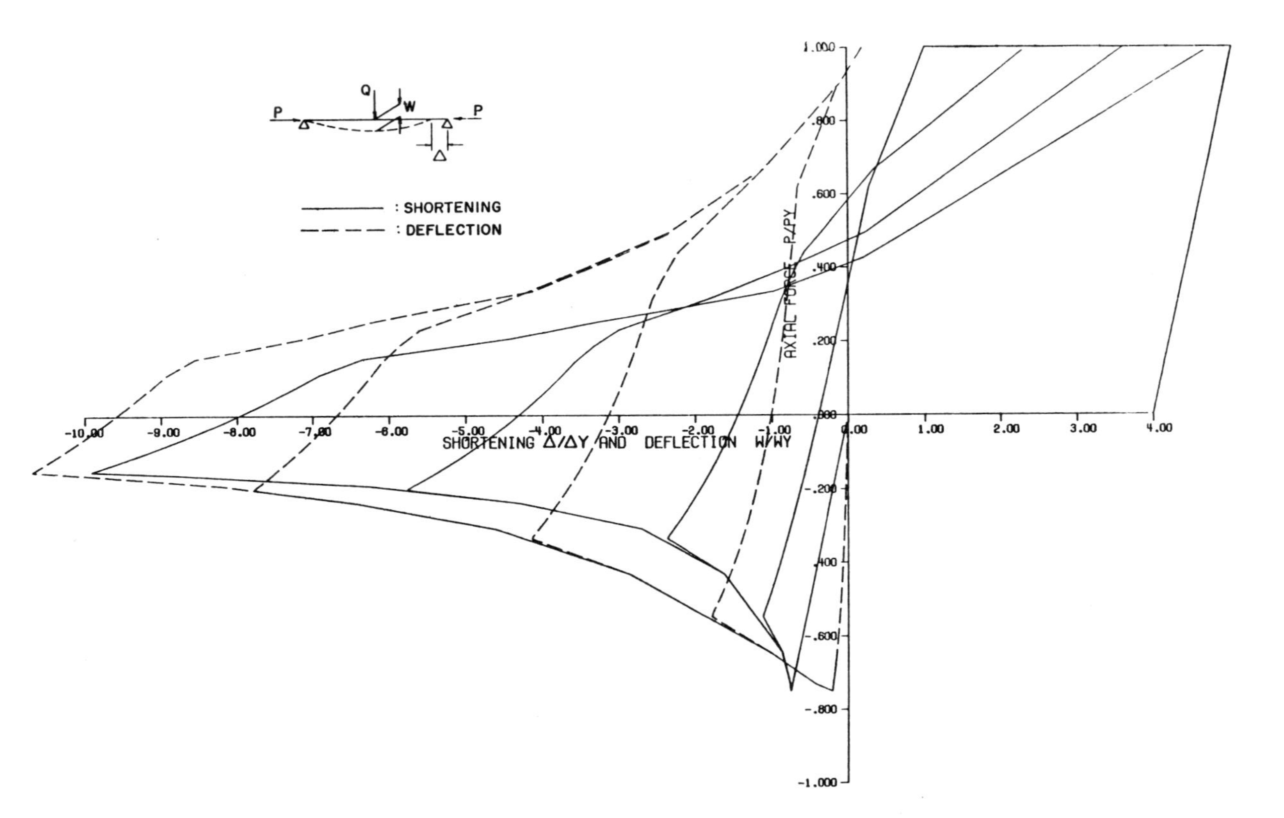

FIG. 26. Cyclic behaviour of pin-ended column, $KL/r=80$.

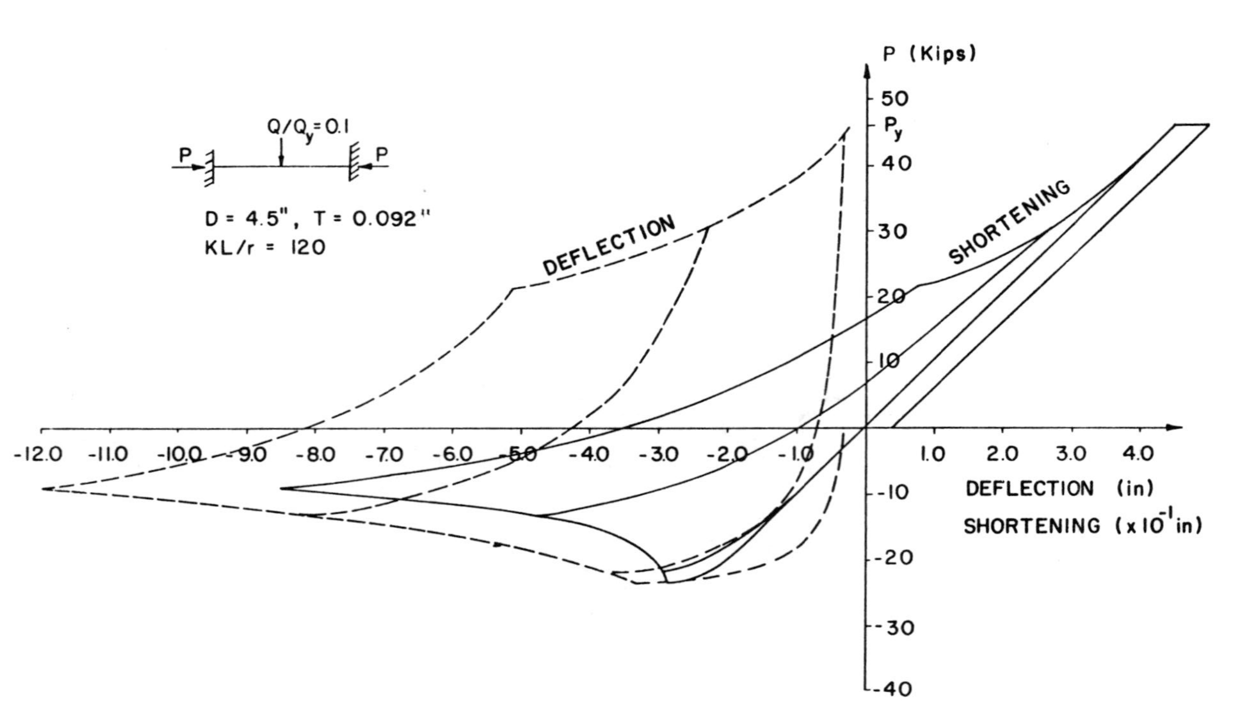

FIG. 27. Cyclic behaviour of fixed-ended column, $KL/r = 120$.

buckling and plastic tension branches are fixed, while the elastic unloading–tensioning branch connects these two fixed envelopes. The sooner the axial load is reversed, the closer the slope of the curve is to the elastic slope. Details of this are given elsewhere (Chen and Toma, 1980).

3.9 CONCLUSIONS

A general understanding of the behaviour and strength of an isolated column and beam-column of welded built-up box and fabricated cylindrical tubular sections subjected to in-plane and biaxial loading has been achieved in recent years through many analytical and experimental investigations. The analytical methods available to date are quite reliable and they can be used to predict accurately the behaviour and strength of any isolated column and beam-column. These predictions have very good accuracy and little variability when applied to column tests for which the required data are carefully determined, which include material properties, initial deflections, residual stresses, and loading and boundary conditions. As a result of this analysis capability, a considerable amount of data has been obtained and critically studied. Against the background of this information, several design formulas have been derived and developed in recent years in the area of fabricated cylindrical tubular members and braces as commonly used in offshore structures, and biaxially loaded columns of welded built-up sections as used in building frames.

The general validity of these proposed interaction formulas has been demonstrated by comparisons of computed loads with test results on non-sway columns and beam-columns. There have been few comparisons with sway columns and beam-columns, particularly in the case of biaxial bending. For further development in this aspect, a systematic check of the proposed design formulas with analytical and experimental results on sway columns and beam-columns is required. Only through this critical evaluation, can the design formula be recommended for general use.

REFERENCES

AISC (1969). *Specifications for the Design, Fabrication and Erection of Structural Steel for Buildings*, American Institute of Steel Construction.

API (1972). *Specifications for Fabricated Structural Steel Pipe*, American Petroleum Institute, API Specification 2B.

ASCE (1971). *Plastic Design in Steel, A Guide and Commentary*, ASCE Manuals and Reports on Engineering Practice No. 41.

ASCE (1978). Inelastic Behavior of Members and Structures, ASCE Chicago Convention, October 16–20, 1978, preprint 3302.

CHEN, W. F. (1977). Theory of Beam-Columns—The State-of-the-Art Review, *Proceedings, International Colloquium on Stability of Structures Under Static and Dynamic Loads*, SSRC/ASCE, Washington, D.C.

CHEN, W. F. (1980). End Restraint and Column Stability, *Journal of the Structural Division, ASCE*, **106** (ST11), November, Proc. Paper 15796, 2279–95.

CHEN, W. F. (1981). Recent Advances on Analysis and Design of Steel Beam-Columns in USA, *Proceedings of the 1981 U.S.-Japan Seminar on Inelastic Instability of Steel Structures and Structural Elements*, May 25–29, Tokyo, Japan.

CHEN, W. F. (1982). *Plasticity in Reinforced Concrete*, McGraw-Hill, New York.

CHEN, W. F. and ATSUTA, T. (1976/77). *Theory of Beam-Columns, In-Plane Behavior and Design*, Vol. 1. *Space Behavior and Design*, Vol. 2. McGraw-Hill Publishing Co., Inc., New York.

CHEN, W. F. and CHEONG-SIAT-MOY, F. (1980). Limit States Design of Steel Beam-Columns, *Solid Mechanics Archives*, **5** (1), 29–74.

CHEN, W. F. and MCGRAW, J. (1977). Behavior and Design of HSS-Columns Under Biaxial Bending, *Proceedings on Advances in Civil Engineering Through Engineering Mechanics*, Second Annual Engineering Mechanics Division Specialty Conference, Raleigh, North Carolina, May 23–25.

CHEN, W. F. and ROSS, D. A. (1978). *Test of Fabricated Tubular Columns*, Fritz Engineering Laboratory Report No. 393.8, Lehigh University, Bethlehem.

CHEN, W. F. and ROSS, D. A. (1977). Tests of Fabricated Tubular Columns, *Journal of Structural Division, ASCE*, **103** (ST3), Proc. Paper 12809, 619–34.

CHEN, W. F., and ROSS, D. A. (1976). The Axial Strength and Behavior of Cylindrical Columns, OTC Paper No. 2683, Eighth Annual Offshore Technology Conference, Houston, Texas, May 3–6, pp. 741–54.

CHEN, W. F. and TOMA, S. (1980). *Inelastic Cyclic Behavior of Tubular Members*, Report No. CE-STR-80-24, Purdue University, West Lafayette.

CHEN, W. F. and TOMA, S. (1979). *Effect of External Pressure on the Axial Capacity of Fabricated Tubular Columns*, Final Report Submitted to Exxon Production Research Co., Houston, TX, Report No. CE-STR-79-1, Purdue University.

CHEN, W. F., TOMA, S., and YUAN, R. L. (1979). Strength of Fabricated Tubular Columns in Offshore Structures, *Proceedings of the International Conference on Thin-Walled Structures*, University of Strathclyde, Glasgow.

GALAMBOS, T. V. (1981). Beam-Columns, Paper prepared for 'The Bruce G. Johnston Session on Stability of Metal Structures: A World View,' 1981 ASCE International Convention at New York.

HIGGINBOTHAN, A. B. (1976). Axial Hysteretic Behavior of Steel Member, *Journal of the Structural Division, ASCE*, **102** (ST7), Proc. Paper 12245, 1365–81.

JOHNSTON, B. G., ed. (1976). *Guide to Stability Design Criteria for Metal Structures*, 3rd ed., John Wiley and Sons, Inc. New York.

JAIN, A. K., GOEL, S. C. and HANSON, R. D. (1978). Inelastic Response of Restrained Steel Tubes, *Journal of the Structural Division, ASCE*, (ST6), Proc. Paper 13832, 887–910.

KAHN, L. K. and HANSON, R. D. (1976). Inelastic Cycles of Axially Loaded Steel Members, *Journal of the Structural Division, ASCE*, **102** (ST5), Proc. Paper 12111, 947–60.

MARSHALL, P. J. and ELLIS, J. S. (1970). Ultimate Biaxial Capacity of Box Steel Columns, *Journal of the Structural Division*, ASCE, **96** (ST9), 1873–87.

PILLAI, S. U. and ELLIS, J. S. (1974). Beam-Columns of Hollow Structural Sections, *Canadian Journal of Civil Engineering*, **1** (4).

POPOV, E. P., ZAYAS, V. A., and MAHIN, S. A. (1978). Cyclic Inelastic Buckling of Thin Tubular Columns, ASCE Preprint, 3302, Inelastic Behavior of Members and Structures, ASCE Chicago Convention.

POPOV, E. P., MAHIN, S. A., and ZAYAS, V. A. (1980). Inelastic Cyclic Buckling of Tubular Braced Frames, Preprint 80–135, ASCE Convention in Portland.

POPOV, E. P., ZAYAS, V. A., and MAHIN, S. A. (1979). Cyclic Inelastic Buckling of Thin Tubular Columns, *Journal of the Structural Division, ASCE*, **105** (ST11), 2261–77.

PRATHUANGSIT, D., GOEL, S. C., and HANSON, R. D. (1978). Axial Hysteresis Behavior with End Restraints, *Journal of the Structural Division ASCE*, (ST6), Proc. Paper 13831, 883–96.

ROSS, D. A., and CHEN, W. F. (1976). Design Criteria for Steel I-Columns Under Axial Load and Biaxial Bending, *Canadian Journal of Civil Engineering*, **3** (3).

SHERMAN, D. R. (1978). Cyclic Inelastic Behavior of Beam-Columns and Struts, ASCE Preprint 3302, Inelastic Behavior of Members and Structures, ASCE, Chicago Convention.

TALL, L. (1961). *The Strength of Welded Built-Up Shapes*, PhD Dissertation, Department of Civil Engineering, Lehigh University, Bethlehem, PA.

TEBEDGE, N. and CHEN, W. F. (1974). Design Criteria for Steel H-Columns Under Biaxial Loading, *Journal of the Structural Division, ASCE*, **100** (ST3), Proc. Paper 10400.

TIMOSHENKO, S. P. and GERE, J. M. (1961). *Theory of Elastic Stability*, Chapter 5, McGraw-Hill, New York.

TOMA, S. (1980). *Analysis of Fabricated Tubular Columns*, PhD Thesis, School of Civil Engineering, Purdue University, West Lafayette, IN.

TOMA, S. and CHEN, W. F. (1979). Analysis of Fabricated Tubular Columns, *Journal of the Structural Division, ASCE*, (ST11), Proc. Paper 14994, 2343–66.

WAGNER, A. L., MUELLER, W. H., and ERZURUMLU, H. (1977). Ultimate Strength of Tubular Beam-Columns, *Journal of the Structural Division, ASCE*, ST1, 9–22.

WAGNER, A. L., MUELLER, W. H., and ERZURUMLU, H. (1976). Design Interaction Curve for Tubular Steel Beam-Columns, OTC Paper No. 2684, Offshore Technology Conference, Houston, TX, 1976, pp. 755–64.

Chapter 4

COMPOSITE COLUMNS IN BIAXIAL LOADING

K. S. Virdi
Department of Civil Engineering,
The City University, London, UK
and
P. J. Dowling
Imperial College of Science and Technology, London, UK

SUMMARY

An analytical method for computing the ultimate failure loads of composite columns, subjected to an axial load and biaxial end moments, is presented. As an integral part of the procedure, a rapid method of establishing the moment–thrust–curvature relations is described. The efficiency of the method stems from the use of Gauss quadrature formulae. With the aid of a simple mapping technique, any quadrilateral component of the cross-section can be accurately considered, including any arbitrary residual stress pattern. This enables the analysis of both concrete encased steel stanchions and concrete filled rectangular hollow sections. The ultimate load is obtained by establishing the highest load for which an equilibrium deflected shape can be obtained. For this latter purpose, a technique based on the generalised Newton–Raphson method is described. The question of convergence of the method is discussed followed by comparison with a series of experimental results.

4.1 INTRODUCTION

The most common form of composite column construction is the steel stanchion encased in concrete. Few designers, however, take adequate account of the contribution made by concrete to the strength of the column. The same applies to another form of composite column, namely concrete filled hollow structural sections whether circular or rectangular. This has been largely due to a lack of knowledge about the behaviour of composite columns. The principal factors that make the analysis of such columns complicated are the nonlinear material characteristics of both concrete and steel, geometrical imperfections in the form of a lack of initial straightness, and residual stresses in the steel section.

The approach adopted in studying the ultimate behaviour of such columns is to calculate the deflected shape of the column first under a small fraction of the applied loading. The process is repeated with higher levels of the loading, until at a certain level, a deflected shape in equilibrium with the applied loading cannot be obtained. The highest load for which a deflected shape can be obtained is then taken as the ultimate load. Beyond this stage, equilibrium deflected shapes with increasing deflections may be obtained by reducing the level of applied loading. In this manner, the complete load versus deflection response of the column can be established.

To determine the equilibrium deflected shape of a column of inelastic materials, it becomes imperative to establish the moment–thrust–curvature relations for the particular cross-section. For an elastic cross-section of a single material subjected to uniaxial bending moments, these can be adequately represented by the well-known equation:

$$\frac{f}{y}=\frac{M}{I}=\frac{E}{R} \tag{1}$$

in which f is the flexural stress only, to which must be added the direct stress due to the applied thrust, to obtain the total stress.

Where the cross-section is composed of two or more materials which may have nonlinear stress–strain characteristics and, in addition, where the shape of the cross-section is not simple (circular or rectangular), a closed-form evaluation of the moment–thrust–curvature relations becomes very involved if not altogether impossible. The problem is made more acute when the column is subjected to biaxial bending. Use of numerical procedures then becomes unavoidable. The method described here is very general. It can be applied to a variety of cross-sections including concrete-encased steel stanchions, concrete-filled rectangular hollow sections, as well as reinforced concrete columns. The method yields results which are sufficiently accurate, and yet the procedure requires a minimum of computations.

4.2 ASSUMPTIONS

It is assumed that there exists a complete interaction between steel and concrete. This is always assumed in the case of reinforced concrete sections. Tests on composite columns have shown (Virdi and Dowling, 1973) that this is a reasonable assumption even for columns in biaxial bending.

The stress–strain curves for concrete and steel are assumed to be reversible. This implies that all plastic strains are fully recoverable. Clearly, this assumption does not agree with the observed behaviour for most materials, but the error introduced by this assumption is not likely to influence the computed values of ultimate collapse load.

Twisting and warping effects on the cross-section are ignored as are the effects of shear stresses. This facilitates the assumption that plane sections before bending remain plane upon flexure. As a corollary, the neutral axis can be assumed to be a straight line. The inclination of the neutral axis with respect to, say, the principal axes of the cross-section depends upon the shape and composition of the section, as well as on the applied loading. Establishing this dependance is the essence of moment–thrust–curvature relationships.

A further assumption made is that the deflections are small, so that curvature can be expressed by the second derivatives of deflections.

4.3 STRESS–STRAIN CHARACTERISTICS

The method described later for the moment–thrust–curvature calculations is so general that virtually any stress–strain characteristic for the composing materials can be adopted. Thus, realistic curves for steel including the plastic plateau and strain-hardening zones (Fig. 1) can be included. For concrete, it becomes feasible to specify a falling branch of the stress–strain curve (Fig. 2) and a crushing strain beyond which the spalling is simulated by assuming that the stress is reduced to zero. Tensile cracking is similarly treated.

In the computer program, several other stress–strain characteristics have been implemented, e.g. a polynomial equation, a multilinear curve, curves recommended in CP 110 (1972) and CEB-FIP Recommendations (1970), etc.

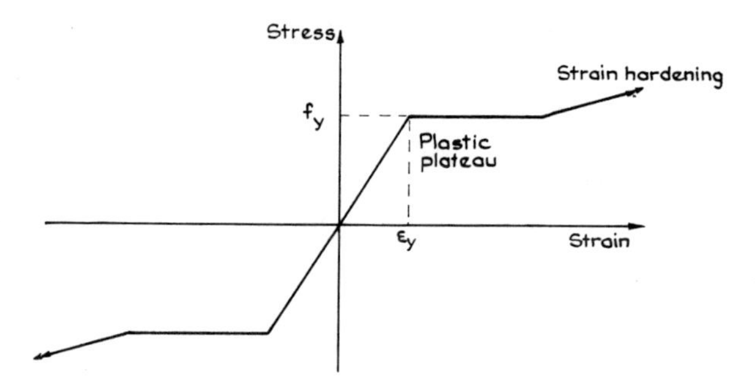

FIG. 1. Typical stress–strain curve for steel.

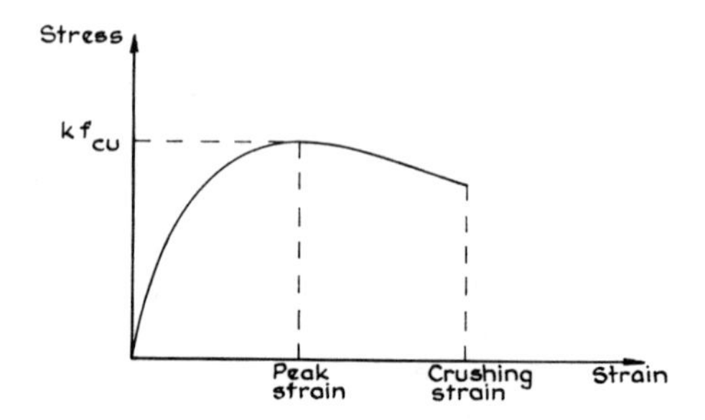

FIG. 2. Typical stress–strain curve for concrete. Ordinarily k has values between 0·6 and 0·7.

4.4 MOMENT–THRUST–CURVATURE CALCULATIONS

The moment–thrust–curvature relations, in the absence of torsional and warping effects, involve six variables: biaxial curvatures ϕ_x and ϕ_y, axial thrust N, biaxial moments M_x and M_y, and a sixth parameter required to locate the position of neutral axis. In the present chapter, this is taken as the distance d_n (Fig. 3) of the neutral axis from the most highly stressed corner in compression. As an alternative, the strain at the centroid of the section can be adopted as the sixth variable. In a similar manner, the biaxial curvatures ϕ_x and ϕ_y can be replaced by the total curvature ϕ and the inclination θ of the neutral axis with respect to some reference axis. By virtue of the assumption regarding plane sections, it can be shown that

$$\phi = \sqrt{\phi_x^2 + \phi_y^2} \tag{2}$$

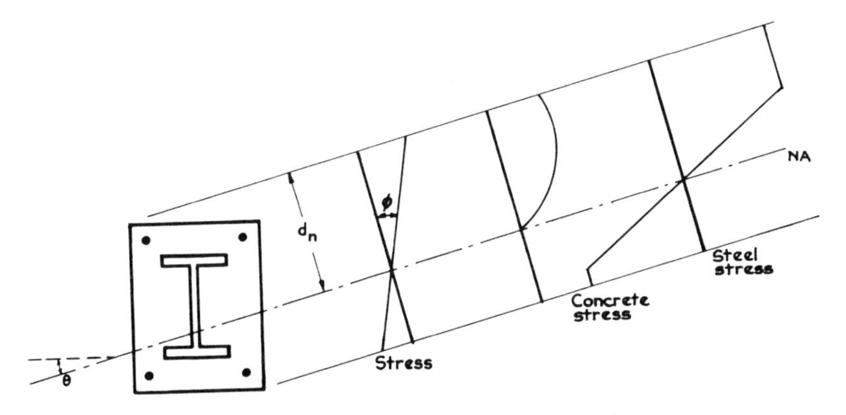

FIG. 3. A typical composite column under biaxial curvature.

and,

$$\theta = \tan^{-1}\ (\phi_y/\phi_x) \tag{3}$$

The strain at any distance d from the neutral axis is given by:

$$\varepsilon = \phi d \tag{4}$$

The corresponding stress σ can be obtained by referring to the material stress–strain curve applicable at the point. The stress resultants acting on the cross-section are then given by:

$$N = \int_A \sigma\ \mathrm{d}A \tag{5}$$

$$M_x = \int_A \sigma x\ \mathrm{d}A \tag{6}$$

$$M_y = \int_A \sigma y\ \mathrm{d}A \tag{7}$$

These three equations, in essence, represent the moment–thrust–curvature relations. By assigning known values to any three of the six variables N, M_x, M_y, ϕ_x, ϕ_y, and d_n, the other three can be determined. In the present analysis, the independent variables chosen are ϕ_x, ϕ_y, and N. This means that d_n has to be varied, so that upon integration, the chosen value of N is obtained, at which stage the biaxial moments M_x and M_y are also calculated. Various iteration schemes are possible, but in the authors' experience, a Newton type convergence technique yields rapid results.

The evaluation of the three integrals involved in eqns. (5)–(7) is not, ordinarily, a simple task, especially when the cross-section is as complicated as an encased stanchion, with the material stress–strain characteristics being highly nonlinear. A closed form solution, which would be most desirable, is virtually impossible to obtain. The simplest approach is to discretise the cross-section into a sufficiently fine grid of elemental areas. The summations are then carried out assuming that the stress over each elemental area is uniform. Typically, a concrete-encased section has to be divided into a 15×15 grid before satisfactory accuracy in integrations is achieved. This approach has been adopted by Gesund (1967), Warner (1969), Sen (1976), and also the authors in their earlier work (Virdi and Dowling, 1973, 1976). An alternative approach is to adopt Gaussian or some other type of quadrature formulae, and this has been shown by Virdi (1976, 1980, 1981a, 1981b) to be highly versatile and yet remarkably rapid.

4.5 GAUSS QUADRATURE FORMULAE

According to these formulae, a definite integral between the limits -1 and $+1$ can be replaced by a weighted sum of the values of the integrand at certain specific points. Thus:

$$\int_{-1}^{1} f(\xi)\,\mathrm{d}\xi = \sum_{i=1}^{m} H_i f(a_i) \tag{8}$$

where H_i are the weighting coefficients and $\xi = a_i$ are the specified Gauss points. The integration is exact if $f(\xi)$ is a polynomial of degree up to $(2m-1)$. In other cases, use of higher values of m results in improved accuracy. Values of H_i and a_i are available in tabular form (Kopal, 1961; Zienkiewicz, 1971) for different values of m.

A double integral, similarly, can be replaced by a double summation:

$$\int_{-1}^{1}\int_{-1}^{1} f(\xi,\eta)\mathrm{d}\xi\mathrm{d}\eta = \sum_{i=1}^{m}\sum_{j=1}^{n} H_i H_j f(a_i, b_j) \tag{9}$$

where H_i and H_j are the weighting coefficients and a_i and b_j are the coordinates of the points where the function $f(\xi,\eta)$ is to be evaluated.

It will be noted that the above equation implies a square area between the limits -1 and 1 for the two axes ξ and η. However, any rectangular, or for that matter a quadrilateral area can be successfully mapped onto the square area between the limits -1 and $+1$, as shown in Fig. 4. The converse mapping, from the so-called natural coordinates (ξ, η) to the Cartesian coordinates (x, y) is readily performed through the following equation:

$$x = [(1-\xi)(1-\eta) \ \ (1+\xi)(1-\eta) \ \ (1-\xi)(1+\eta) \ \ (1+\xi)(1+\eta)] \begin{bmatrix} x_p \\ x_q \\ x_r \\ x_s \end{bmatrix} \tag{10}$$

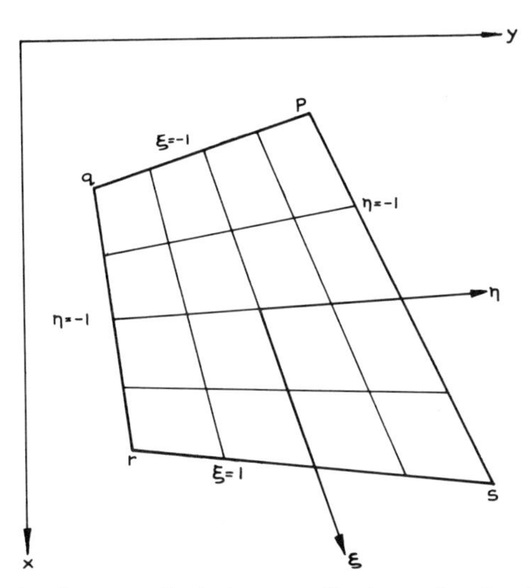

FIG. 4. Mapping between Cartesian coordinates and natural coordinates.

The y-ordinate of the mapped point is obtained by replacing the vector of x-ordinates of the four vertices by their y-ordinates. A close inspection of eqn. (10) would reveal that the function of (ξ, η) in that equation is in fact an interpolation function. Thus, providing the values of a given function, varying over the quadrilateral, are known at its four vertices, its value at a point (ξ, η), which is notionally the same as the point (x, y), can be obtained by replacing the vector of x-ordinates in eqn. (10) by the values of the function at the four vertices. This facility lends itself neatly to the treatment of residual strains (or stresses) acting within a section.

The elemental area $d\xi d\eta$ gets transformed to $dxdy$ thus:

$$dxdy = |J|\ d\xi d\eta \tag{11}$$

where,

$$[J] = \frac{1}{4}\begin{bmatrix} -(1-\eta) & (1-\eta) & -(1+\eta) & (1+\eta) \\ -(1-\xi) & -(1+\xi) & (1-\xi) & (1+\xi) \end{bmatrix} \begin{bmatrix} x_p & y_p \\ x_q & y_q \\ x_r & y_r \\ x_s & y_s \end{bmatrix}$$

Thus an integral in Cartesian coordinates can be evaluated as follows:

$$\int_A g(x, y) dxdy = \int_{-1}^{1}\int_{-1}^{1} g(x, y)|J|\, d\xi d\eta$$

$$= \sum_{i=1}^{m}\sum_{j=1}^{m} H_i H_i g(x_i, y_j)|J| \tag{13}$$

where, (x_i, x_j) are the Gauss points in Cartesian coordinates.

4.6 APPLICATION OF GAUSS QUADRATURE

It will now be shown how Gauss quadrature formulae can be used to evaluate the integrals in eqns. (5)–(7), representing the moment–thrust–curvature relations. The concrete-encased steel stanchion is first idealised by a series of rectangular blocks as shown in Fig. 5. For the three rectangles representing the steel section, integrations have to be performed twice, once for the steel area, and once for the concrete area that

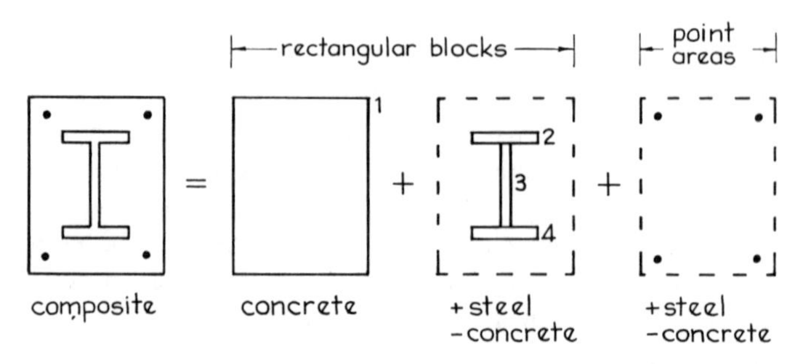

FIG. 5. Idealisation of a composite column section.

the steel has replaced. Longitudinal reinforcement bars, usually placed near the corners, are treated as point areas, and are similarly compensated for the concrete area displaced.

The application of Gauss quadrature can be illustrated for a typical rectangle. Assuming that the two biaxial curvatures ϕ_x and ϕ_y are known, and that the integration is to be performed for a chosen value of d_n, the first step would be to determine the Cartesian coordinates, using eqn. (10), for each of the Gauss points. This would lead to the calculation of distance d to the neutral axis (Fig. 6) for all the Gauss points involved.

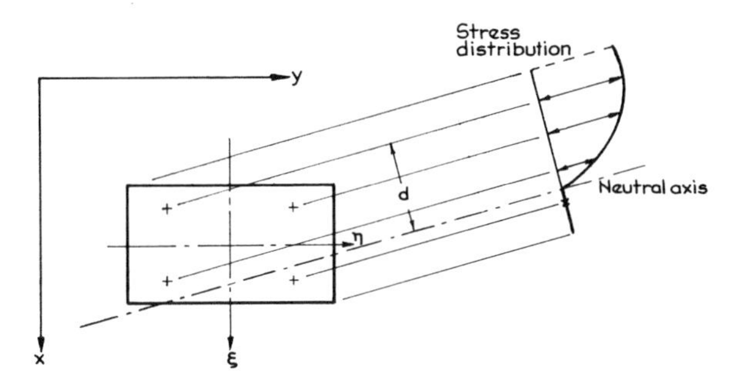

FIG. 6. A typical rectangle shown with 2 × 2 Gauss points.

The imposed strains at the Gauss points can then be calculated using eqn. (4). Where there exist residual strains in the rectangle, the strain obtained from eqn. (4) has to be added to the residual strain, interpolated for the particular Gauss point, to obtain the total strain at the point. Using the material stress–strain curve applicable, the stresses at the Gauss points can be calculated. With the known natural coordinates for each Gauss point, and the Cartesian coordinates of the vertices, the value of the determinant $|J|$ at each point is easily evaluated. The remaining step is to apply the weighting coefficients to obtain the integrals for N, M_x, and M_y.

4.7 EQUILIBRIUM DEFLECTED SHAPE USING SECOND-ORDER ITERATION

A column subjected to biaxial moments, which may be different at the two ends, is shown in Fig. 7. For simplicity, deflections and other quantities are shown only for one axis. Under the action of all end forces and moments, the biaxial curvatures at different points along the length

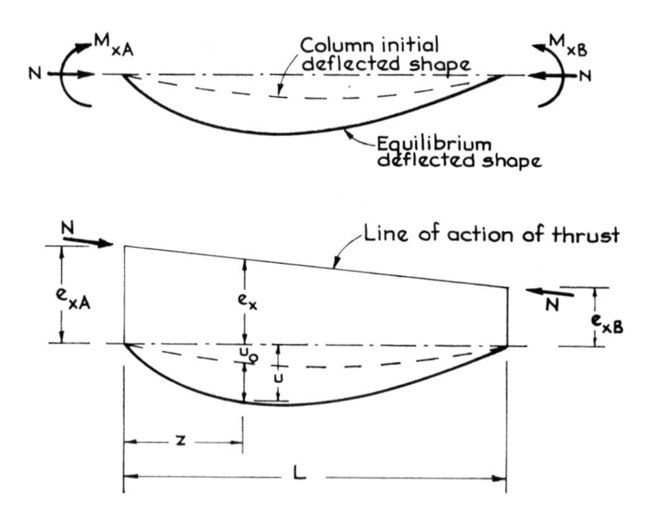

FIG. 7. Deflected shape of a column under an axial force and end moments.

would vary, giving rise to varying internal moments which have to be in equilibrium with the external forces. Thus, if the total deflections of a point on the column in the x–z and y–z planes are denoted by u and v, and the biaxial moments by M_x and M_y, one obtains:

$$M_x = N(e_x + u) \text{ or } u = (M_x/N) - e_x \tag{14a}$$

and,

$$M_y = N(e_y + v) \text{ or } v = (M_y/N) - e_y \tag{14b}$$

where, e_x and e_y are the biaxial eccentricities of the effective line of action of the thrust N with respect to the originally straight column axis, at the point in question. The internal stress resultants exist due to the induced biaxial curvatures given by:

$$\phi_x = \frac{\partial^2 u}{\partial z^2} - \frac{\partial^2 u_0}{\partial z^2} \tag{15a}$$

and

$$\phi_y = \frac{\partial^2 v}{\partial z^2} - \frac{\partial^2 v_0}{\partial z^2} \tag{15b}$$

where u_0 and v_0 represent the initial deflections representing a lack of straightness in the unloaded column.

Recalling the fact that the nonlinear material characteristics result in nonlinear moment–thrust–curvature relationships, it can be stated that M_x and M_y are nonlinear functions of u and v. The solution to the problem now requires a simultaneous solution of eqns. (14) at all points along the column length.

Adopting an approach similar to the finite difference method for elastic problems, the column length is divided into n equal parts, each of length h. The resulting nodes are labelled 1, 2, ... $(n-1)$. The equilibrium is then satisfied at the $(n-1)$ internal nodes only. Clearly, the larger the value of n, the more accurate the solution obtained will be.

At each node s, the biaxial curvatures can be calculated from the finite difference approximations:

$$\phi_{xs} = (-u_{s-1} + 2u_s - u_{s+1})/h^2 \tag{16a}$$

and

$$\phi_{ys} = (-v_{s-1} + 2v_s - v_{s+1})/h^2 \tag{16b}$$

Since M_x, M_y, and N depend upon the values of ϕ_x and ϕ_y, as implied by the moment–thrust–curvature relations, it follows that eqns. (15) can be restated thus:

$$u_s = U(u_{s-1}, u_s, u_{s+1}, v_{s-1}, v_s, v_{s+1}) \tag{17a}$$

and

$$v_s = V(u_{s-1}, u_s, u_{s+1}, v_{s-1}, v_s, v_{s+1}) \tag{17b}$$

Equations (17) can be further generalised by replacing u_s by w_{2s-1}, v_s by w_{2s}, etc. Thus:

$$\{w\} = \{W(w_1, w_2, \ldots w_{2n-3}, w_{2n-2})\} \tag{18}$$

The solution to problems represented by equations of this type can be obtained using several alternative methods. The solution technique adopted here is based on the generalised Newton–Raphson method for a system of nonlinear equations. The method suggests that if $\{w^k\}$ represents an approximate solution to eqn. (18), then a better approximation can be obtained by the formula:

$$\{w^{k+1}\} = \{w^k\} - [I - K]^{-1} \{w^k - W^k\} \tag{19}$$

where $[I]$ is an identity matrix, and $[K]$ is a Jacobian matrix, the elements of which are

$$K_{ij}=\frac{\partial W_i}{\partial w_j}=\frac{\Delta W_i}{\Delta w_j} \tag{20}$$

The incremental definition is particularly relevant here since the derivatives have to be calculated numerically.

When eqn. (20) is viewed in the light of eqns. (17), it becomes clear that the evaluation of the elements of the Jacobian matrix $[K]$ requires one basic and six incremental computations of the moment–thrust–curvature relations for each nodal point along the column length. These can be reduced to one basic and only two incremental computations of the moment–thrust–curvature relations, providing the magnitude of the increment in all u's and v's is kept the same, say Δ. In this event, it will be noted that

$$\frac{\Delta\phi_{xs}}{\Delta u_{s-1}}=\frac{\Delta\phi_{xs}}{\Delta u_{s+1}}=-\frac{1}{2}\frac{\Delta\phi_{xs}}{\Delta u_s}=-\frac{\Delta}{h^2} \tag{21}$$

Similar equations would be obtained for the other bending plane. Since the computed W_i depend directly on the curvature values, it follows that similar correlation can be obtained between the elements of the Jacobian matrix. Thus, it will be found that,

$$\frac{\partial W_{2s-1}}{\partial w_{2s-3}}=-\frac{1}{2}\frac{\partial W_{2s-1}}{\partial w_{2s-1}}=\frac{\partial W_{2s-1}}{\partial w_{2s+1}}=\frac{U'_s-U_s}{\Delta} \tag{22a}$$

and

$$\frac{\partial W_{2s}}{\partial w_{2s-3}}=-\frac{1}{2}\frac{\partial W_{2s}}{\partial w_{2s-1}}=\frac{\partial W_{2s}}{\partial w_{2s+1}}=\frac{V'_s-V_s}{\Delta} \tag{22b}$$

in which U_s and V_s are the computed deflections at nodes based on assumed deflected shape, and U'_s and V'_s are the computed deflections when the x-curvature at node s is incremented by an amount Δ/h^2. Equations (22) show how with only one incremental computation, corresponding to increments in x-deflections, at each node s, all the Jacobian coefficients relating to x-deflections can be obtained. The second incremental computation with respect to a y-curvature increment of the amount Δ/h^2 would, similarly, yield the remaining required coefficients of the Jacobian matrix.

The matrix eqns. (19) can now be formed and solved for $\{w^{k+1}\}$. The process can be repeated until some suitable criterion for convergence is satisfied. At this stage the equilibrium deflected shape corresponding to the chosen level of applied loading would be obtained.

4.8 ANALYSIS OF INSTABILITY

The procedure described above for the determination of the deflected shape of the column can be applied, in general, for the nonlinear analysis of any structure subjected to a given loading. Analysis of instability further requires that the deflection response of the structure be monitored for increasing loads until, at some stage, equilibrium is no longer possible. Thus, if v is defined as a load factor on the initial loading $\{F_0\}$, the structure is analysed for varying end loads $\{F_v\}$ given by:

$$\{F_v\}=[(v-1)G+I]\ \{F_0\} \tag{23}$$

where $[I]$ is an identity matrix and $[G]$ is a diagonal matrix the elements of which are either 1 or 0, depending upon whether the corresponding load component varies with v or not. The highest value of v so obtained would correspond to the load factor for the limit state of collapse.

In the case of a biaxially loaded column, the vector $\{F_v\}$ consists of five components, namely the applied thrust and the biaxial moments at the two ends.

4.9 ACCURACY OF THE PROPOSED METHOD

The accuracy of the method outlined above depends upon two factors: the number of Gauss points chosen for integration across the section, and the number of nodal points along the column length. The sensitivity of the computed ultimate loads to variation in these two parameters is now explained by analysing a typical composite column cross-section in two modes of bending. The section chosen is a 250 UB 37 encased in 52 mm concrete all round. Additionally, there are 4 bars of 12 mm diameter, one at each corner as shown in Fig. 8. It is assumed that the steel has a yield stress of 250 MPa, and the concrete characteristic strength is 10 MPa. The stress–strain curves for the two materials are shown in Fig. 9. In both modes of bending, the column length adopted was 5 m, with a sinusoidal initial bow in the minor bending plane of $L/1000$. In the first mode of bending (S), the column is bent in single curvature. There are equal end eccentricities at both ends in the major

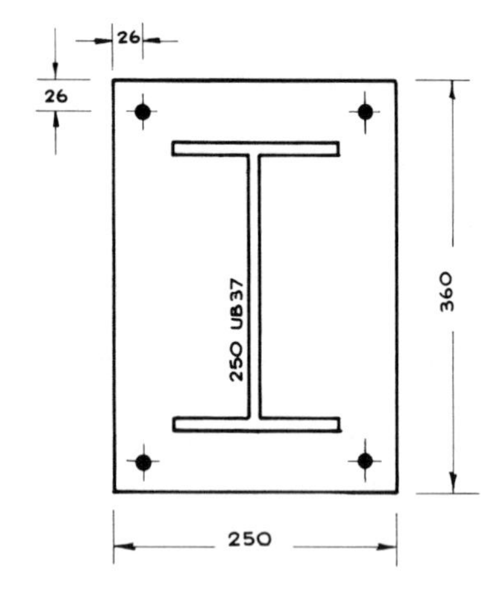

FIG. 8. 250 UB37 section encased in concrete, 52 mm cover all round. Four bars 12 mm diameter at corners.

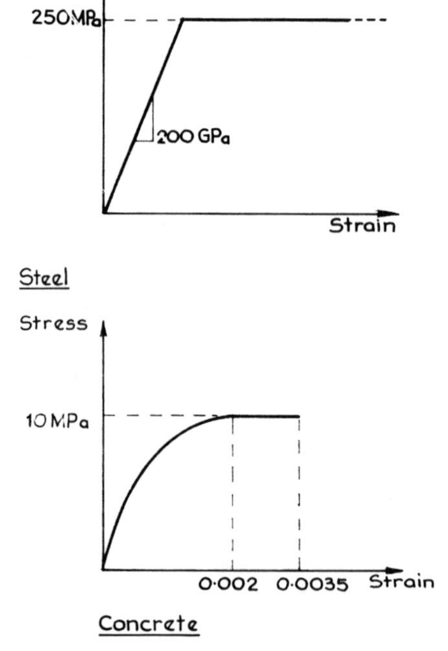

FIG. 9. Stress–strain curves for steel and concrete.

bending plane only. Because of the interaction between the eccentricities in the major bending plane and the initial lack of straightness in the minor bending plane, the column fails in biaxial bending. In the second mode of bending (D), the biaxial eccentricities are 75 mm each, but are of opposite sense at the two ends, causing the column to bend in double curvature in both planes of bending.

Both cases were analysed for a range of m, the number of Gauss integration points, varying from 2 to 6 for each rectangle. The number of segments along the column length was varied from 8 to 28. The results are given in Table 1.

TABLE 1
ACCURACY OF THE PROPOSED METHOD

Case	*Number of column segments*	*Failure load (kN) for Number of Gauss Points*				
		2	3	4	5	6
[S]	8	1392·7	1380·18	1377·4	1377·2	1376·7
	10	1395·2	1383·5	1380·9	1380·4	1379·9
	12	1396·5	1384·8	1382·2	1381·8	1381·3
	14	1397·3	1385·6	1383·2	1382·7	1382·2
	16	1398·1	1386·5	1383·8	1383·4	1382·8
[D]	8	873·4	786·2	797·0	787·7	787·6
	12	846·4	765·6	770·0	762·0	759·9
	16	834·8	753·4	755·9	747·9	745·6
	20	829·8	752·0	747·5	739·7	737·4
	24	822·5	741·0	742·0	734·1	732·1
	28	818·9	744·0	738·1	730·3	728·5

It will be noted that, for a constant number of column segments along the length, use of $m=3$ Gauss points for each rectangle yields results that are within 0·25% of the convergence value. This should be so, as long as the material stress–strain curve is a polynomial of degree less than $(2m-1)$. Of course, some allowance must be made for the fact that the biaxial nature of the integration renders the formulae slightly less accurate. It is suggested that if r is the highest degree of a polynomial among the stress–strain curves in a given case, a suitable value of m, the number of Gauss points for each rectangle in each axis, would be $m=1+(r+1)/2$ rounded up to the next integer.

The results given in Table 1 also show that for a given value of the Gauss point index m, the convergence in single and double curvature cases is obtained in different ways. For single-curvature bending, the results with fewer segments tend to be conservative, whereas for the double-curvature bending these could be non-conservative. Furthermore, the number of column segments required for satisfactory levels of error is also much larger. For the example chosen, it will be noted that with 16 column segments along the length, results are obtained within 2% of the convergence value for a constant Gauss point index. As a rule of thumb, between every pair of points of contraflexure expected, 8–12 column segments should be allowed.

4.10 COMPARISON WITH EXPERIMENTAL RESULTS

Test results on nine composite columns reported by Virdi and Dowling (1973) are now compared with computer results based on the method described above. The data relating to the nine columns are given in Table 2. The comparative ultimate load values are given in Table 3.

TABLE 2
DATA FOR TESTED COLUMNS

Column Label	*Length (in)*	*Eccentricities*		*Cube Strength* (lbf/in^2)
		x-Axis (in)	*y-Axis (in)*	
A	72	2·50	1·45	8957
B	72	5·00	2·90	8576
C	72	7·50	4·35	8957
D	144	2·50	1·45	9489
E	144	5·00	2·90	8957
F	144	7·50	4·35	9489
G	288	2·50	1·45	8234
H	288	5·00	2·90	8975
I	288	7·50	4·35	9758

In calculating the ultimate loads given in Table 3, it was assumed that the columns had a minor axis initial sinusoidal bow of amplitude 0·001 times the column height. Also, based on experimental evidence reported by Virdi and Dowling (1973) it is assumed that the maximum compressive strength of concrete in the column is only 0·64 times the observed cube strength. No additional safety factor on material strengths was applied. For steel, a bilinear elastic–plastic stress–strain characteristic

TABLE 3
COMPARISON WITH EXPERIMENTAL RESULTS

Column	*Experimental Ultimate load (tonf)*	*Theoretical Ultimate load (tonf)*	*Ratio (3)/(2)*
(1)	(2)	(3)	(4)
A	126	137·66	1·093
B	65	71·67	1·103
C	47·5	46·95	0·988
D	93	110·43	1·187
E	57·5	61·23	1·065
F	42	42·75	1·018
G	67	56·10	0·837
H	35·5	37·76	1·064
I	29·5	29·36	0·995
		Mean	1·039
		Standard Deviation	9·7%

was adopted with yield stress of 45 642 lbf/in^2 and a related yield strain of 0·001 55. For concrete, a stress–strain curve in line with CEB–FIP Recommendations was adopted. It will be noted that, except for column G the agreement between the experimental and analytical ultimate loads is very good indeed.

4.11 APPLICATIONS

The method outlined above offers a tool for parametric studies on a range of column sections in biaxial bending. The computer programs developed have been used to validate an interaction formula for composite columns (Virdi and Dowling, 1973). It is not difficult to see that the procedure outlined above can be readily applied to concrete-filled rectangular hollow steel sections, reinforced concrete sections, and torsionally stiff bare steel sections. Some of these applications have been described by Virdi (1980, 1981a, 1981b).

4.12 CONCLUSION

A general method for a rapid ultimate analysis of biaxially loaded rectangular composite columns has been described. The efficiency of the method lies in two advanced numerical techniques adopted. First, Gauss quadrature formulae have been used to speed up the integrations involved in the essential phase of moment–thrust–curvature calculations. Second, a second-order iteration technique, the generalised Newton–Raphson method, has been adopted to evaluate the deflections of the column at evenly spaced points along the column length.

The influence of two parameters on the accuracy of the results has been explored. It is shown that for columns having materials with a quadratic stress–strain curve, it is sufficient to have 3×3 Gauss points for each rectangle forming the section. It is suggested that where the degree of the highest polynomial among the material stress–strain curves is r, it would be sufficient to adopt the Gauss first index $m = 1 + (r + 1)/2$. Also investigated was the number of column segments required to achieve a satisfactory level of accuracy. It has been shown that for columns in single-curvature bending, 8 column segments along the length are sufficient, whereas for columns in double-curvature bending, the minimum number of column segments required seems to be 16. Further, for double-curvature bending fewer than the required number of segments may lead to unacceptable errors on the non-conservative side.

REFERENCES

CEB–FIP (1970). *International Recommendations for the design and construction of concrete structures. Principles and Recommendations*, Cement and Concrete Association.

CP110 (1972). *The Structural Use of Concrete*, British Standards Institution.

GESUND, H. (1967). Stress and moment distributions in three dimensional frames composed of non prismatic members made of nonlinear material. Chapter 13 in *Space Structures*, pp. 145–153, Blackwell, Oxford and Edinburgh.

KOPAL, Z. (1961). *Numerical Analysis*, Chapman and Hall, London.

SEN, T. K. (1976). *Inelastic H-Column performance at high axial loads*, PhD Thesis, University of London.

VIRDI, K. S. (1976). A new technique for moment–thrust–curvature calculations for columns in biaxial bending, *Sixth Australian Conference on Mechanics and Strength of Materials*, Canterbury, New Zealand, pp. 307–313.

VIRDI, K. S. (1980). Variable cross-section columns loaded up to failure, *International Conference on Numerical Methods for Nonlinear Problems in Engineering*, pp. 553–564, Pineridge Press, Swansea.

VIRDI, K. S. (1981a). Biaxially loaded slender reinforced concrete columns. *Advanced Mechanics of Reinforced Concrete Colloquium*, Delft.

VIRDI, K. S. (1981b). Design of Circular and Rectangular Hollow Section Columns, *Journal of Constructional Steel Research*, September, 35–45.

VIRDI, K. S. and DOWLING, P. J. (1973). The ultimate strength of composite columns in biaxial bending, *Proc. Instn Civ. Engrs*, March, 251–272.

VIRDI, K. S. and DOWLING, P. J. (1976). The ultimate strength of biaxially restrained columns, *Proc. Instn Civ. Engrs*, March, 41–68.

WARNER, R. F. (1969). Biaxial moment–thrust–curvature relations. *Journal of the Structural Division, Proc. Am. Soc. Civ. Engrs*, May.

ZIENKIEWICZ, O. C. (1971). *The Finite Element Method in Engineering Science*, McGraw Hill, London.

Chapter 5

COLD-FORMED WELDED STEEL TUBULAR MEMBERS

BEN KATO
Department of Architecture, University of Tokyo, Japan

SUMMARY

Buckling behaviour of cold-formed structural hollow section columns is discussed. Local buckling strength and flexural buckling strength of these columns are evaluated taking the deformational residual stresses and strain-hardening of material due to cold-forming into consideration. The effect of stress-relieving on the improvement of buckling behaviour is assessed. Finally, preliminary recommendations are presented for the design of this type of column in association with the SSRC column curves.

5.1 INTRODUCTION

Throughout the last decade, the use of manufactured tubular sections has increased significantly, a recognition of the inherent advantages of such shapes for a variety of structural purposes. Since the manufacturing methods may vary from one producer to the next, depending on equipment, technical preference, and the intended applications, a host of products called 'manufactured tube' is at present available.

Typical manufactured tubes, classified according to the forming processes and the heating conditions used in manufacture, are as follows:

1. *Seamless*: A heated bar is pierced one or more times, rolled with an internal mandrel.
2. *Casting*: Usually, centrifugal force is applied during the casting by rotating the tube.

3. *Continuous butt weld*: A continuous strip is heated and fed through forming and welding rolls where edges are forged together by pressure at a high temperature, and sometimes referred to as 'hot-formed'.
4. *Electric resistance welding*: The plate is cold-formed by rolls into a circular shape, and the edges are heated to welding temperature by resistance to the flow of an electric current. This is referred to as 'cold-formed and electric resistance welded'.
5. *Fusion welding*: After cold-forming to a circular shape, the edges are fusion welded (usually by submerged-arc method), either longitudinally or spirally, depending on the type of forming.
6. *Hot-finished*: If heat is applied before the final sizing in forming by method 4, the tube is referred to as 'hot-finished'.
7. *Cold-formed stress relieved*: If heat is applied after forming by method 4, the product is referred to as 'cold-formed stress relieved'.

The structural hollow sections produced by methods 1, 2, 3, and 6 are deemed to be residual stress free, and to have no change in mechanical properties from base material. In cold-formed and electric resistance welded structural hollow sections, residual stresses are introduced by cold-forming, and at the same time, material strength is increased by the strain-hardening. These two effects will strongly influence the buckling behaviour of this type of column.

This chapter describes the analysis and measurement of residual stresses, local buckling behaviour, and flexural buckling behaviour of cold-formed and electric resistance welded tubular columns, and attempts to improve the predictability of the performance of these products as structural compression members. The structural performance of cold-formed stress-relieved tubular members is surveyed briefly in order to assess the effect of stress-relieving. Very little information about the structural behaviour of tubular members made by fusion welding is available at the present time.

5.2 MANUFACTURING PROCESS OF COLD-FORMED ELECTRIC RESISTANCE WELDED TUBES

The standard manufacturing process of cold-formed and electric resistance welded tubes is illustrated in Fig. 1. In the forming steps, the predominant cold-working which affects the mechanical properties of

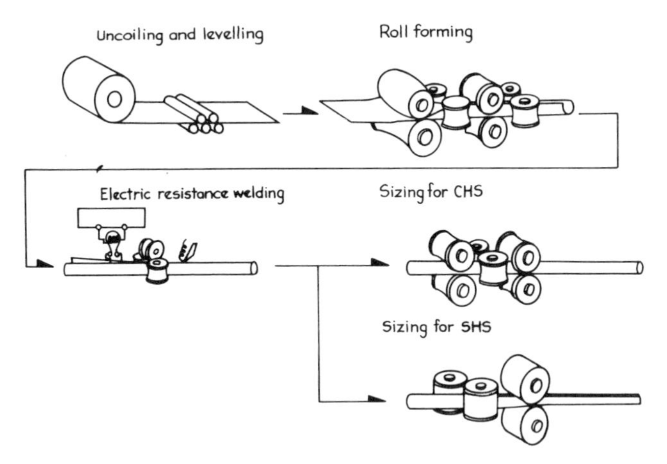

FIG. 1. Manufacturing processes of cold-formed steel tubes.

steel tubes are (1) uncoiling and levelling, (2) roll forming, and (3) sizing. These are characterised as follows;

(1) Uncoiling and levelling
It is assumed that no cooling residual stresses are contained in a sheet coil because the sheet coil is kept in the tempering furnace. The sheet coil is flattened in this process. The plate can be considered to be bent plastically in the plane-strain condition in the longitudinal direction. When the plate has passed through the leveller, it is released from the bending moment applied by the leveller. The plate may be unloaded elastically and thus deformational residual stresses will be left in the plate.

(2) Roll-forming
The plate is bent along the direction of the width to form a tube and then welded. Also in this process, the plate is bent in the plane-strain condition. Elastic and plastic bending stresses are not released this time and will be locked in the tube. The cooling residual stress due to the electric resistance welding, the heat input of which is very small compared with that in arc-welding, is negligibly small as will be demonstrated later.

(3) Sizing
For circular hollow section (CHS), the tube is compressed uniformly in the radial direction to reduce the diameter to a certain extent. This

process is necessary to ensure a proper profile. In the case of square hollow section (SHS), the additional rollers force the circular cross-section into a square shape. When a tube has passed through the sizer, it is released from bending and/or radial compression, and is unloaded elastically. Thus another set of deformational residual stresses are induced in the tube.

5.3 ANALYSIS AND MEASUREMENT OF RESIDUAL STRESSES

Kato and Aoki (1978) analysed strains and stresses in CHS tubes at each step of the forming process mentioned above and the analytical prediction thus obtained was compared with the experimental measurements.

The basic assumptions made for this analysis are: (1) Plane sections normal to the geometric axis of the plate before deformation remain plane and normal to this axis after deformation (Navier's assumption); (2) The normal stress in the thickness direction can be ignored; (3) The stress–strain relationship is defined by the Prandtl–Reuss equations applicable to a strain-hardening material; (4) The yield locus moves and changes its shape depending on the strain hysteresis of the material and on the change of direction of the principal axis. Among the rules controlling this phenomenon, the hypothesis of isotropic hardening after Prandtl–Reuss and that of independent hardening based on a slip theory (Batdorf and Budianski, 1949) are assumed as the upper and lower bound respectively, (5) The Bauschinger effect is ignored. An example of the analytical results is shown in Fig. 2. Stress distributions in an arbitrary point of a tube wall are shown both in circumferential and longitudinal directions for each step of the forming process. Considerable plastic bending stresses are locked in along both circumferential and longitudinal directions at the finished state.

To examine the accuracy of the prediction, cold-formed and welded steel tubes were cut into small coupons and the released strains were measured. An example of measured values is shown in Fig. 3; these are compared with the theoretical prediction in the figure. Though the measured strains fluctuate along the circumference, the theory reasonably predicts the average magnitude of the released strains. It can be observed that the cooling residual strain due to electric resistance welding is quite insignificant.

A measurement of the residual strain distribution in a cold-formed

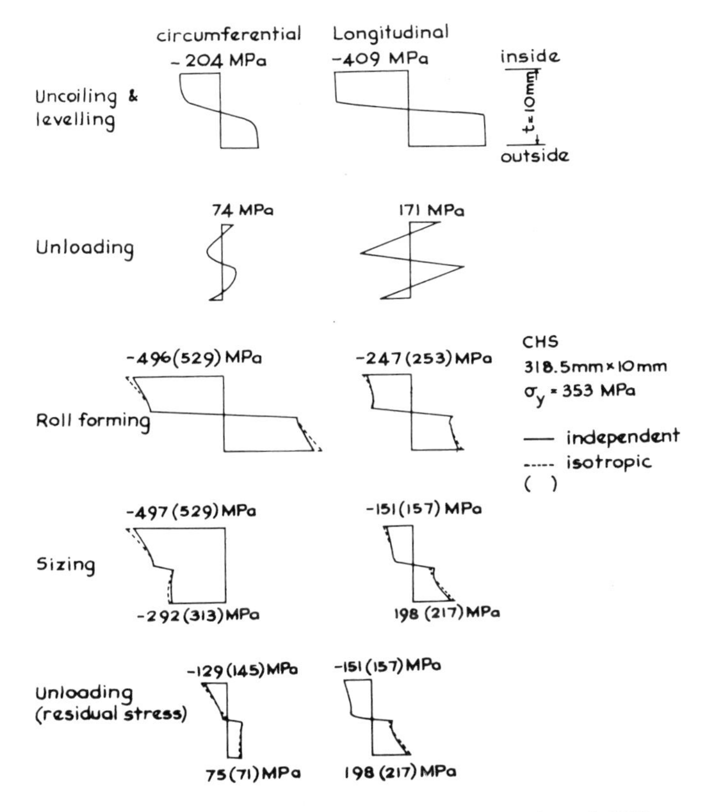

FIG. 2. Stress distributions at each step of forming of CHS.

SHS was carried out in the similar way (Katsurai, 1980), and the result is shown in Fig. 4. Sharp strain concentration was observed at the inner surface of the four corners; another interesting point is that the sign of residual bending strains along the circumference of SHS and CHS are opposite each other. This characteristic difference is illustrated in Fig. 5 schematically.

5.3.1 Load-Deformation Relationship of the CHS Stub-Column under Axial Compression

The compressive behaviour of a short cold-formed tube containing the complex residual stress patterns as described in the foregoing was analysed and compared with a test result as shown in Fig. 6 (Kato and Aoki, 1978), in which the result is expressed in terms of average stress–

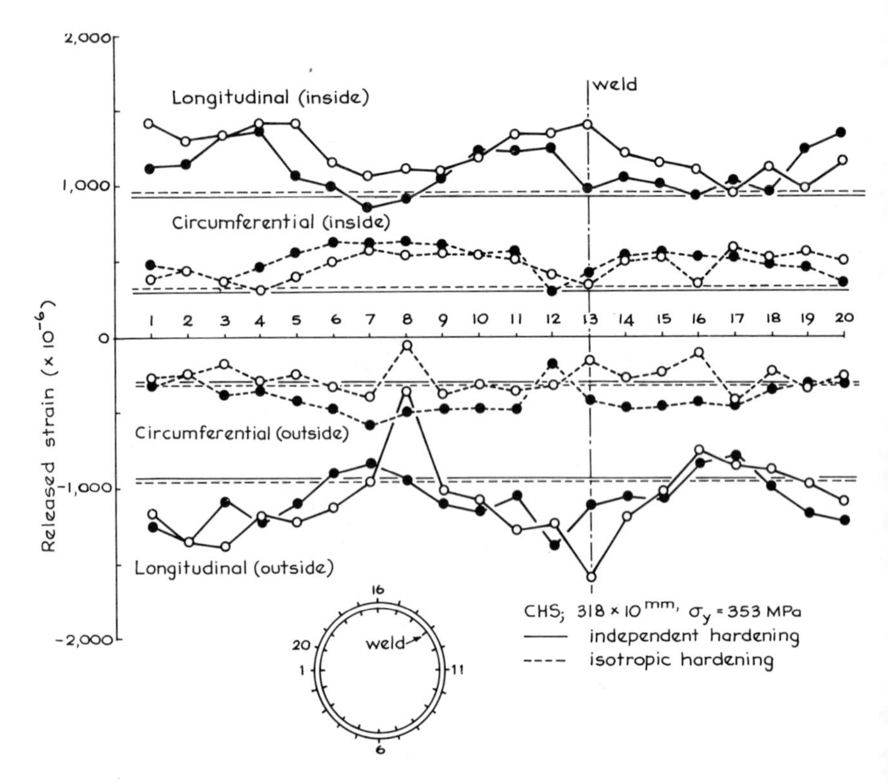

FIG. 3. Measured released strains and theoretical prediction for CHS.

strain relationship. The prediction based on the assumption of the independent hardening appears to be the lower bound and the prediction based on the assumption of isotropic hardening to be the upper bound of the actual behaviour. The proportional limit of the actual stress–strain relationship is slightly lower than those of the theoretical curves. This may be attributed to the Bauschinger effect and/or to an inadequacy of the modelling of forming processes.

5.4 LOCAL BUCKLING

The diameter-to-thickness ratios of most of the manufactured structural tubes listed in material catalogues are relatively small so that local buckling will take place only in the inelastic region. The inelastic

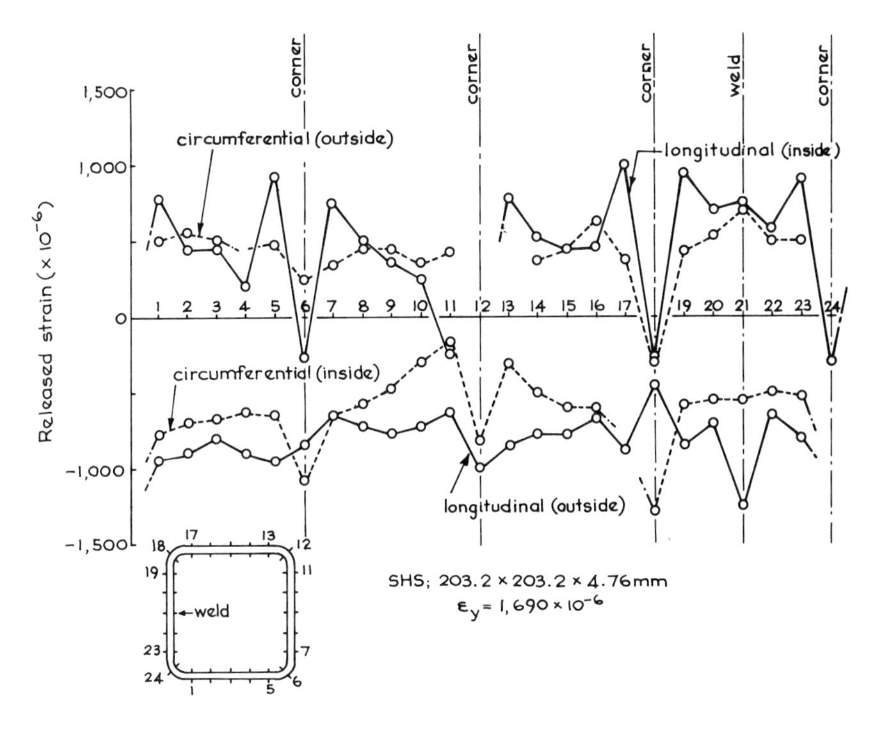

FIG. 4. Measured released strains for SHS.

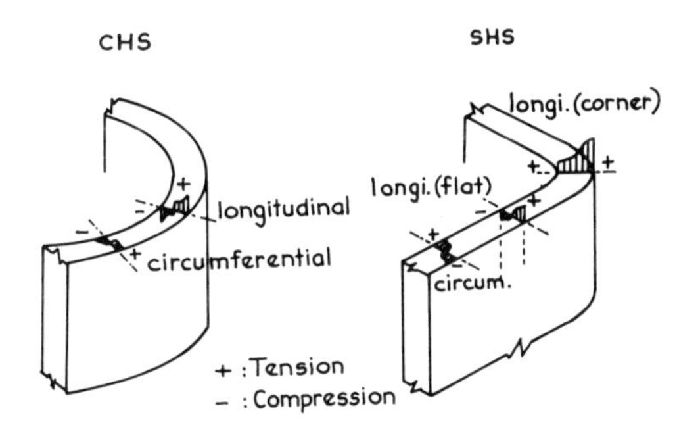

FIG. 5. Difference of residual stresses between CHS and SHS.

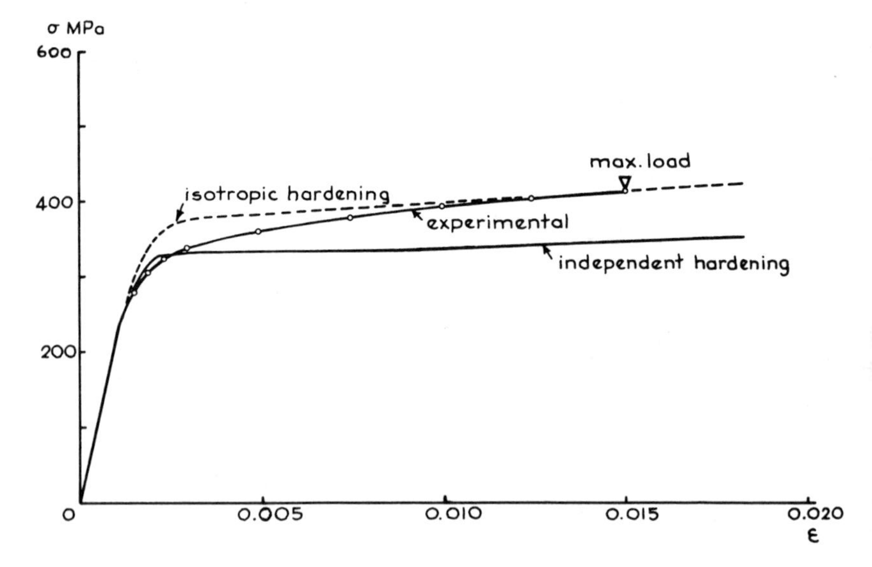

FIG. 6. Stress–strain relationships of stub-column, experimental and theoretical.

buckling stress of tubes is usually obtained in one of two ways. Either the elastic formula is used with an effective modulus in place of the elastic modulus, or empirical relations are developed for specific classes of materials. The inelastic buckling of cold-formed steel tubes, which are nonhomogeneous due to the presence of residual stresses, is more conveniently handled with an empirical formula.

Information on the inelastic local buckling of structural steel tubes is very limited. Plantema equation is probably the most commonly used formula for manufactured tubes (Plantema, 1946). This relation is given by:

$$\left.\begin{aligned} \frac{\sigma_{cr}}{\sigma_y} &= 1{\cdot}0 \quad \text{for } \alpha_c \geq 8 \\ \frac{\sigma_{cr}}{\sigma_y} &= 0{\cdot}75 + 0{\cdot}03\alpha_c \quad \text{for } 2{\cdot}5 \leq \alpha_c < 8 \\ \frac{\sigma_{cr}}{\sigma_y} &= 0{\cdot}33\alpha_c \quad \text{for } \alpha_c \leq 2{\cdot}5 \end{aligned}\right\} \tag{1}$$

in which, σ_{cr} = critical buckling stress, σ_y = yield stress of the material, and α_c is a nondimensional local buckling parameter for CHS and is expressed as

$$\alpha_c = \left(\frac{E}{\sigma_y}\right)\left(\frac{1}{D_c/t_c}\right) \tag{2}$$

in which, D_c = diameter of CHS, t_c = wall thickness of CHS, and E = Young's modulus. No studies on the local buckling behaviour of manufactured cold-formed steel tubes are available at present except for those by the writer. Hence the following descriptions are all based on the writer's studies (Kato, 1977; Kato and Nishiyama, 1981). Tests were carried out using stub-columns according to the procedure recommended by the SSRC (Johnston, 1976).

In addition to the basic information on the maximum stress and corresponding maximum strain of tubes governed by the local buckling under axial compression, the test results of stub-columns will be utilised for the assessment of (1) rotation capacity of beams and beam-columns, (2) interaction between column and local buckling, and (3) general yield stress of tubes which is one of the basic parameters in the evaluation of the flexural buckling strength of columns. These will be discussed at relevant places in this chapter.

Tests were carried out on cold-formed electric resistance welded CHS and SHS made from SM41 steel, which is a typical mild steel with a minimum specified yield stress of 235 MPa. Geometrical properties of the cross-sections are defined by the symbols shown in Fig. 7. The

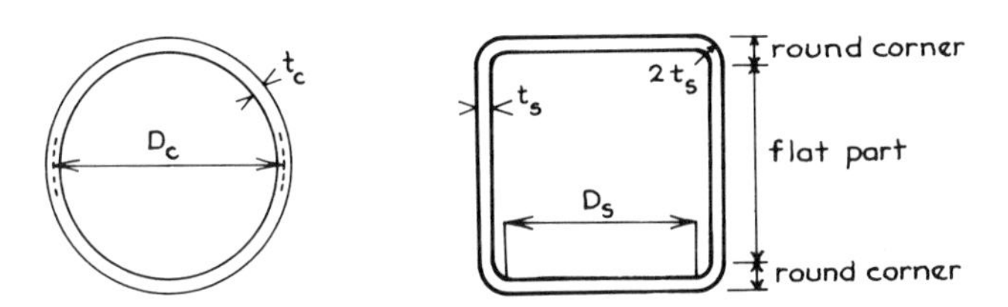

FIG. 7. Geometrical parameters of CHS and SHS.

external corner radius of SHS is twice the wall thickness, and the width of the flat portion D_s is used in the definition of width-to-thickness ratio.

In the evaluation of the local buckling strength of cold-formed structural tubes, both the adverse effect of residual stresses as was discussed in the preceding section and the advantageous effect of the increase of yield stress due to work-hardening must be taken into consideration, and keeping this in mind, the following discussions will be developed.

5.4.1 Change of Yield Stress by Cold Forming

Tensile yield stress of base material before being formed (${}_{b}\sigma_{yt}$), that of cold-formed CHS in longitudinal direction (${}_{c}\sigma_{yt}$) and that of stress-relieved CHS in longitudinal direction (${}_{a}\sigma_{yt}$) were measured by using coupons. The cut coupons from cold-formed CHS had shown a gradual yielding characteristic and their yield stresses were evaluated on the 0·2% offset basis. The heating condition for the stress-relieving was by keeping the tube at 560°C for one hour and then cooling gradually in the furnace. The results are summarised in Fig. 8, in which the ratio of yield stress of

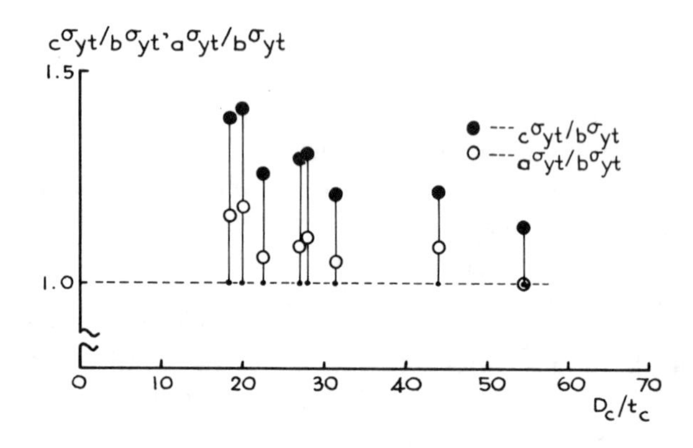

FIG. 8. Change of yield stress by cold-forming, stress relieving.

cold-formed tube to that of base material ${}_{c}\sigma_{yt}/{}_{b}\sigma_{yt}$, and the ratio of yield stress of stress-relieved tube to that of base material ${}_{a}\sigma_{yt}/{}_{b}\sigma_{yt}$ are taken in vertical axis, and the diameter-to-thickness ratio D_c/t_c is taken in horizontal axis. Note that the inverse of the diameter-to-thickness ratio t_c/D_c is the approximate measure of the extreme fibre strain in the tube wall due to cold forming. It can be observed that the yield stress of cold-formed tube increases with the increase of bending strain $\varepsilon = t_c/D_c$, and that the effect of work hardening was not completely removed by this level of heat treatment.

The mean yield stress of coupons taken from the flat parts of SHS was 386 MPa, and the mean yield stress of CHS coupons was 395 MPa. There was no substantial difference between them, but the yield stress of corner parts of SHS must be much higher than these values. The mean yield stress of base material was 310 MPa, which was much higher than its minimum specified value of 235 MPa.

5.4.2 Stub-Column Behaviour

In the stub-column tests of cold-formed tubes, the stress–strain relationship does not exhibit a sharp yielding point, but shows gradual yielding when it exceeds the proportional limit. On the other hand, the stress–strain relationship of cold-formed stress-relieved tubes shows a sharp yielding point with subsequent plastic flow. (Here, the stress means the working load divided by sectional area of tube and the strain means the axial deformation of a stub-column divided by its initial length.) The basic properties which characterise the stress–strain relationship were evaluated according to the following definitions:

Yield stress σ_y: The 0·2% offset yield stress is adopted for tubes which show gradual yielding, and the lower yield point is adopted for tubes which show a sharp yielding point.

Yield strain ε_y: Strain at yield, and $\varepsilon_y = \sigma_y/E + 0{\cdot}002$ for cold-formed tubes.

Maximum stress σ_m: The peak in a stress–strain curve is defined as the maximum stress and this peak is governed by the local buckling of tube wall.

Maximum strain ε_m: The strain at maximum stress. The average shortening of a stub-column, Δ, was measured by four deformation dials and Δ/l (l is the original length of a stub-column) was assumed as the average axial strain.

(a) Yield Stress

Yield stresses of CHS ${}_c\sigma_y$ are plotted in Fig. 9 in relation to their

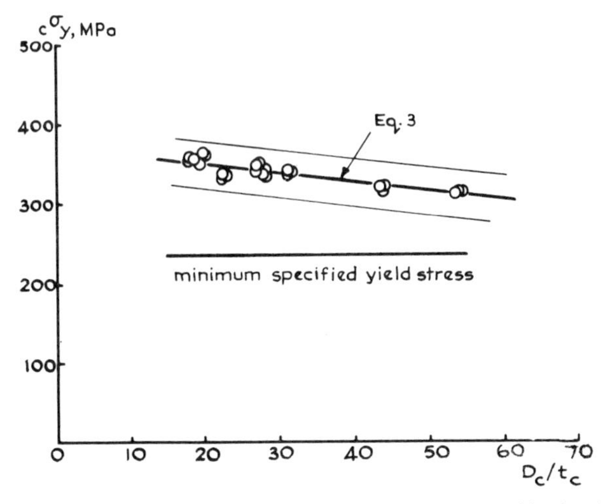

FIG. 9. Stub-column yield stress versus D_c/t_c relationship for CHS.

diameter-to-thickness ratio D_c/t_c. A regression line with 95% confidence limits is obtained as eqn. (3), and is shown in the figure.

$$ {}_c\sigma_y = 375 - 0{\cdot}114 \frac{D_c}{t_c} \pm 30{\cdot}4 \text{ (in MPa)} \tag{3} $$

Yield stress decreases proportionally with D_c/t_c-ratio. Yield stresses of all CHSs tested are higher than the minimum specified yield value (about 50% higher on average).

Yield stresses of SHS ${}_s\sigma_y$ are plotted in Fig. 10 in relation to the width-to-thickness ratio D_s/t_s. A regression line with 95% confidence limits is

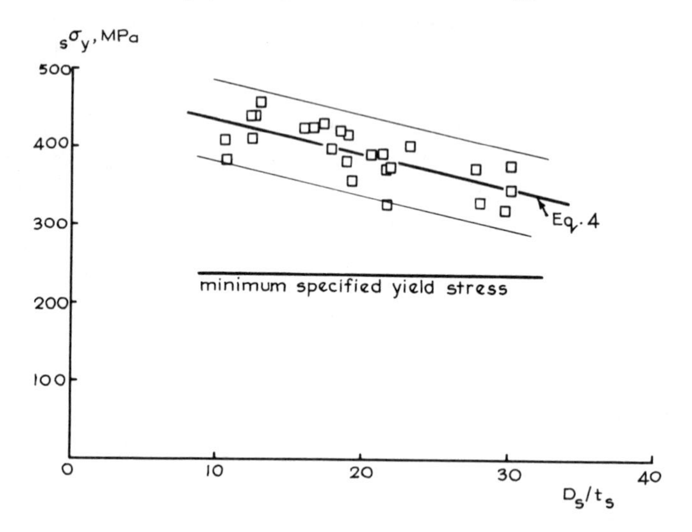

FIG. 10. Stub-column yield stress versus D_c/t_c relationship for SHS.

obtained as eqn. (4), and is shown in the figure.

$$ {}_s\sigma_y = 477 - 4{\cdot}35 \frac{D_s}{t_s} \pm 51{\cdot}5 \text{ (in MPa)} \tag{4} $$

Similarly, the yield stress decreases proportionally with D_s/t_s-ratio. In general, yield stress of SHS is about 60% higher than the minimum specified yield stress.

(b) Maximum Stress

The maximum stress of CHS ${}_c\sigma_m$ is expressed in terms of a maximum stress index defined by $S_c = {}_c\sigma_m/{}_c\sigma_y$ which is a convenient parameter in the evaluation of the rotation capacity of members subject to bending with or without axial force. The relation between $1/S_c$ and $1/\alpha_c$ (α_c is the non-dimensional buckling parameter for CHS, and is defined by eqn. (2)) is shown in Fig. 11. Equation (5) is a regression line with 95% confidence

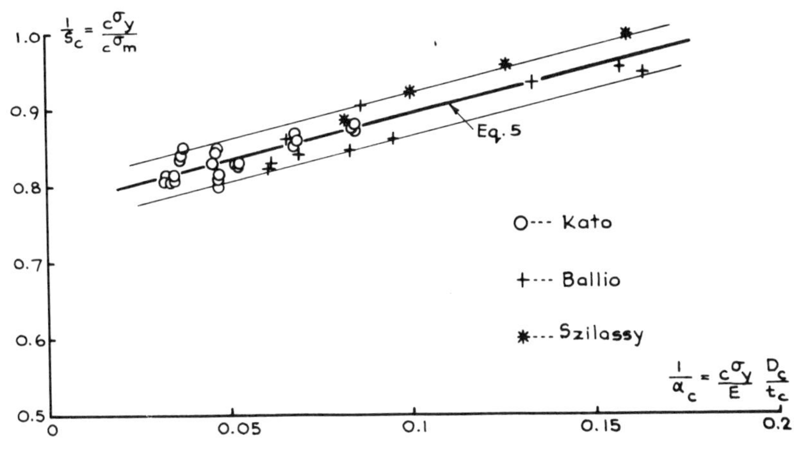

FIG. 11. $1/S_c$ versus $1/\alpha_c$ relationship (CHS).

limits and is shown in the figure.

$$1/S_c = 0{\cdot}772 + 1{\cdot}21\ \frac{1}{\alpha_c} \pm 0{\cdot}029 \qquad (5)$$

In the figure, the test results by Ballio, *et al.* (1977) and Szilassy (1977), where CHS of high strength steel were tested, are also plotted. It seems that eqn. (5) is also applicable for CHS of higher yield stresses. Alternatively, test results and eqn. (5) are plotted in terms of S_c versus α_c relationship in Fig. 12. Plantema equation (eqn. (1)) is also plotted for comparison showing that, though the Plantema equation is rather conservative, the correlation between both formulae is reasonable.

The maximum stress of SHS ${}_s\sigma_m$ is expressed in terms of another maximum stress index $S_s = {}_s\sigma_m / {}_s\sigma_y$, and is evaluated as a function of non-dimensional buckling parameter α_s for SHS which is defined as

$$\alpha_s = \frac{E}{{}_s\sigma_y}\left(\frac{t_s}{D_s}\right)^2 \qquad (6)$$

The relation between $1/S_s$ and $1/\alpha_s$ is shown in Fig. 13, and the following regression line with 95% confidence limits is obtained and shown in the

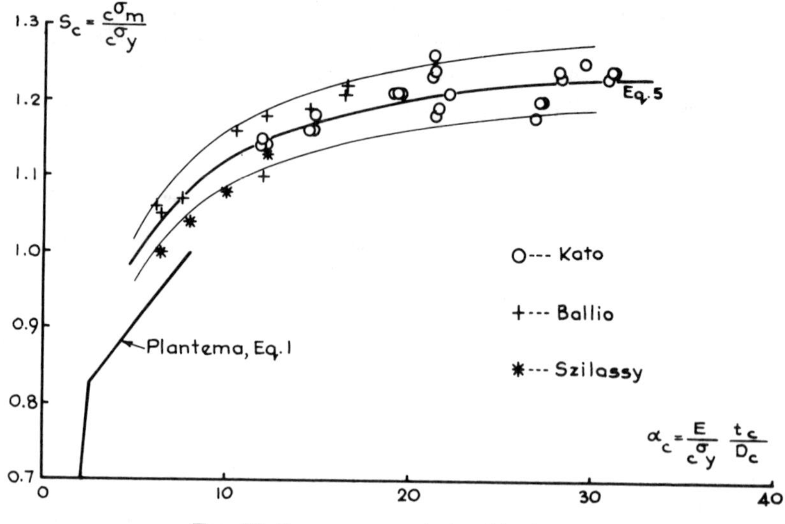

FIG. 12. S_c versus α_c relationship (CHS).

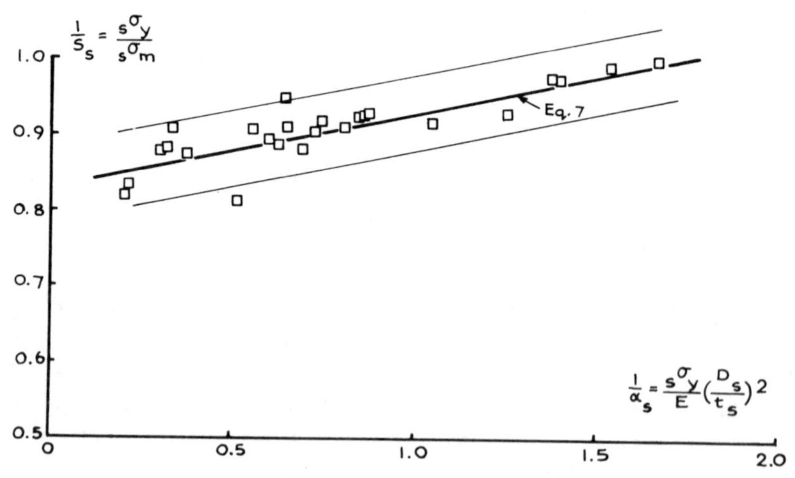

FIG. 13. $1/S_s$ versus $1/\alpha_s$ relationship (SHS).

figure,

$$1/S_s = 0{\cdot}831 + 0{\cdot}10\,\frac{1}{\alpha_s} \pm 0{\cdot}049 \tag{7}$$

In Fig. 14 the experimental results, the curve obtained from eqn. (7) and the theoretical elastic line are depicted as the S_s versus α_s relationship.

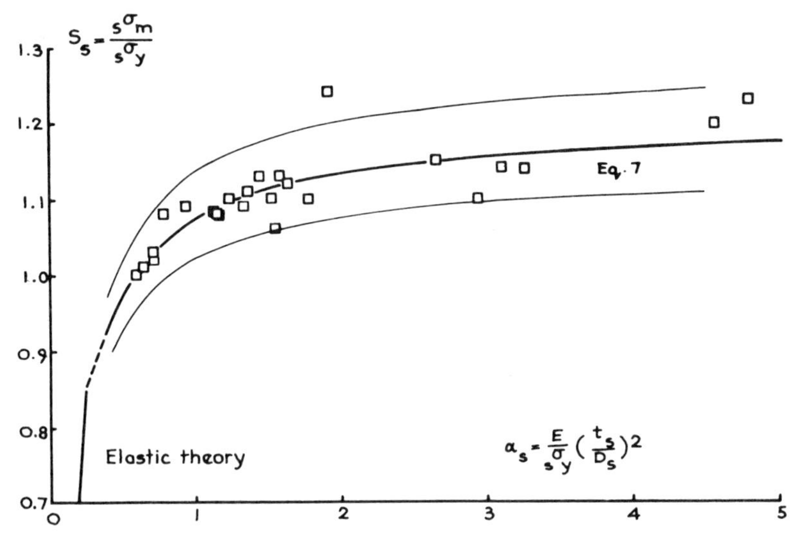

FIG. 14. S_s versus α_s relationship (SHS).

The continuity between elastic line and inelastic line given by eqn. (7) appears to be good.

(*c*) *Maximum strain*

The maximum strain of CHS ${}_c\varepsilon_m$ is expressed in terms of the ductility index $\Xi_c = {}_c\varepsilon_m / {}_c\varepsilon_y$. The relation between Ξ_c and α_c is plotted in Fig. 15. Equation (8) is a regression line with 95% confidence limits and shown in the figure.

$$\Xi_c = (0{\cdot}293 \pm 0{\cdot}051)\alpha_c \tag{8}$$

The test results on higher strength tubes by Ballio and Szilassy are also plotted in the figure. Again, eqn. (8) seems to be applicable for CHS with higher yield stresses as well.

The result of similar investigation on SHS is shown in Fig. 16. The corresponding regression line is

$$\Xi_s = (2{\cdot}183 \pm 0{\cdot}735)\alpha_s \tag{9}$$

5.4.3 Comparison of CHS and SHS

As CHS and SHS differ from each other in geometrical shape and in the forming process, it is of interest to compare their structural performances

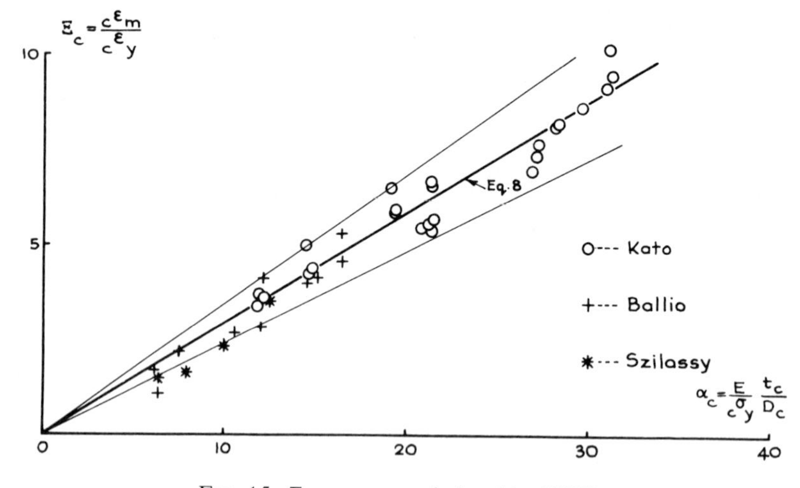

FIG. 15. Ξ_c versus α_c relationship (CHS).

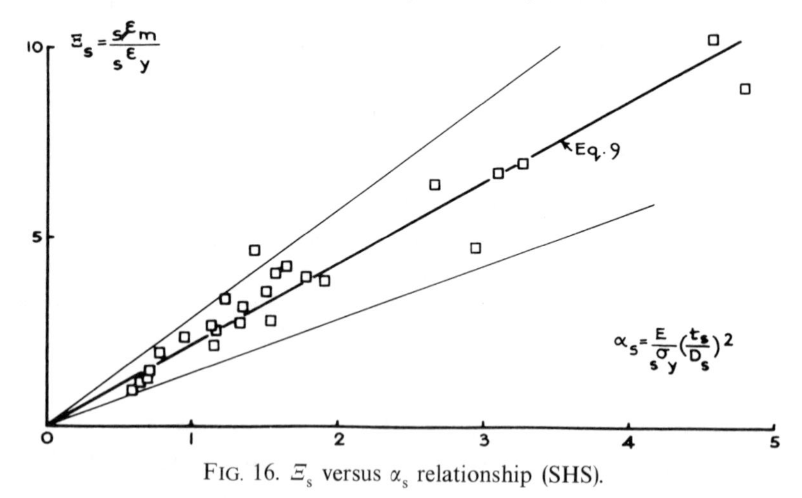

FIG. 16. Ξ_s versus α_s relationship (SHS).

associated with local buckling. The comparison was made for CHS and SHS with identical ratio of circumferential length to thickness. The circumferential length–to–thickness ratio β is defined as

$$\left.\begin{aligned} \beta &= \pi D_c/t_c \quad \text{for CHS} \\ \beta &= 4D_s/t_s + 3\pi \ \text{for SHS} \end{aligned}\right\} \tag{10}$$

Figure 17 is the comparison of yield stresses. The yield stress of the corner parts of SHS is much increased due to excessive cold bending, and

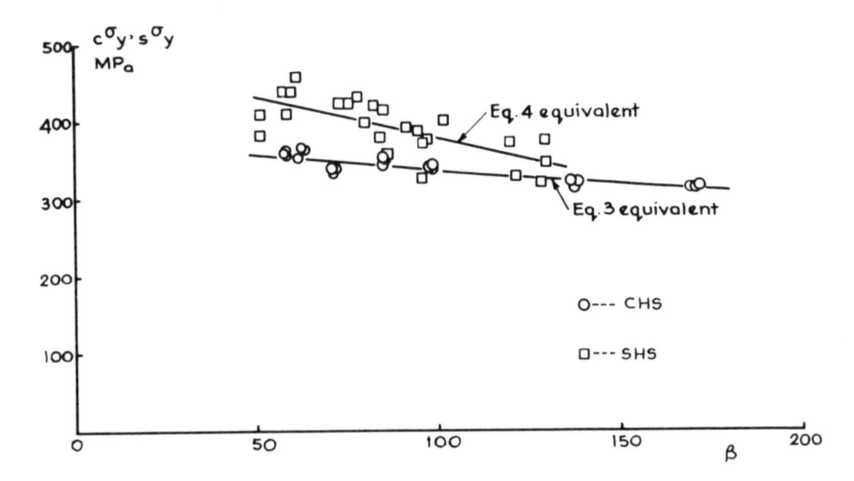

FIG. 17. Comparison of yield stresses between CHS and SHS.

the ratio of the area of corner parts to the whole sectional area becomes higher when D_s/t_s or β decreases. This fact will explain the feature observed in the figure.

Figure 18 compares the maximum stresses. Lines corresponding to

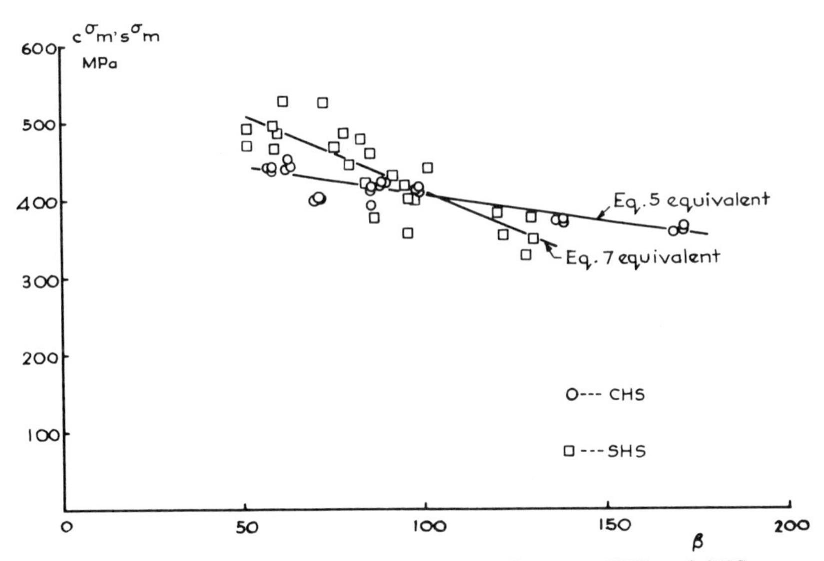

FIG. 18. Comparison of maximum stresses between CHS and SHS.

eqn. (5) and eqn. (7) are seen to be crossing, since in lower ranges of β the member is insensitive to buckling and the maximum stress is largely affected by material strength, while, in higher ranges of β the difference of buckling characteristics between shell (CHS) and plate (SHS) will dominate. If a comparison is made on the maximum stress index, SHS values are located in lower level through the whole range tested as seen in Fig. 19. This is because the yield stress of SHS increases more steeply

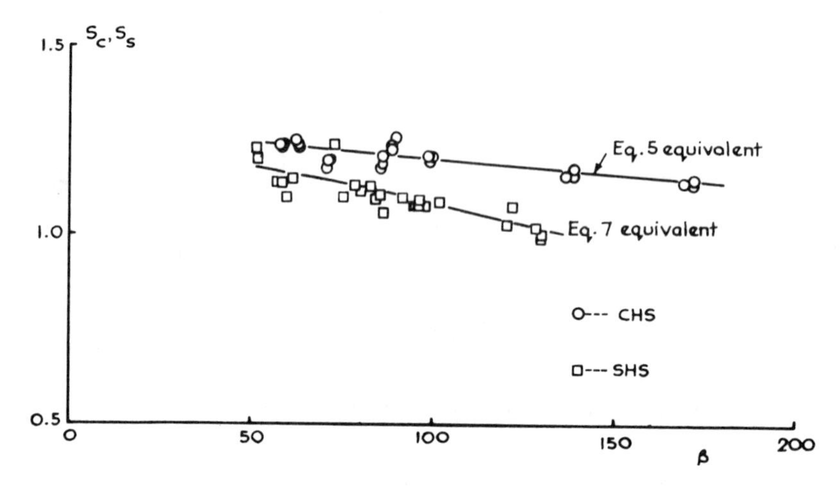

FIG. 19. Comparison of maximum stress indices between CHS and SHS.

with lower values of β (see Fig. 17). Figure 20 compares the maximum strains, and shows that CHS is more ductile than SHS through the whole range of β. The difference diminishes toward the lower value of β and will reach the same value at about $\beta = 50$. This tendency does not change even if the comparison is made on the ductility indices.

5.4.4 Effect of Stress-Relieving

Cold-formed CHS and SHS are subject to work-hardening and deformational residual stresses as well. The work-hardening makes the yield stress higher and the residual stresses reduce the stiffness in the earlier stage of loading. The integrated outcome of these opposing effects is investigated by comparing the structural performances of as-formed and stress-relieved tubes. Note that in the so-called stress-relieved tubes studied herein, the effect of work-hardening is not completely removed as has been observed in Fig. 8.

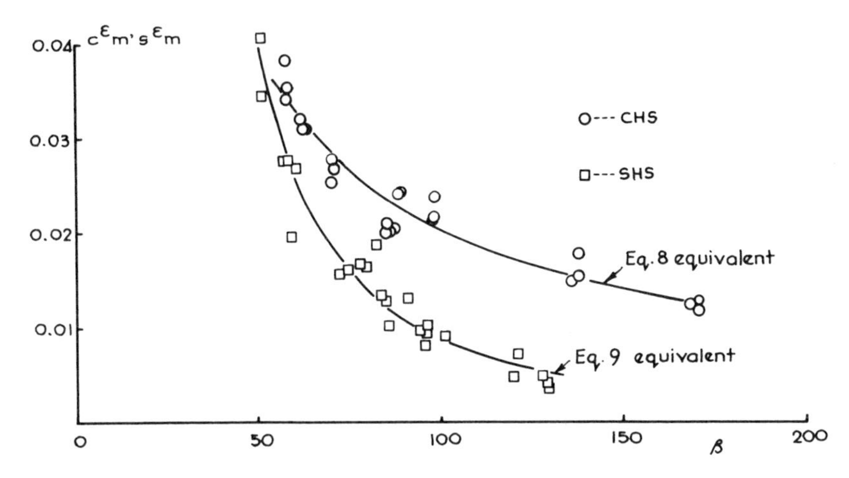

FIG. 20. Comparison of maximum strains between CHS and SHS.

The maximum stresses of as-formed and stress-relieved tubes with identical β-ratio are compared in Fig. 21, in which the ratios of maximum stresses of as-formed tubes ($_c\sigma_m$, $_s\sigma_m$) to those of stress-relieved tubes ($_c\sigma_{ma}$, $_s\sigma_{ma}$) are taken in vertical axis. Though there is a slight undulation for SHS, and though the test data are limited to a short range of β for CHS, roughly speaking, there is no substantial difference in the maximum stress. In other words, the above-mentioned opposing influences seem to be compensating each other.

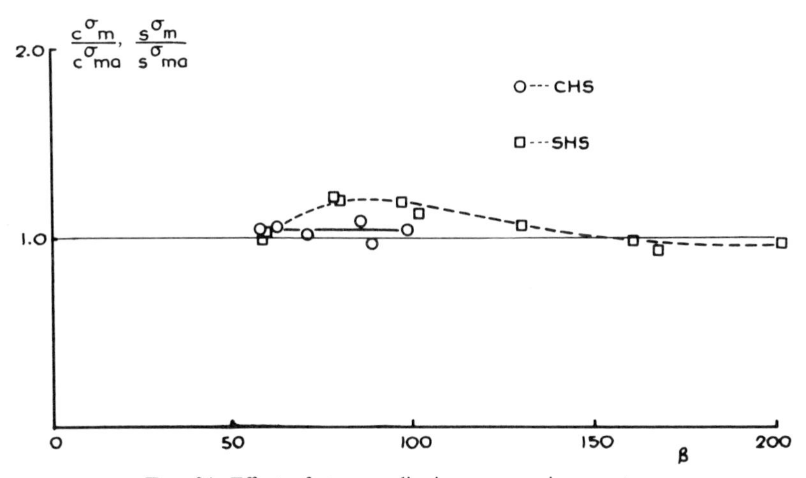

FIG. 21. Effect of stress relieving on maximum stress.

A similar comparison is made for the maximum strains in Fig. 22, in which the ratios of maximum strains of as-formed tubes (${}_c\varepsilon_m$, ${}_s\varepsilon_m$) to those of stress-relieved tubes (${}_c\varepsilon_{ma}$, ${}_s\varepsilon_{ma}$) are taken in vertical axis. This shows an interesting configuration, but no reasonable explanation for this can be given at this stage.

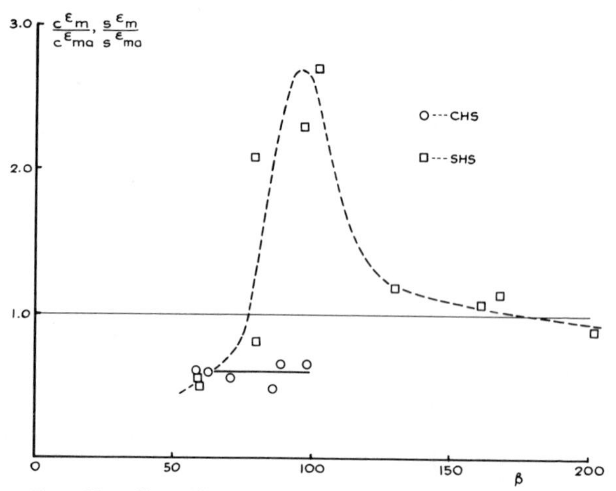

FIG. 22. Effect of stress relieving on maximum strain.

5.5 FLEXURAL BUCKLING

Similar to the local buckling behaviour, the flexural buckling strength is influenced by residual stresses, variation of yield properties, initial crookedness, end eccentricities, and the geometrical accuracy of cross-section. Among these, geometrical properties such as straightness and the accuracy of cross-sectional dimensions of cold-formed tubes are believed to be superior to those obtained in hot-rolled open-section members, while the distribution of residual stresses is unique compared to the typical one produced by cooling in hot-rolled sections.

The inelastic flexural buckling stress of tubular members is usually obtained in one of two ways; the first is the analytical approach based on the tangent modulus concept and the other is the statistical evaluation based on a large number of test results and/or numerical analyses.

5.5.1 Tangent Modulus Approach

In CHS members, residual stress is uniformly distributed along the perimeter of a section as was observed in the previous section. When a

CHS member with this type of residual stress is compressed axially and is partially yielded, it behaves as though the wall thickness is reduced as illustrated schematically in Fig. 23. In such a case, the tangent modulus obtained from the stress–strain curve in a stub-column test can be applied directly to the evaluation of the flexural buckling strength.

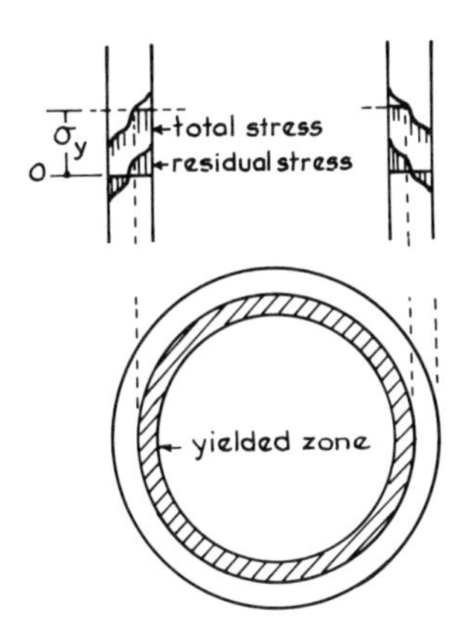

FIG. 23. Yielding process of cold-formed CHS.

Unfortunately, the theoretical approach described in the previous section (Fig. 6) could not predict the stress–strain relationship of stub-columns accurately enough for evaluating the tangent modulus. In practice, it is found to be better to develop an empirical stress–strain relation on the basis of test data. The following is an example of this approach carried out by Kato (1977*a*).

(*a*) *Stress–strain relationship*

Using the test data on stub-columns discussed in the preceding section, the following empirical stress–strain relations are derived. Since the configuration of stress–strain relationship of cold-formed CHS is influenced by the D_c/t_c-ratio, experimental curves are subdivided into three groups, and were best fitted by the following three formulae:

A group ($20 \leq D_c/t_c \leq 32$, $336 \leq \sigma_y \leq 359$ MPa):

$$\bar{\varepsilon} = \bar{\sigma}[1 - 0{\cdot}52(1 - \bar{\sigma}^{11})] \quad (11)$$

B group ($D_c/t_c \approx 45$, $\sigma_y \approx 319$ MPa):

$$\bar{\varepsilon} = \bar{\sigma}[1 - 0{\cdot}56(1 - \bar{\sigma}^{9})] \quad (12)$$

C group ($D_c/t_c \approx 56$, $\sigma_y \approx 317$ MPa):

$$\bar{\varepsilon}=\bar{\sigma}[1-0{\cdot}57(1-\bar{\sigma}^8)] \tag{13}$$

in which, $\bar{\varepsilon}=\varepsilon/{}_c\varepsilon_y$ and $\bar{\sigma}=\sigma/{}_c\sigma_y$, ${}_c\sigma_y$ is given in eqn. (3) and ${}_c\varepsilon_y={}_c\sigma_y/E+0{\cdot}002$. The above formulae are applicable between the proportional limit ${}_c\sigma_p$ and the maximum stress ${}_c\sigma_m$ given by eqn. (5). The proportional limit is given by

$$ {}_c\sigma_p=0{\cdot}726\ {}_c\sigma_y. \tag{14}$$

Since the proportional limit depends on the sensitivity of instrumentation and personal judgement, it was decided to define proportional limit as the stress at which the permanent strain is 0·01%. These formulae can trace the actual stress–strain curves quite well. An example from A group is shown in Fig. 24.

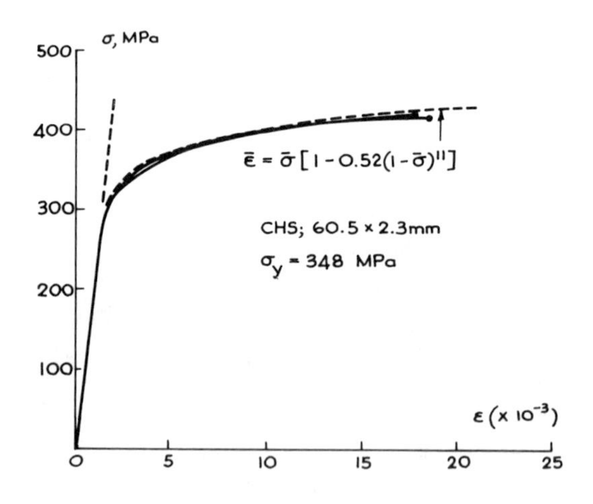

FIG. 24. Stress–strain relationship of CHS stub-column.

(b) Tangent moduli

Equations (15) to (17) are tangent moduli $E_t=d\bar{\sigma}/d\bar{\varepsilon}$ derived from eqns. (11) to (13) respectively, and they are shown plotted in Fig. 25.

$$\text{A group: } E_t/E=0{\cdot}48\ (\bar{\sigma}/\bar{\varepsilon}) \tag{15}$$

$$\text{B group: } E_t/E=0{\cdot}44\ (\bar{\sigma}/\bar{\varepsilon}) \tag{16}$$

$$\text{C group: } E_t/E=0{\cdot}43\ (\bar{\sigma}/\bar{\varepsilon}) \tag{17}$$

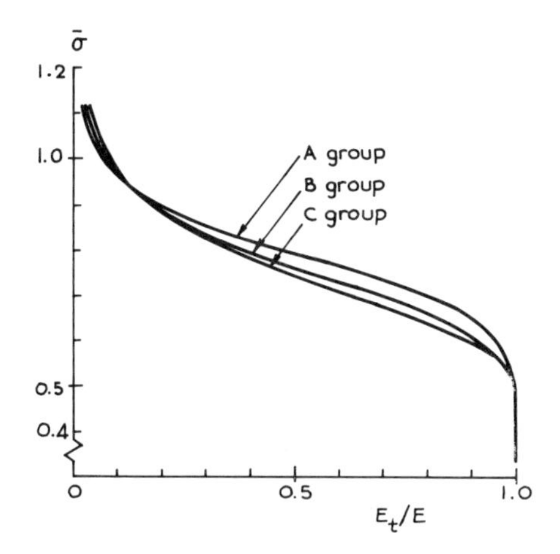

FIG. 25. Tangent modulus of cold-formed CHS.

(*c*) *Column curves*

The non-dimensional expression of the column buckling curve on the basis of tangent modulus concept is given as

$$\bar{\sigma}_{cr} = (E_t/E)\,\frac{1}{\lambda^2} \tag{18}$$

in which, $\bar{\sigma}_{cr} = \sigma_{cr}/{}_c\sigma_y$ and $\lambda = (1/\pi)\sqrt{{}_c\sigma_y/E}\,Kl/r$ = non-dimensional slenderness ratio.

Column curves thus obtained for each group are depicted in Fig. 26 and Fig. 27. These compare the test results (Suzuki, 1964; Wakabayashi, 1968; Nishida, 1971) with the prediction for group A members demonstrating a good correlation with each other. Most of the test results exceed the prediction since the geometrical imperfections such as eccentricity and initial crookedness are not taken into account in this prediction. The CHS members used for stub-column tests and for flexural buckling tests were taken from different lots, and there might be some difference in tangent moduli.

In SHS members, residual stresses are not uniformly distributed along the perimeter of a section as has been shown in Fig. 4. For such a situation, the effective modulus for inelastic flexural buckling should be derived as a function of the tangent modulus obtained from the stub-column test, and therefore exact information about the distribution of residual stresses along the perimeter of the section is necessary in order to predict the critical load of a column.

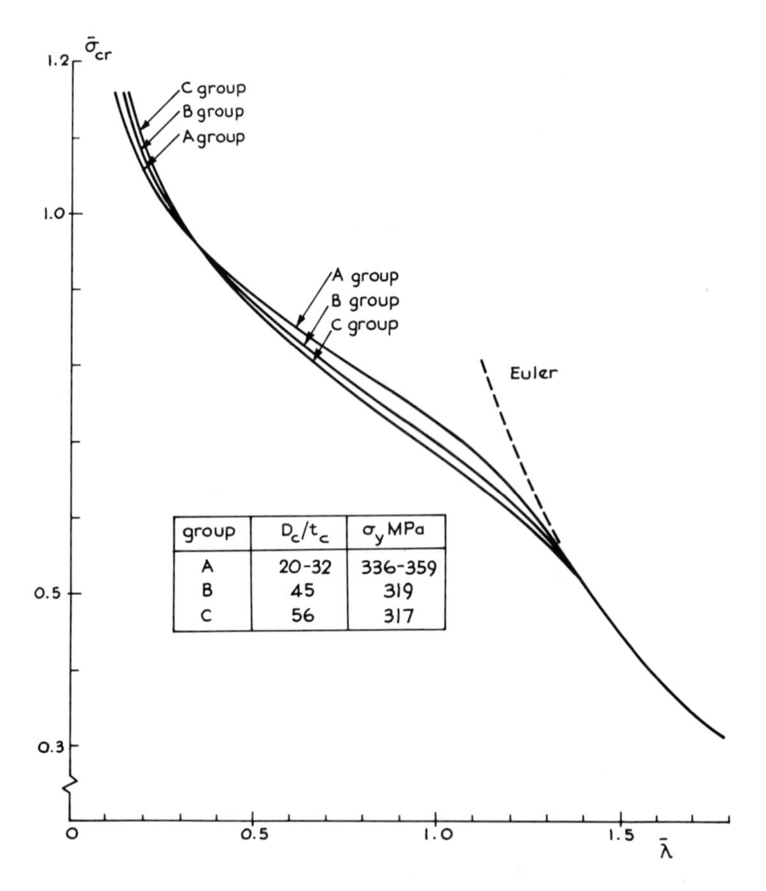

group	D_c/t_c	σ_y MPa
A	20-32	336-359
B	45	319
C	56	317

FIG. 26. Critical stress curves for cold-formed CHS columns.

5.5.2 Maximum Strength, related to Column Strength Curves

The tangent modulus theory accounts for residual stresses and material-related non-linearities, but imperfections such as initial crookedness and eccentricity of load are not incorporated. Such imperfections are always present in real columns, and to assess the combined effect of residual stresses and geometric imperfections, the maximum strength of columns must be evaluated by means of numerical analysis.

The maximum strength curves were developed by numerical analysis using measured values of residual stress, yield properties, and the other significant column strength parameters (Bjorhovde, 1972; Johnston, 1976). A total of 112 column curves for various types of cross-sections

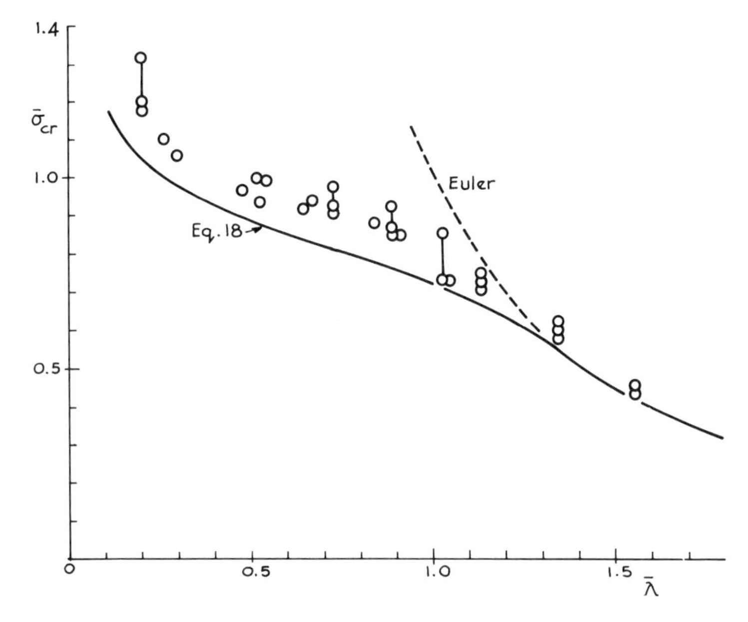

FIG. 27. Test results and theoretical prediction of CHS columns.

were developed in the investigation and these show a considerable scatter band. Such uncertainties as to column strength can be reduced by defining subgroups, each of which is represented by a single 'average' curve. The family of average curves so established, is termed 'multiple column curves'. Research basic to the evaluation of multiple column curves has been carried out since 1960 at Lehigh University (Bjorhovde, 1972). The results are presented as SSRC curves 1, 2, 3 (Johnston, 1976). At the same time that the SSRC multiple column curves were being developed, a similar study was conducted in Europe under the auspices of the European Convention for Constructional Steelwork (ECCS). To verify their validity, extensive series of full-size long column tests were carried out (Sfintesco, 1970). However, hollow structural section columns did not form part of the family of SSRC curves, due to the lack of residual stress data at that time. In Canada, studies on column behaviour of hollow structural sections are under way (Birkemoe, 1977; Bjorhovde, 1979). In Europe, buckling tests on hollow structural section columns were carried out under CIDECT (International Committee for the Research and Development of Tubular Construction) sponsorship, but the major part of the test specimens were of hot-formed shapes and no definite evaluation for the cold-formed section members has been achieved as yet.

In this context, the following is a review of the extensive data on the maximum strength of cold-formed and cold-formed stress-relieved hollow structural section columns, and an evaluation of these test results in relation to the SSRC column curves.

(*a*) *Outline of the available test data*

CHS columns: Test results used in the evaluation were obtained by Japanese researchers (Wakabayashi, 1968; Nishida, 1971), and by CIDECT (Guiaux). Steel grades of the Japanese specimens were SM41 (minimum specified yield stress = 235 MPa) and SM50 (minimum specified yield stress = 323 MPa); a total 32 columns were tested (4 specimens of which were stress relieved). In all, 40 specimens (10 identical specimens for each slenderness ratio) were tested in the CIDECT project, and they came from five producers in four countries (Germany, Belgium, Spain, and Italy); no information about their steel grades is available. The maximum D_c/t_c-ratio of all specimens tested is 37, and the maximum stress index S_c calculated by eqn. (5) is 1·17, which means that there is no possibility of local buckling occurring or flexural–local buckling interaction.

SHS columns: Test results by research workers in Japan, Canada, and CIDECT were used in this evaluation. As seen in the table (p. 175), the initial crookedness of all specimens was very small compared with that assumed in developing a design formula (1/1000). In fact, analysis of test results has shown that inevitable load eccentricities were more influential than such a level of crookedness. It is plausible that a smaller tolerance can be maintained for these shapes, considering the continuous nature of process and the inherent multi-axial strength and stiffness of hollow structural section members.

Heat conditions for stress relieving are somewhat different from each other for the different series of specimens. In the Japanese project, columns were kept at 560°C for one hour as reported in the preceding section. The products used in the Toronto tests were heated to about 455°C. In the Alberta tests, it was reported that the tubes were 'fully stress relieved (annealed)'. Though the temperature was not reported, it must have been much higher than that for Japanese specimens, and it was also reported that, by annealing, they exhibited some initial twist as well as out-of-square cross-sections (slightly rhomboid, or outward 'bow' of the sides), and resulted in dubious test results. In view of these differences of heat conditions, the effect of stress-relieving must have been somewhat different in each group of tests.

THE OUTLINE OF SHS COLUMNS TESTED IS SHOWN IN THE FOLLOWING TABLE

	Univ. of Tokyo	*Univ. of Toronto*	*Univ. of Alberta*	*CIDECT*
Minimum specified yield stress	235 MPa	345 MPa	345 MPa	—[1]
No. of specimens	25 (incl. 2 stress-relieved)	20	30 (incl. 5 annealed)	40 (10 for each slenderness)
mean crookedness	1/7500	1/6500	1/6384	—
D_s/t_s	10–30	16–30	6–30	19–28 (41)[2]
references	Katsurai (1980)	Salvarinass (1978)	Bjorhovde (1977)	Guiaux

[1] products of five producers of four countries (Germany, Belgium, Spain, and Italy), no information about their steel grades.
[2] D_s/t_s-ratio of some specimens of CIDECT project are about 40 to 41 (In shorter columns, the maximum strength of these columns must have been governed by local buckling, and they were discarded.)

(b) Evaluation of column strength

The available column test data can be compared by using the common non-dimensionalised plot of load (stress) versus slenderness ratio. Non-dimensional load (stress) is defined as $P_u/P_y = \sigma_m/\sigma_y$, in which $P_u = P_m$ = ultimate static load obtained from full-size column tests, $\sigma_m = P_u/A$ = maximum stress, P_y = stub-column yield load, and $\sigma_y = P_y/A$ = stub-column yield stress as defined in the preceding section.

Stub-column yield loads (stresses) provide the best comparison of column behaviour, since they demonstrate total column performance at zero slenderness, i.e., in the absence of overall stability effects.

Non-dimensional slenderness ratio is defined as $\lambda = \sqrt{\sigma_y/E}\ Kl/r$, in which Kl/r = effective slenderness ratio, l = column length, and r = radius of gyration.

End fixtures and geometric alignment procedure for column testing of all these tests met the prescription of SSRC column testing procedure (Johnston, 1976).

Test results on CHS columns are shown in Fig. 28, in which the SSRC column curves 1, 2, 3 are also plotted. Though the number of test specimens is not sufficient for statistical evaluation, the SSRC curve 2 seems to be applicable for cold-formed CHS columns. Column strength may be improved somewhat by stress relieving.

Test results on SHS columns are shown in Fig. 29, in which the SSRC column curves 1, 2, 3 are also shown. As with the CHS columns, the SSRC

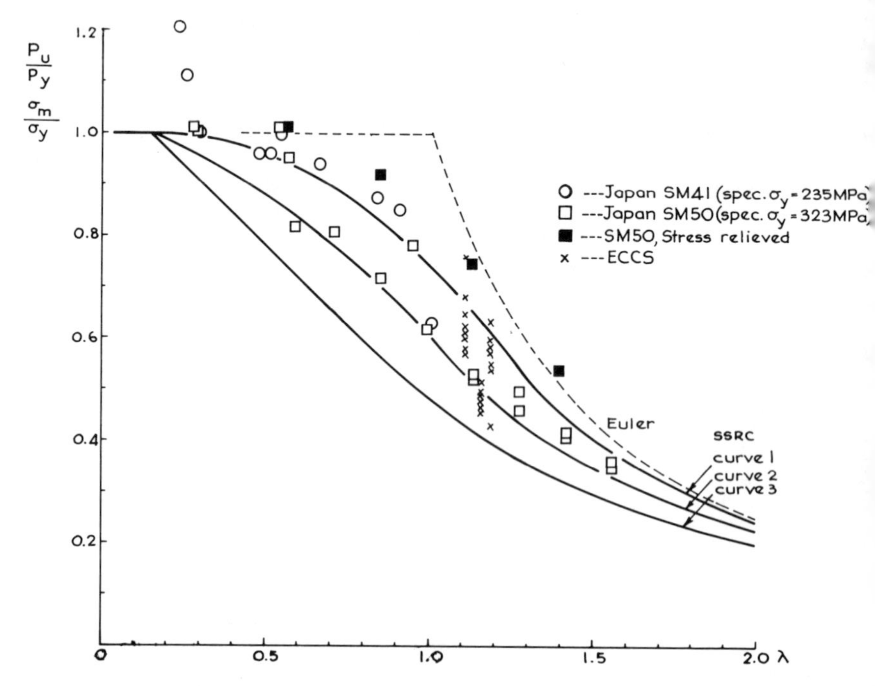

FIG. 28. Comparison of CHS column data with SSRC curves (stub-column strength normalisation).

curve 2 seems to be applicable for cold-formed SHS members, and column strength may be improved by stress relieving. Flexural and local buckling interaction was observed for columns with D_s/t_s-ratio of 30, which could be anticipated from eqn. (7). Even for such a situation, column behaviour could be reasonably explained if the effect of local buckling is reflected by the stub-column tests, but the configuration of the column curve might be somewhat distorted by this interaction.

Although the use of the stub-column test for mechanical property evaluation is theoretically reasonable, it is tedious and expensive. As a practical tool the stub-column test therefore leaves much to be desired. As an alternative approach, the use of the yield strength of a tensile coupon taken from the structural hollow section as the basis of normalisation may be suggested, but since the yield strength of a tensile coupon will vary for each combination of diameter (width) and wall thickness, one must provide the list of yield strength for all profiles of structural

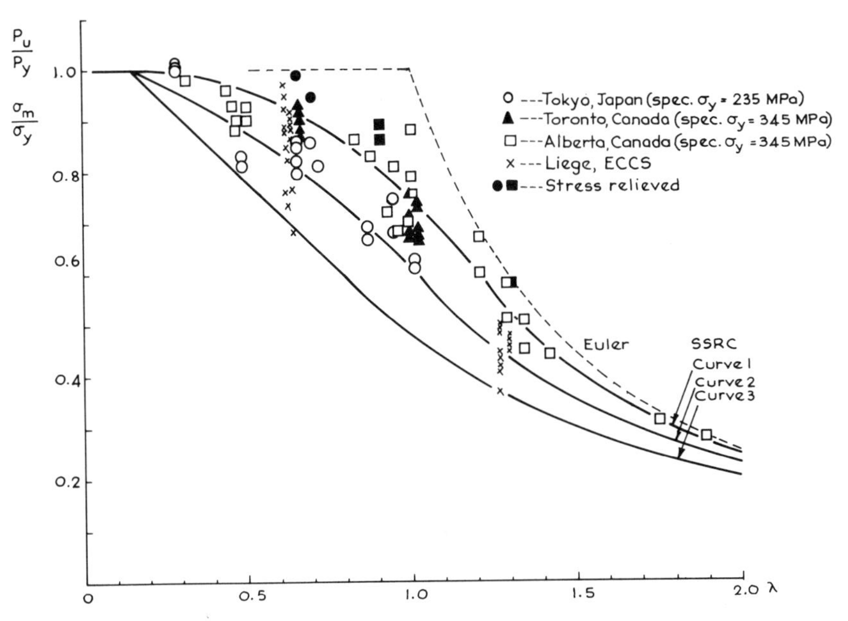

FIG. 29. Comparison of SHS column data with SSRC curves (stub-column strength normalisation).

hollow sections in order to give the designer the necessary information for proportioning his member. This is still tedious for both users and producers.

The strength parameter which is readily available to a designer is the minimum specified yield stress, and this is therefore the most likely candidate for incorporation into calculations of resistance in practical design. Figures 30 and 31 are plots of test data normalised by the minimum specified yield stress, i.e., the ratio of the maximum stress of a column to its minimum specified yield stress is taken in vertical axis for CHS and SHS columns respectively. Since the minimum specified yield stress of base material is always lower than the stub-column yield stress as was demonstrated in the preceding section, test points shift upwards and to the left. From these figures, it may be suggested that SSRC curve 1 is applicable for cold-formed SHS columns if the minimum specified yield stress is taken as the strength parameter. For CHS columns, however, more test data are needed before any recommendations can be made. At this stage, the test data are insufficient to apply the ECCS method of interpretation.

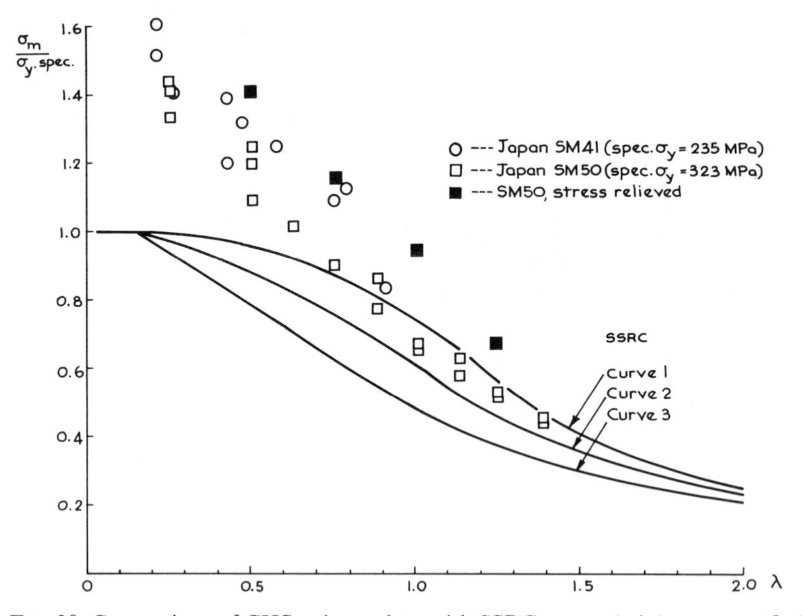

FIG. 30. Comparison of CHS column data with SSRC curves (minimum specified yield stress normalisation).

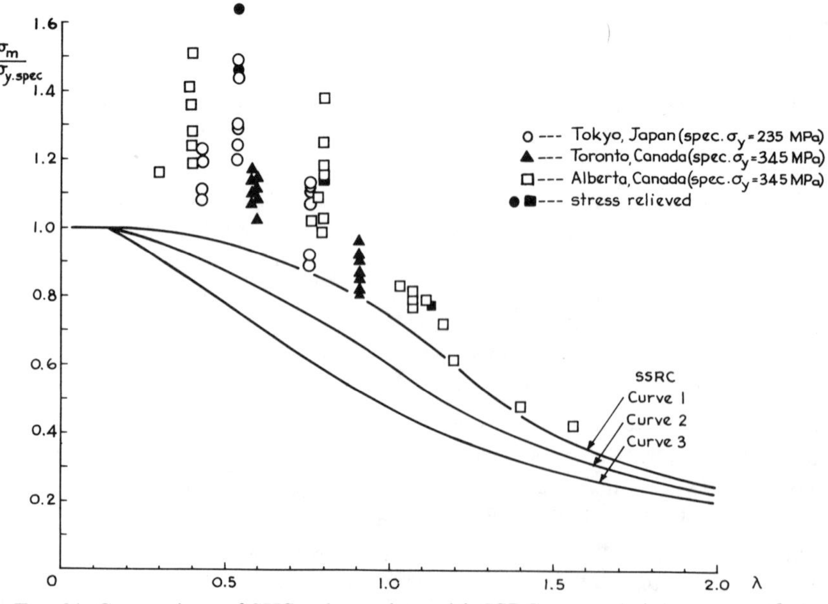

FIG. 31. Comparison of SHS column data with SSRC curves (minimum specified yield stress normalisation).

5.6 CONCLUSIONS

1. Deformational residual stresses in a cold-formed resistance welded structural hollow section were analysed, and the predicted residual strains showed good agreement with measured strains. The gross behaviour of a stub-column subjected to axial compression was reasonably explained by the analysis.
2. Inelastic local buckling behaviour of cold-formed CHS and SHS was investigated experimentally. Simple empirical formulae which predict the maximum stresses and the maximum strains in terms of non-dimensional local buckling parameters were presented. The effect of stress relieving on the improvement of local buckling behaviour was shown to be insignificant.
3. Flexural buckling strength of cold-formed CHS and SHS was evaluated on the basis of tangent modulus concept and of maximum strength concept. It was suggested that SSRC curve 2 may be applicable to cold-formed CHS and SHS columns when the stub-column yield stress is taken as the strength parameter, and that SSRC curve 1 may be applicable to cold-formed SHS columns when the minimum specified yield value is taken as the strength parameter. The column strength may be improved by stress relieving.

REFERENCES

BALLIO, G., FINZI, L., and URBANS, C. (1977). Centrally compressed high-strength steel round and square tubes, *2nd International Colloquium on Stability of Steel Structures*, Liege.

BATDORF, S. B. and BUDIANSKY, B. (1949). A mathematical theory of plasticity based on the concept of slip, NACA Technical Note, No. 1871.

BIRKEMOE, P. C. (1977). Development of column curves for HSS, *International Symposium on Hollow Structural Sections*, Toronto.

BJORHOVDE, R. (1972). *Deterministic and probabilistic approaches to the strength of steel columns*, PhD thesis, Lehigh University, Bethlehem, Pa.

BJORHOVDE, R. (1977). *Strength and behaviour of cold-formed HSS columns*, Department of Civil Engineering, University of Alberta.

BJORHOVDE, R. and BIRKEMOE, P. C. (1979). Limit states design of HSS columns, *Canadian Journal of Civil Engineering*, **6**, No. 2.

GUIAUX, P. (1974) *Essais de Flambement sur Profiles Creux Formes à Froid, Carres et Circulaires*, Executes à la Demande du CIDECT, Universite de Liege, Prog. 2 Document No. 74/18.

JOHNSTON, B. G. (1976). *Guide to stability design criteria for metal structures*, Third edition, Structural Stability Research Council, John Wiley & Sons.

KATO, B. (1977a). Column curve for cold-formed and welded steel tubular members. *2nd International Colloquium on Stability of Steel Structures, Preliminary report*, ECCS, Liege.

KATO, B. (1977b). Local buckling of steel circular tubes in plastic region, *International Colloquium on Stability of structures under static and dynamic loads*, ASCE, Washington, D.C.

KATO, B. and AOKI, H. (1978), Residual stresses in cold-formed tubes, *Journal of Strain Analysis*, **13** (4), 193–204.

KATO, B. and NISHIYAMA, I. (1981). Inelastic local buckling of cold-formed circular hollow section and square hollow section members, *Japan-US seminar on inelastic instability of steel structures and structural elements*, Tokyo.

KATSURAI, S. (1980), *Maximum strength of cold-formed square hollow section columns*, Master's thesis, University of Tokyo (in Japanese).

NISHIDA, Y. *et al.* (1971), Plastic design of steel tubular columns, *Sumitomo Metals*, **23**, No. 4 (in Japanese).

PLANTEMA, F. J. (1946). Collapsing stress of circular cylinders and round tubes, *Nat. Luchtraart Lab. Rep.*, Amsterdam.

SALVARINASS, J. J., BARBER, J. D., and BIRKEMOE, P. C. (1978). An experimental investigation of the column behaviour of cold-formed stress-relieved hollow structural steel sections, University of Toronto, ISSN 0316–7968, Publication 78–02.

SFINTESCO, D. (1970). Fondement Experimental des Courbes Europeennes de Flambement, *Construction Metallique*, No. 3.

SZILASSY, K. (1977), Design of cold-formed steel tubular compression members on the basis of their stress-strain curve, *Regional Colloquium on Stability of Steel Structures*, Budapest, Hungary.

SUZUKI, T. and FUJIMOTO, M. (1964). *NKK design manual of tubular structure*, Nippon Kokan K. K. (in Japanese).

WAKABAYASHI, M. *et al.* (1968), Column buckling test on welded steel tubes, *Trans. Architectural Inst. of Japan*, Oct. (in Japanese).

Chapter 6

BUCKLING OF SINGLE AND COMPOUND ANGLES

J. B. Kennedy and M. K. S. Madugula
Department of Civil Engineering, University of Windsor, Ontario, Canada

SUMMARY

A state-of-the-art review is presented on the analysis as well as on the design of single and compound angle columns; both concentric and eccentric axial loads are treated for columns with singly-symmetric, doubly-symmetric, and unsymmetric cross-sections. The current European and North American design practices are presented and compared, together with the available design aids. Scope for future work is also provided.

NOMENCLATURE

A	= cross-sectional area;
b	= width of leg of angle;
C_w	= warping constant of cross-section;
E	= modulus of elasticity;
E_t	= tangent modulus of elasticity;
e_x, e_y	= x and y coordinates of point of application of axial load with reference to the centroidal principal axes;
G	= modulus of rigidity $= E/2(1+\nu)$;
I_{pc}	= polar moment of inertia of cross-section about the centroid;

I_{ps} = polar moment of inertia of cross-section about the shear centre;
I_x = moment of inertia of cross-section about the x–x axis;
I_y = moment of inertia of cross-section about the y–y axis;
J = torsional constant of the cross-section;
K = effective length factor;
K_t, K_x, K_y = effective length factors for torsional buckling, flexural buckling about the x–x axis and flexural buckling about the y–y axis, respectively;
k = plate buckling coefficient;
L = length of column between supports;
P_{cr} = critical load;
P_t = torsional buckling load $= \dfrac{A}{I_{ps}}\left[GJ + \dfrac{\pi^2 EC_w}{(K_t L)^2}\right]$;
P_{ty} = torsional–flexural buckling load for simultaneous twisting about the centre of rotation and bending about the symmetric y-y axis;
P_x = flexural buckling load about the x–x axis $= \pi^2 EI_x/(K_x L)^2$;
P_y = flexural buckling load about the y–y axis $= \pi^2 EI_y/(K_y L)^2$;
P_{yield} = squash load on specimen $= A\sigma_y$;
r = radius of gyration;
r_t, r_x, r_y = equivalent radius of gyration for torsional buckling about the shear centre, radius of gyration about the x–x axis and radius of gyration about the y–y axis, respectively;
t = thickness of leg of angle;
x_s, y_s = x and y coordinates of the shear centre with reference to the centroidal principal axes;
η $= [(\sigma_y - \sigma_{cr})\sigma_{cr}]/[(\sigma_y - \sigma_p)\sigma_p]$;
λ_y $= \pi\sqrt{E/\sigma_y}$;
$\dfrac{\lambda}{\lambda_y}$ = non-dimensional slenderness ratio $= \dfrac{KL/r}{\pi\sqrt{E/\sigma_y}}$;
ν = Poisson's ratio;
σ_a = ultimate maximum stress according to ASCE Manual No. 52 (N/mm^2);
σ_{cr} = critical stress (N/mm^2);
σ_p = proportional limit of material (N/mm^2);
σ_y = yield stress of material (N/mm^2).

6.1 INTRODUCTION

Angles are the most common structural shapes used in latticed electrical transmission towers and antenna-supporting towers; they are easy to fabricate and erect because of the basic simplicity of their cross-section. Angles are also used as chord and web members of trusses, as web members of long-span open-web steel joists, and as bracing members to provide lateral support to steel structures; they are classified as equal or unequal angles, single or compound (i.e., built-up) angles, hot-rolled or cold-formed angles. Members of angle construction are loaded concentrically or eccentrically, axially or transversely, inducing stresses below or above the proportional limit of the material. In this chapter the buckling of members of angle construction is discussed by reviewing the literature and the classical theoretical analyses and focusing on North American and European design practices as well as on the available design aids. The treatment of the subject will be limited to the behaviour of single and compound angles under concentric or eccentric axial compressive loading; members subjected to transverse loads or pure end moments are thus outside the scope of the chapter. Figure 1(a) shows some possible arrangements for singly-symmetric compound angles while Fig. 1(b) shows doubly-symmetric star-shaped and cruciform compound sections.

6.2 LITERATURE REVIEW

The earliest tests on plain and flanged angles, made of aluminium, were carried out by Wagner and Pretschner (1934) to confirm Wagner's theoretical results about the torsional buckling of open thin-walled sections. Tests on thin angle struts were made by Thomas (1941); and Kollbrunner (1935, 1946) tested more than 500 steel and aluminium equal-leg angles of various cross-sectional dimensions and lengths to check the theoretical results on torsional buckling. Mackey and Williamson (1953) conducted tests on two steel lattice girders consisting of double-angle chord members and single-angle web members; one of the aims of their investigation was to obtain information about the effect of adjacent members and joints on the actual end restraint of compression members in lattice girders. More recently, Marshall, Nelson, and Smith (1963) carried out tests on aluminium alloy equal-leg, unequal-leg, and bulb angles; they found that for slenderness ratios greater than 60, failure was primarily due to excessive deformation; angles with smaller

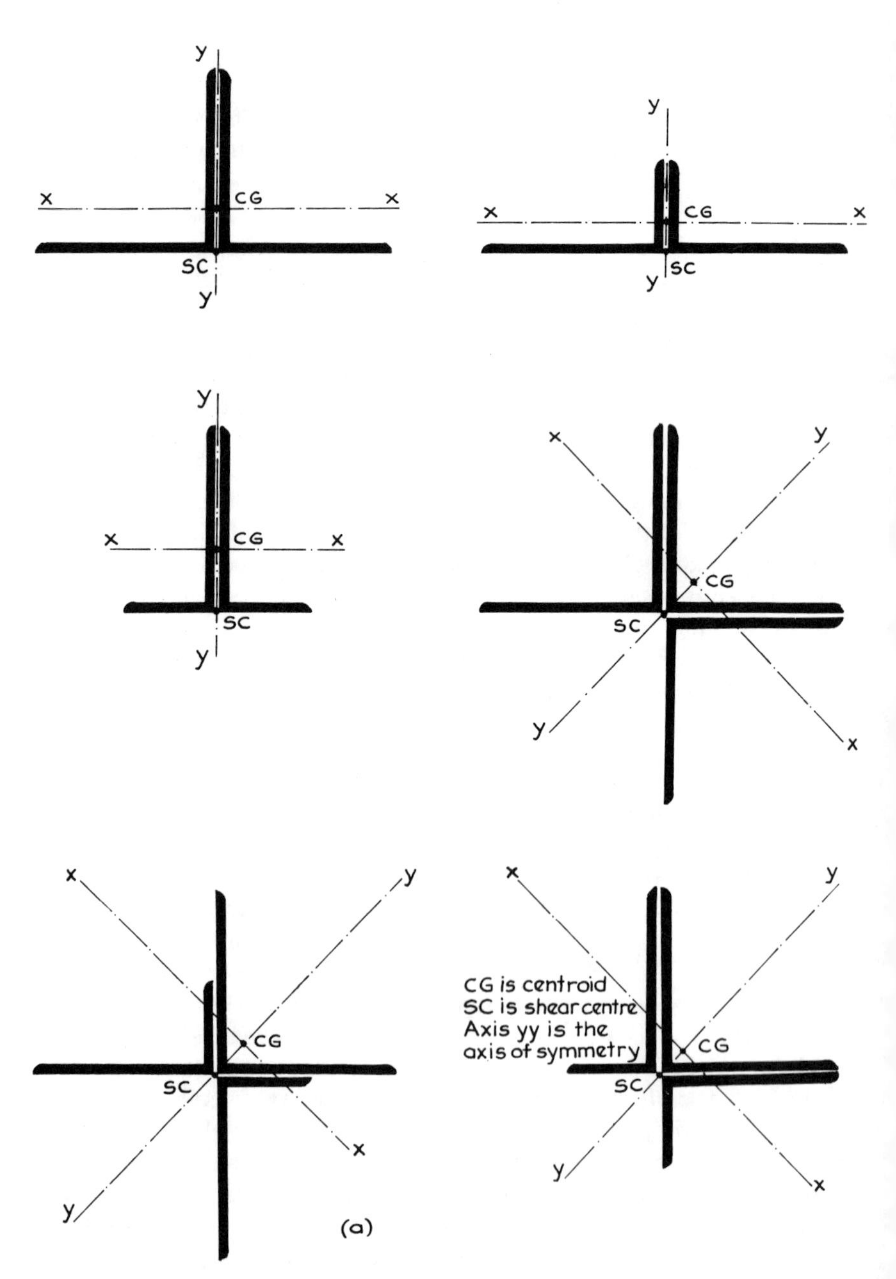

FIG. 1(a). Compound angles—singly symmetric.

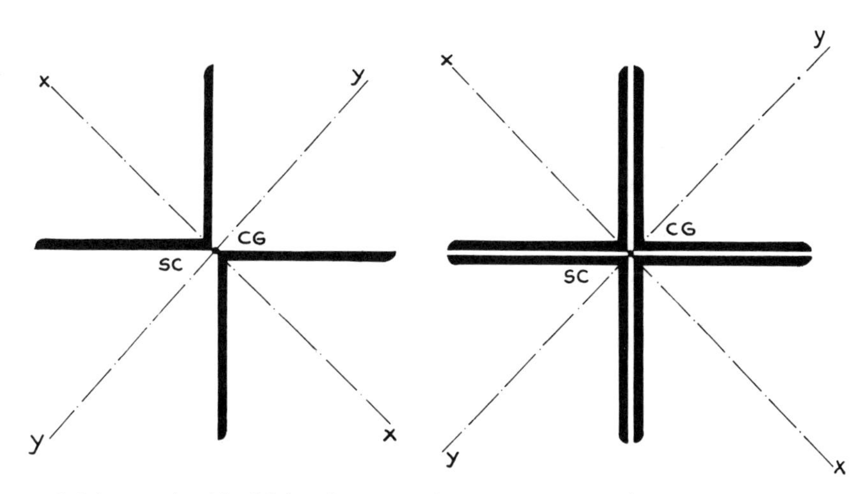

FIG. 1(b). Compound angles—doubly symmetric.

slenderness ratios failed by either overstressing or local buckling. No significant difference was observed in the load-carrying capacities of multiple-bolted connections as compared to single-bolted connections; they concluded that unequal-leg angles connected by the short leg were stronger than the angles connected by the long leg. Wakabayashi and Nonaka (1965) studied experimentally the buckling strength of $90 \times 90 \times 7$ mm structural steel angles; various eccentricities and slenderness ratios were included in the test programme; test results agreed with the theoretical predictions. Fifty-seven mild steel equal-leg angles of $90 \times 90 \times 7$ mm size were also tested by Yokoo, Wakabayashi and Nonaka (1968); concentric as well as eccentric loading tests (including eccentricity about the weak axis, eccentricity about the strong axis, and eccentricity about both axes) were conducted; both positive and negative eccentricities (Fig. 2) were included in the experimental programme. Tests on concentrically loaded angles showed that more torsional deformation occurred at the middle region of the member than at the end regions, confirming that the buckling strength of equal-leg single angles is not significantly affected by the boundary conditions for twisting. For all slenderness ratios, angle specimens loaded eccentrically with respect

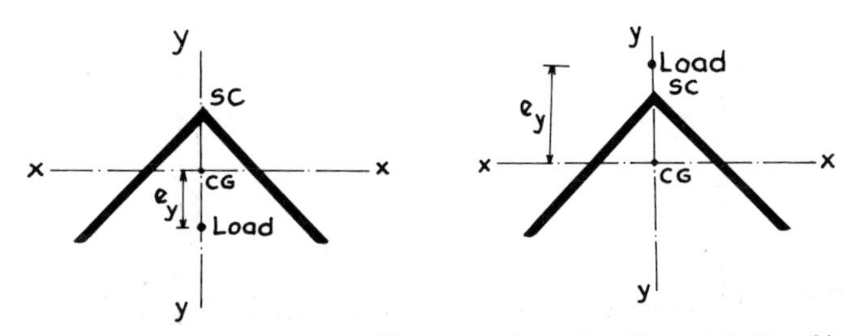

FIG. 2. Definition of positive and negative eccentricities of load.

to their weak axis showed predominantly flexural deformations with a considerable reduction in the load-carrying capacity as compared to concentrically loaded specimens. Angle specimens loaded eccentrically with respect to their strong axis showed considerable torsional deformation at early stages of loading; the decrease in the load-carrying capacity was not appreciable for slender angles which failed mainly by bending about the weak axis; on the other hand, the load-carrying capacity of angles with smaller slenderness ratios was considerably reduced (as compared to concentrically loaded angles), owing to failure caused by local plate buckling.

Tests were conducted by Ishida (1968) on semi-killed high-strength 'SHY' steel angles of size $75 \times 75 \times 6$ mm and $65 \times 65 \times 6$ mm. Both concentric and eccentric loading (eccentricity about the weak axis) tests were carried out. Because of higher residual stresses found in the test specimens, the behaviour of rolled high-strength steel angles was found to be different from that of rolled mild-steel angles; the load-carrying capacity of the former, when concentrically loaded, was generally less than that of mild-steel angles; however, for eccentric loads, the residual stresses did not have any significant effect on the load-carrying capacity.

Marsh (1969) conducted tests on $25 \times 25 \times 2{\cdot}5$ mm and $38 \times 38 \times 2{\cdot}5$ mm aluminium equal-leg single angles; for single-bolt connection, no difference was found between slack and tight bolts. The use of an effective length of $0{\cdot}8L$ for single-bolt connection and $0{\cdot}7L$ for double-bolt connection provided good agreement between the experimental results and the results obtained by the use of the following analytical expression:

$$\sigma_{cr} = 0{\cdot}9\ \pi^2 \frac{E}{\lambda_e^2} \tag{1}$$

in which

$$\lambda_e = \sqrt{\left(5\,\frac{b}{t}\right)^2 + (K\lambda_{max})^2} \tag{2}$$

b = leg width measured from the root fillet; t = thickness of leg; K = end-fixity factor, 0·8 for single-bolt connection and 0·7 for double-bolt connection; and, $\lambda_{max} = KL/r_{min}$. It was suggested that eqn. (1) be used up to $\sigma_{cr} = 0{\cdot}5\sigma_y$ for angles with single-bolt end connections, and up to $\sigma_{cr} = 0{\cdot}67\sigma_y$ for angles with double-bolt end connections.

A total of 721 single angle single-bolted connections were tested by Kennedy and Sinclair (1969) and empirical formulas were developed for the ultimate load-carrying capacities of bolted connections for both end-type and edge-type failures. Trahair, Usami, and Galambos (1969) and Usami and Galambos (1971a, 1971b) investigated, theoretically and experimentally, single angle columns under biaxially eccentric loading; the test specimens were representative of the web members used in standard long-span steel joists and with their ends welded to structural T-sections to simulate the chords of such joists. It was found that the method of loading the T-section end blocks had an effect on the load-carrying capacity of angles; an analytical investigation was carried out by assuming that the column is made of an elastic–perfectly plastic material, and representing the out-of-plane end restraint by an elastic–plastic rotational spring and the in-plane end restraint by an elastic rotational spring; good correlation between theory and experimental results was found. Usami and Fukumoto (1972) studied the behaviour of bracing members of steel bridges; it was found that the effect of residual stresses was not significant, a conclusion which agrees with the earlier finding of Ishida (1968). It was also revealed that the cross-sectional dimensions did not have any marked influence on the non-dimensional column curves, i.e. P_{cr}/P_{yield} versus $(L/r_x)\sqrt{\sigma_y/\pi^2 E}$, where r_x is the radius of gyration about the centroidal axis parallel to the gusset plate. It was recommended that the maximum load of an eccentrically loaded angle bracing member can be taken as 58% of the load of a corresponding concentrically loaded member.

Tests were carried out by Kennedy and Murty (1972) on 72 single and double angle struts with hinged and fixed end conditions, under concentric axial loading. The results from the experimental investigation provided verification of the established theoretical solutions for inelastic flexural, torsional–flexural, and plate buckling. Klöppel and Ramm

(1972) conducted tests on lattice columns to find the effective lengths of leg members. The influence of end connections on the load-carrying capacity of web members was experimentally studied by Lorin and Cuille (1977). They found that increasing the yield strength of the gusset plate did not increase the load-carrying capacity of the web member; however, doubling of the thickness of the gusset plate from 10 mm to 20 mm increased the buckling load by approximately 40%; the tests showed that if the crossing web members are continuous at their intersection (Fig. 3(a)), their capacity is 40% more than the capacity of crossing web

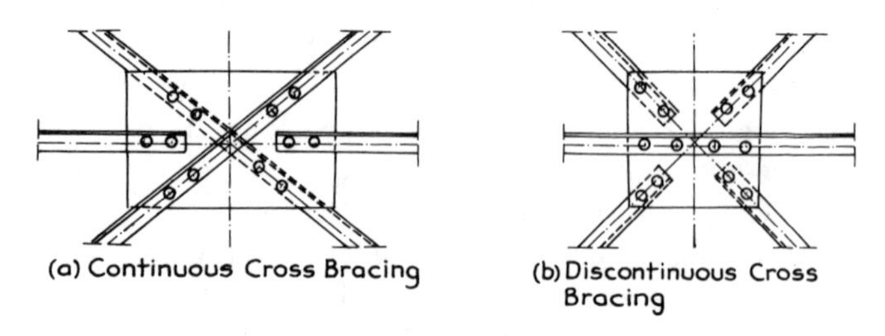

FIG. 3. Continuous and discontinuous cross bracing in latticed towers.

members which are discontinuous at the intersection (Fig. 3(b)). The use of a double angle compound section instead of a larger single angle, at approximately the same cost, increased the strength by 30%. The use of a thicker gusset plate with the compound section increased the load-carrying capacity by 70%. In addition, Lorin and Cuille (1977) also derived empirical formulae for the strength of single-bolt and two-bolt connections.

Fourteen tests were carried out by Short (1977a) on compound angles to investigate their buckling about the x–x axis (Fig. 4); while thirteen tests were done to investigate buckling of compound angles about the symmetric y–y axis. The effect of varying the gap between the angles was also studied by Short (1977b); it was found that for compound angles having two bolts at each end, the effect of end eccentricity must be considered for buckling about the x–x axis. Slotting of end connection holes took place in a direction transverse to the axial load thus providing very little end fixity in the plane of the connected legs; it was suggested that for buckling about the symmetric y–y axis, the effect of spacing between the stitch bolts and the effect of gap between the angles should be considered by computing the effective slenderness ratio λ_{eff} as

$$\lambda_{\text{eff}}^2 = A^2 + 0{\cdot}5B^2 \tag{3}$$

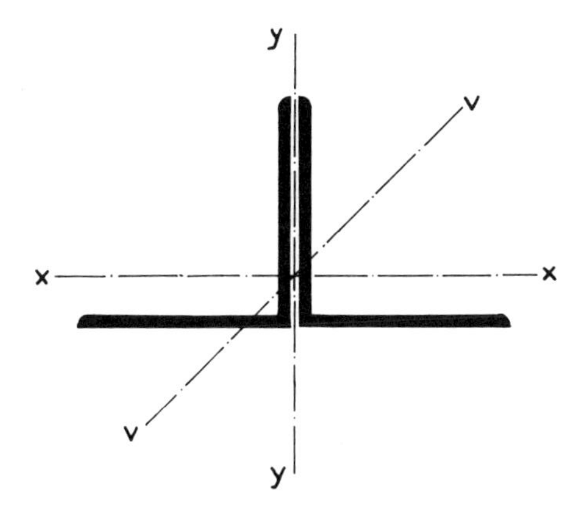

FIG. 4. Definition of x–x, y–y, and v–v axes for compound angle.

in which A = geometric slenderness ratio based on a gap equal to 1/5 of the true gap; and B = slenderness ratio of single angle between stitch bolts. Tests were also carried out by Short (1977c) on single angles to study their buckling about the x–x axis (Fig. 7) as well as buckling about the weak axis. Results for the weak-axis buckling agreed well with curves recommended by the European Convention for Constructional Steelwork (ECCS) (1976a). However, angles which were prevented from failing about their weak axis failed about the x–x axis at considerably less load than the load obtained by using the recommended ECCS curves; this point should be kept in mind when designing transmission towers using single angle shapes (Short and Morse, 1979). Reasonably good agreement with the experimental results can be expected if the buckling load for buckling about the x–x axis is computed by considering the angle as a beam column with a bending stress equal to 70% of the axial stress.

Requirements regarding interconnection of starred angles (Fig. 1(b)) vary from code to code: Canadian Standards Association CAN3-S16.1-M78 (1978), clause 18.1.3(a); American Institute of Steel Construction Specifications (1978), clause 1.18.2.4; German Buckling Specifications DIN 4114 (1952), clauses 8.22 and 8.36; and British Standards Institution specifications BS 449: Part 1: 1970, clauses 30, 36, and 37. To study the problem of interconnection, series of tests were conducted by Temple and Schepers (1980) with the number of interconnectors varying from zero to five. From the study it was found that the interconnection requirements of North American codes are adequate if the buckling load is calculated as twice the buckling load of one angle buckling about its weak axis, using an effective length factor of 0·6. To utilise the load tables for starred angles given in the *Limit States Design Steel Manual* published by the Canadian Institute of Steel Construction (1977), two interconnectors, i.e., at third points, are required.

Some specifications deem that if the eccentricity of load is positive, Fig. 2(a), it is more unfavourable than the case of negative eccentricity, Fig. 2(b). However, based on analytical study, Shao-Fan Chen (1980) concluded that the lateral–torsional buckling load of double angle sections under load with positive eccentricity is larger than the failure load with a negative eccentricity; these theoretical results were confirmed by tests. Angle specimens were tested under static and slow dynamic loading by Jain, Goel, and Hanson (1980) to quantify residual elongation and reduction in maximum compressive strength with number of cycles. It was concluded that the effective slenderness ratio of a member is the most influential parameter on its hysteresis behaviour, and that the total energy dissipation through hysteretic cycles is independent of the direction of loading. Tests were also conducted by Wakabayashi, Nakamura, and Yoshida (1980) on the elastic–plastic behaviour of angle-braced frames under repeated horizontal loading. Experimental and analytical investigations are currently being carried out by Massonnet and Plumier (1981) to ascertain the effectiveness of reinforcing the relatively thin leg angles of transmission towers.

Torsional–flexural buckling of cold-formed thin-walled columns both in the elastic and inelastic range under concentric loading was investigated by Chajes and Winter (1965) and Chajes, Fang, and Winter (1966); a simple method of calculating the torsional–flexural buckling load of singly-symmetric sections was presented. Since the buckling of singly-symmetric sections can occur either in flexural or torsional–flexural mode, curves for determining the critical modes were presented; for single equal-leg angles, it was shown that torsional–flexural buckling occurs when

$$\left(\frac{\text{thickness} \times \text{effective length}}{(\text{width of leg})^2}\right) < 1{\cdot}1$$

Tests on cold-formed angles, as well as other shapes, were carried out to check the accuracy of the analytical procedures developed. The behaviour of singly-symmetric thin-walled open sections under eccentric axial load acting in the plane of symmetry was investigated by Peköz and Celebi (1969). Interaction-type equations were suggested for eccentrically loaded sections, for both positive and negative eccentricities (Fig. 2(a) and 2(b)). In order to check the behaviour of cold-formed equal-leg angles when used as diagonal members in transmission towers, a series of buckling tests were carried out by Carpena, Cauzillo, and Nicolini (1976). For comparison, tests on hot-rolled sections were also made using the same test set-up; the results showed that cold-formed angles had an average buckling strength greater than that of hot-rolled angles. The specifications for the design of cold-formed steel structural members published by the American Iron and Steel Institute (1980) are based mostly on the work mentioned above of Professor Winter and his associates at Cornell University; section 3.7.4 of the specifications states that the strength of singly-symmetric shapes subjected to both axial compression and bending applied out of plane of symmetry must be determined by tests. An exploratory study was conducted recently by Haaijer, Carskaddan, and Grubb (1981) to investigate the feasibility of using a finite element analysis in lieu of a physical test; only the elastic behaviour was considered so that the results are applicable to relatively slender members.

6.3 CLASSICAL THEORETICAL ANALYSES

6.3.1 Elastic Buckling—Concentric Loading

Concentrically loaded columns of angle construction can buckle in one of the following three modes:

(i) flexural buckling—by bending about the weaker principal axis;
(ii) torsional buckling—by twisting about the shear centre; or
(iii) torsional–flexural buckling—by simultaneous bending and twisting.

Concentrically loaded single *equal-leg* angles fail either by flexural buckling about the asymmetric minor principal axis of inertia or by simultaneous bending and twisting about the symmetric major principal axis of inertia, depending upon the length, cross-sectional dimensions, and end-conditions of the member; whereas, concentrically loaded single *unequal-leg* angles always fail by torsional–flexural buckling. Concentrically loaded compound angles which have one axis of symmetry (Fig. 1(a)) fail either by flexural buckling about the asymmetric axis or by torsional–flexural buckling about the symmetric axis (similar to equal-leg single angles). Concentrically loaded compound angles which have two axes of symmetry (Fig. 1(b)) fail either by flexural buckling about the weak axis or by torsional buckling about the shear centre, which in this case coincides with the centroid of the cross-section. The theory of elastic flexural buckling of concentrically loaded columns was formulated by Euler (1759). Wagner (1929) was the first to investigate torsional buckling of open thin-walled sections; however, in his theory, Wagner assumed arbitrarily that the centre of rotation coincides with the shear centre, which, in general, is not the case. In fact, Ostenfeld (1931) was the first to present the exact solutions for the torsional–flexural buckling of angle sections. Among the early investigators who studied the problem of torsional–flexural buckling were Bleich (1936), Kappus (1937), Lundquist and Fligg (1937), Vlasov (1940), Goodier (1941, 1942), and Timoshenko (1945). Since the theories of flexural buckling and torsional–flexural buckling are readily available in the well-known textbooks of Bleich (1952) and Timoshenko and Gere (1961), only the final results of the theoretical analyses are given herein.

(*a*) *Critical Load for Columns with Doubly-symmetric Cross-sections* (Fig. 1(b))

The buckling load in this case is the least of the following three critical loads:

$$P_x = \text{Euler load for flexural buckling about the } x\text{–}x \text{ axis} = \frac{\pi^2 EI_x}{(K_x L)^2} \tag{4}$$

$$P_y = \text{Euler load for flexural buckling about the } y\text{–}y \text{ axis} = \frac{\pi^2 EI_y}{(K_y L)^2} \tag{5}$$

P_t = load for torsional buckling about the shear centre

$$= \frac{A}{I_{ps}}\left[GJ + \frac{\pi^2 EC_w}{(K_t L)^2}\right] \tag{6}$$

in which the various symbols are defined under 'Nomenclature', (page 181).

The torsional constant J of angle sections can be approximately taken as $\Sigma\, bt^3/3$, where b is the centre-line width and t is the thickness of each leg of the angle. A more accurate value for J of hot-rolled angle sections can be obtained from the following expressions developed by El Darwish and Johnston (1965):

Referring to Fig. 5, for a single angle,

$$J=\frac{1}{3}b_1t^3+\frac{1}{3}b_2t^3+\alpha D^4-0{\cdot}315t^4 \tag{7}$$

in which D = juncture diameter parameter $=0{\cdot}343r+1{\cdot}172t$ (8)

$$\alpha=\text{juncture coefficient}=0{\cdot}0728+0{\cdot}0571\left(\frac{r}{t}\right)-0{\cdot}0049\left(\frac{r}{t}\right)^2 \tag{9}$$

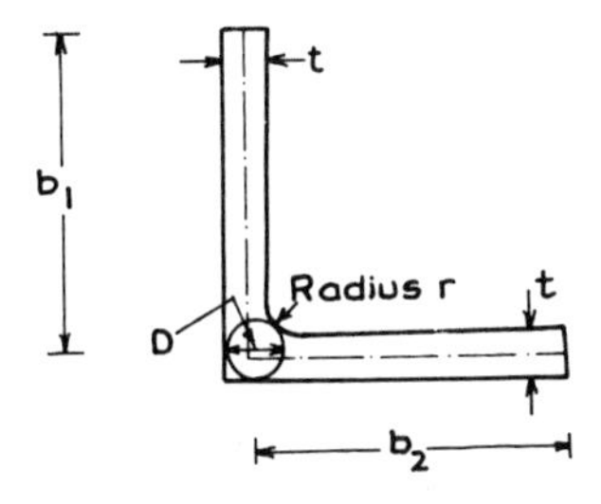

FIG. 5. Geometry of hot-rolled angle section for evaluation of torsional constant J.

The torsional constant J of compound angle is obtained by summing up the J values of the component angles. The warping constant C_w is computed as $\Sigma\, b^3t^3/36$; the summation is carried out over the legs of all the angles in the compound section. Quite often the concept of an equivalent radius of gyration r_t is useful in the analysis of torsional buckling; thus eqn. (6) can be written as

$$P_t=\frac{\pi^2E(Ar_t^2)}{(K_tL)^2} \tag{10}$$

in which

$$r_t=\sqrt{\frac{1}{I_{ps}}\left[\frac{J(K_tL)^2}{2\pi^2(1+v)}+C_w\right]} \tag{11}$$

after substituting $G=E/[2(1+v)]$. It can be readily observed that the two flexural modes (eqns. (4) and (5)) and the torsional mode (eqn. (6)) are independent and therefore there is no coupling between flexural and torsional buckling for such columns.

(b) Critical Load for Columns with Singly-symmetric Cross-sections

For such sections (Fig. 1(a)) the centroid and shear centre do not coincide and therefore, as stated earlier, failure will be either by

(i) Euler flexural buckling about the asymmetric x–x axis; or by
(ii) flexural buckling about the symmetric y–y axis and simultaneous twisting about the centre of rotation, which is a point lying on the symmetric axis close to the shear centre.

The Euler flexural buckling load can be calculated from eqn. (4). For the same end conditions for flexure and torsion, the critical torsional–flexural buckling load, P_{cr}, can be deduced from the following equation (Timoshenko and Gere, 1961, p. 236, eqn. 5–39):

$$\frac{I_{pc}}{I_{ps}} P_{cr}^2 - (P_y + P_t)P_{cr} + P_y P_t = 0 \tag{12}$$

This quadratic equation has two values for P_{cr}, one of which is smaller and the other larger than P_y and P_t; the smaller root is the torsional–flexural buckling load P_{ty}. A study of eqn. (12) shows that when the ratio P_t/P_y is small, the critical load P_{ty} is very close to P_t and buckling is essentially torsional. For larger values of P_t/P_y, the critical load P_{ty} is close to P_y pertaining to flexural buckling. For an equal-leg single angle, the ratio $I_{pc}/I_{ps} = 0{\cdot}625$. Substituting this value in eqn. (12) the critical load P_{ty} for equal-leg single angles is

$$P_{ty} = 0{\cdot}8\,[(P_y + P_t) - \sqrt{(P_y + P_t)^2 - 2{\cdot}5\,P_y P_t}\,] \tag{13}$$

(c) Critical Load for Columns with Unsymmetric Cross-sections

When the end conditions are identical for both torsion and flexure, the critical buckling load for unequal-leg angles is given by the smallest root of the following cubic equation (Timoshenko and Gere, 1961, p. 233, eqn. 5–32):

$$\frac{I_{pc}}{I_{ps}} P_{cr}^3 + \left[\frac{A}{I_{ps}}(P_x y_s^2 + P_y x_s^2) - (P_x + P_y + P_t)\right] P_{cr}^2 + (P_x P_y + P_x P_t + P_y P_t)P_{cr} - P_x P_y P_t = 0 \tag{14}$$

where (x_s, y_s) are the coordinates of the shear centre with reference to the centroidal principal axes. The smallest root for P_{cr} is less than P_x, P_y, and P_t which indicates that unequal-leg single angles always fail in a torsional–flexural mode when subjected to concentric axial load.

(d) Local Plate Buckling

For ideal single equal-leg angles, there is no distinction between the torsional–flexural buckling mode and the local plate buckling mode.

However, for single unequal-leg angles and compound angles, the torsional–flexural buckling mode is distinctly different from the local plate buckling mode. Each leg of an ideal equal-leg angle behaves as a plate simply supported on three sides and free on the fourth side; since both legs have equal width, they buckle at the same time and neither leg can provide restraint to the other one. The local buckling stress σ_{cr} can be computed from the following equation (Bleich, 1952, p. 343, eqn. 670):

$$\frac{\sigma_{cr}}{\sqrt{\eta}} = \frac{\pi^2 E}{12(1-\nu^2)} \left(\frac{t}{b}\right)^2 \cdot k \tag{15}$$

in which $\eta = E_t/E$ = ratio of tangent modulus to elastic modulus; and k = plate buckling coefficient, which approaches 0·425 for long equal-leg angles. In the case of unequal-leg angles, the shorter leg exerts a certain restraining effect upon the longer leg, depending on the ratio of the leg widths; if the ratio of widths (of short leg to long leg) is 2 to 3, the plate buckling coefficient for the long leg is 0·504. If the ratio of widths is 1 to 2, the plate buckling coefficient for the long leg is 0·568 (Bleich, 1952, p. 342).

6.3.2 Elastic Buckling—Eccentric Axial Load

Differential equations of equilibrium for the general case of biaxial eccentricities have been solved by Vlasov (1940), Thürlimann (1953), Dabrowski (1961), Prawel and Lee (1964), Culver (1966), and Peköz and Winter (1969) using different procedures.

(a) Critical Load for Columns with Doubly-symmetric Cross-sections

If the cross-section of the member has two axes of symmetry (e.g., compound angles shown in Fig. 1(b)), the shear centre coincides with the centroid. If the coordinates of the point of application of the axial load are e_x and e_y with reference to the centroidal principal axes, the critical load can be computed by solving the cubic equation obtained by expanding the following determinant:

$$\begin{vmatrix} (P_y - P_{cr}) & 0 & P_{cr} \cdot e_y \\ 0 & (P_x - P_c) & -P_{cr} \cdot e_x \\ P_{cr} \cdot e_y & -P_{cr} \cdot e_x & \frac{I_{ps}}{A}(P_t - P_{cr}) \end{vmatrix} = 0 \tag{16}$$

The smallest root of the cubic equation is the critical buckling load, caused by combined bending and torsion. If the eccentric load acts along one of the axes of symmetry, say, along the y–y axis, (i.e., eccentricity about the x–x axis), then $e_x=0$; whence buckling about the x–x axis occurs independently and the corresponding critical load is given by the Euler load P_x (eqn. (4)). Flexural buckling about the y–y axis and torsional buckling are coupled, and in this case the buckling load is given by the following quadratic equation (Timoshenko and Gere, 1961, p. 248, eqn. 5–77):

$$P_{\mathrm{cr}}^2\left(1-\frac{Ae_y^2}{I_{\mathrm{ps}}}\right)-P_{\mathrm{cr}}(P_y+P_{\mathrm{t}})+P_yP_{\mathrm{t}}=0 \qquad (17)$$

A study of eqn. (17) shows that one value of P_{cr} is less than P_y and P_{t}, corresponding to the torsional–flexural buckling load $P_{\mathrm{t}y}$. The failure load is then the smaller value of either P_x or $P_{\mathrm{t}y}$.

(b) Critical Load for Columns with Singly-symmetric Cross-sections

Assuming that the y–y axis is the axis of symmetry then for the compound angles shown in Fig. 1(a) and for single equal-leg angles, the critical load is computed by solving the cubic equation obtained by expanding the following determinant:

$$\begin{vmatrix} (P_y-P_{\mathrm{cr}}) & 0 & -P_{\mathrm{cr}}(y_{\mathrm{s}}-e_y) \\ 0 & (P_x-P_{\mathrm{cr}}) & -P_{\mathrm{cr}}\cdot e_x \\ -P_{\mathrm{cr}}(y_{\mathrm{s}}-e_y) & -P_{\mathrm{cr}}\cdot e_x & \left(P_{\mathrm{t}}\cdot\frac{I_{\mathrm{ps}}}{A}-P_{\mathrm{cr}}\cdot e_y\cdot\beta_1-P_{\mathrm{cr}}\cdot\frac{I_{\mathrm{ps}}}{A}\right) \end{vmatrix}=0 \qquad (18)$$

in which

$$\beta_1=\frac{1}{I_x}\left(\int_A y^3\mathrm{d}A+\int_A x^2y\mathrm{d}A\right)-2y_{\mathrm{s}} \qquad (19)$$

After expanding the determinant, it is found that the buckling of such columns occurs by combined bending and torsion, with the smallest root of the cubic equation being the critical buckling load. For the special case, when the load acts on the y–y axis of symmetry ($e_x=0$), eqn. (18) yields the critical load for buckling about the x–x axis equal to the Euler buckling load P_x. However, flexural buckling about the y–y axis is coupled with torsional buckling and the critical load is given by the following quadratic equation (Timoshenko and Gere, 1961, p. 247, eqn.

5–76):

$$(P_y - P_{cr})\left[P_t \cdot \frac{I_{ps}}{A} - P_{cr}\left(e_y\beta_1 + \frac{I_{ps}}{A}\right)\right] - P_{cr}^2(y_s - e_y)^2 = 0 \tag{20}$$

The smaller root of eqn. (20), i.e. $P_{cr} = P_{ty}$, is less than the Euler buckling load P_y about the y–y axis, and therefore the failure load will be the lesser of either P_x or P_{ty}. For the special case when the line of action of the load passes through the shear centre, $e_y = y_s$. Equation (20) reduces to

$$(P_y - P_{cr})\left[P_t \cdot \frac{I_{ps}}{A} - P_{cr}\left(e_y\beta_1 + \frac{I_{ps}}{A}\right)\right] = 0 \tag{21}$$

with the two solutions:

$$P_{cr} = P_y$$

and

$$P_{cr} = \frac{P_t}{1 + e_y\beta_1(A/I_{ps})} = \frac{P_t}{1 + y_s\beta_1(A/I_{ps})} \tag{21a}$$

Thus when an axial load is applied through the shear centre for a section having one axis of symmetry, the two flexural buckling modes (about the principal axes x–x and y–y) and the torsional buckling mode are uncoupled. The failure load is then the smallest of P_x, P_y or

$$\frac{P_t}{1 + y_s\beta_1(A/I_{ps})}$$

(c) Critical Load for Columns with Unsymmetric Cross-sections

For an unequal-leg angle having no axis of symmetry, the critical load is computed from the cubic equation obtained by expanding the following determinant:

$$\begin{vmatrix} (P_y - P_{cr}) & 0 & -P_{cr}(y_s - e_y) \\ 0 & (P_x - P_{cr}) & P_{cr}(x_s - e_x) \\ -P_{cr}(y_s - e_y) & P_{cr}(x_s - e_x) & \left(P_t \cdot \frac{I_{ps}}{A} - P_{cr} \cdot e_y \cdot \beta_1 - P_{cr} \cdot e_x \cdot \beta_2 - P_{cr} \cdot \frac{I_{ps}}{A}\right) \end{vmatrix} = 0 \tag{22}$$

in which

$$\beta_2 = \frac{1}{I_y}\left(\int_A x^3\,\mathrm{d}A + \int_A xy^2\,\mathrm{d}A\right) - 2x_s \tag{23}$$

Such angles fail in the torsional–flexural buckling mode at a load corresponding to the smallest root of eqn. (22). Even for an unequal-leg angle, the three types of buckling modes become uncoupled if the load passes through the shear centre; thus, substituting $e_x = x_s$ and $e_y = y_s$, the solutions to eqn. (22) reduce to

$$\left.\begin{aligned} P_{cr} &= P_y \\ P_{cr} &= P_x \\ P_{cr} &= \frac{P_t}{1 + e_y \cdot \beta_1 \cdot A/I_{ps} + e_x \cdot \beta_2 \cdot A/I_{ps}} = \\ &\quad \frac{P_t}{1 + y_s \cdot \beta_1 \cdot A/I_{ps} + x_s \cdot \beta_2 \cdot A/I_{ps}} \end{aligned}\right\} \tag{24}$$

The failure load is the smallest of the three buckling loads given by eqn. (24).

6.3.3 Inelastic Buckling

For stocky and intermediate columns (i.e., for members having small or moderate slenderness ratios), the stress at buckling may exceed the proportional limit of the material, and hence inelastic buckling pertains. In the inelastic range, for flexural buckling, the elastic modulus E is replaced by $E\eta$ (where η is the ratio of the tangent modulus to the elastic modulus and is dependent on the stress level) and the equations developed in the preceding sections can be used. However, for torsional and torsional–flexural buckling, there is no consensus on the value to be used for the modulus of rigidity G in the inelastic range. For the sake of simplicity, Bleich (1952) suggested the use of $G\eta$ for the tangent modulus of rigidity. This approach leads to conservative results and also makes it possible to use readily the equations of elastic torsional and torsional–flexural buckling for inelastic buckling, by replacing G with $G\eta$. The same procedure was also followed by Chajes, Fang, and Winter (1966).

6.4 DESIGN PRACTICES

6.4.1 North American Practice

The American Society of Civil Engineers Manual No. 52, entitled *Guide for Design of Steel Transmission Towers* (1971) has been the basis for the design of axially loaded columns of angle shapes in transmission towers. The recommendations in the 'Guide' are not intended to be used when the width to thickness ratio, b/t, exceeds 20; the width 'b' is not the outside width of the leg but is the width measured from the root of the fillet. The pertinent requirements of the 'Guide' will now be discussed:

(i) The L/r ratio of leg members cannot exceed 150; that of other load-carrying members cannot exceed 200; and that of non-load-carrying bracing members cannot exceed 250. It is to be noted that these slenderness ratios are calculated on the basis of length between panel points without any regard to end fixity.

(ii) The limiting b/t ratio, $(b/t)_{\text{limit}} = 208/\sqrt{\sigma_y}$, where σ_y is the yield stress of the material in N/mm^2. If the b/t ratio does not exceed this value, the member is sufficiently compact to develop its material yield stress at small values of L/r without local buckling. If the b/t ratio exceeds the limiting value, σ_{cr} values given by eqns. (25) and (26) should be substituted for σ_y in eqns. (27) and (29). If $(b/t)_{\text{limit}} \leqslant b/t \leqslant 311/\sqrt{\sigma_y}$,

$$\sigma_{cr} = \left[1{\cdot}8 - \frac{0{\cdot}8\, b/t}{(b/t)_{\text{limit}}}\right]\sigma_y \tag{25}$$

For $b/t > 311/\sqrt{\sigma_y}$,

$$\sigma_{cr} = \frac{57\,900}{(b/t)^2} \tag{26}$$

(iii) For $KL/r \leqslant C_c$, the ultimate maximum stress, σ_a, is obtained from

$$\sigma_a = \left[1 - \frac{(KL/r)^2}{2C_c^2}\right]\sigma_y \tag{27}$$

When $KL/r > C_c$,

$$\sigma_a = \frac{1\,971\,900}{(KL/r)^2} \tag{28}$$

in which

$$C_c = \pi \sqrt{\frac{2E}{\sigma_y}} \tag{29}$$

The formulae for the ultimate maximum stress are based on the Column Research Council's Column Strength Curve (Structural Stability Research Council, 1976, p. 64) in the inelastic range and Euler's formula in the elastic range.

(iv) The effective length factor K is taken as 1·0 for leg sections or post members bolted at connections in both faces. For all other compression members, the following adjusted slenderness ratios, KL/r, shall be used:

For members with concentric loading at both ends, and $L/r \leqslant 120$,

$$\frac{KL}{r} = \frac{L}{r} \tag{30}$$

For members with concentric loading at one end and normal framing eccentricities at the other end, and $L/r \leqslant 120$,

$$\frac{KL}{r} = 30 + 0{\cdot}75\frac{L}{r} \tag{31}$$

For members with normal framing eccentricities at both ends, and $L/r \leqslant 120$,

$$\frac{KL}{r} = 60 + 0{\cdot}50\frac{L}{r} \tag{32}$$

For members unrestrained against rotation at both ends, and $120 < L/r \leqslant 200$,

$$\frac{KL}{r} = \frac{L}{r} \tag{33}$$

For members partially restrained against rotation at one end, and $120 < L/r \leqslant 225$,

$$\frac{KL}{r} = 28{\cdot}6 + \frac{0{\cdot}762L}{r} \tag{34}$$

For members partially restrained against rotation at both ends, and $120 < L/r \leqslant 250$,

$$\frac{KL}{r} = 46{\cdot}2 + \frac{0{\cdot}615L}{r} \tag{35}$$

Furthermore, single-bolt connection shall not be considered as offering restraint against rotation, whereas a multiple-bolt connection properly detailed to minimise eccentricities shall be considered to offer partial restraint if the connection is made to a sufficiently strong member.

6.4.2 European Practice

The European practice for the design of angle members is found in the Introductory Report of the European Convention for Constructional Steel Work (ECCS), published in connection with the Second International Colloquium on Stability; this Introductory Report is often referred to as the 'Stability Manual'. Sub-chapter 3.1.5 of the Stability Manual deals with single angles for use in structures other than transmission towers, while sub-chapter 9.2 deals specifically with angles in latticed transmission towers. These form the bases for the ECCS European Recommendations for Steel Construction (1976a). Sub-chapter 3.1.5 deals with concentrically as well as with eccentrically loaded angles connected by one leg. To eliminate the problem of torsional–flexural buckling, the width-thickness ratios of legs are limited to the following: $b/t = 15$ for steel Fe 510; $b/t = 17$ for steel Fe 430; and $b/t = 18$ for steel Fe 360. Here b is the nominal width of leg, unlike ASCE Manual No. 52 (1971) in which b is taken as the width measured from the root of the fillet.

The ECCS Recommendations give five non-dimensional buckling curves designated a_0, a, b, c, d, as shown in Fig. 6; this figure also shows calculated buckling load curves obtained by numerical simulation for angles with and without residual stresses. The experimental results of centrally loaded angles agreed well with curve (b) for non-dimensional slenderness ratios

$$\frac{\lambda}{\lambda_y} = \frac{KL/r}{\pi\sqrt{E/\sigma_y}}$$

greater than 1; for $\lambda/\lambda_y < 1$, the experimental results were lower than curve (b). Therefore, ECCS recommended that the yield stress be lowered

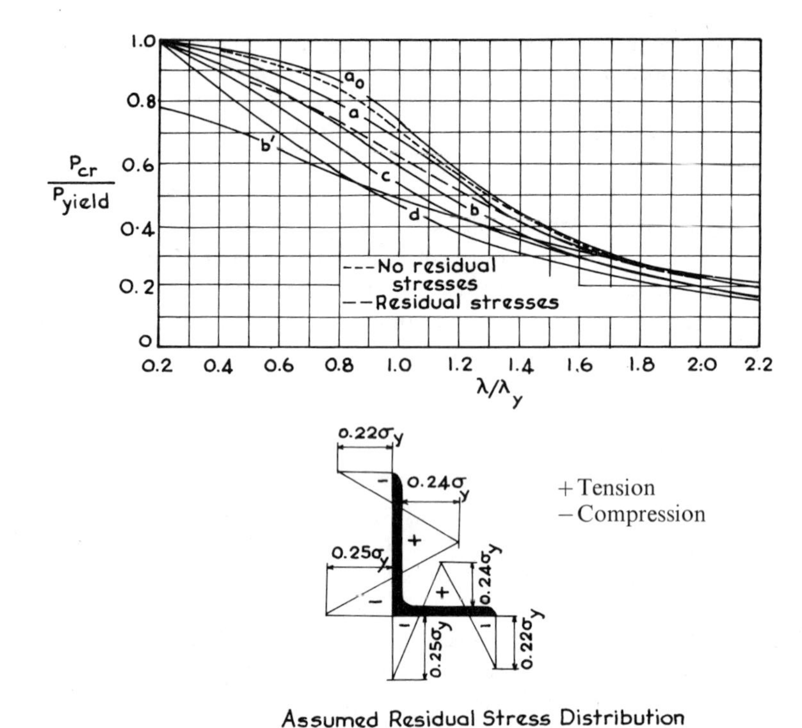

FIG. 6. ECCS buckling curves and numerical simulation curves for angles (initial out-of-straightness $= L/1000$).

and the dimensional buckling curves B_3 and B_4 be used; curves B_3 and B_4, not shown herein, are given in the ECCS Recommendations (1976a).

For eccentrically loaded columns, for the sake of simplicity, a slenderness ratio different from the geometric one is assumed and the same buckling curve (b) is used as a starting point. For non-dimensional slenderness ratio $\lambda/\lambda_y < \sqrt{2}$,

$$\left(\frac{\lambda}{\lambda_y}\right)_{\text{modified}} = 0{\cdot}60 + 0{\cdot}5757\frac{\lambda}{\lambda_y} \tag{36}$$

For a ratio $\lambda/\lambda_y > \sqrt{2}$, no adjustment was found necessary. This is shown as curve (b') in Fig. 6. However, if the angle member is connected to a stiffer main member, the end restraint has a significant beneficial effect as compared to the detrimental effect of eccentricity of the connection. This fact was observed in tests carried out by Working Group 08 of Study Committee 22 of the International Conference on Large High Voltage Electric Systems (CIGRÉ, 1975). In such cases, when the eccentrically loaded angle member has at least two bolts or rivets the equivalent welding), the effective slenderness ratio is less than the actual slenderness ratio and is computed as follows: For $\sqrt{2} < \lambda/\lambda_y < 3{\cdot}5$,

$$\left(\frac{\lambda}{\lambda_y}\right)_{\text{modified}} = 0{\cdot}35 + 0{\cdot}7525\frac{\lambda}{\lambda_y} \tag{37}$$

where λ is calculated from the actual length between the points of intersection of the member centre-line without taking into account any end restraint.

For angles in latticed transmission towers, ECCS adopted the non-dimensional a_0 curve (Fig. 6) and the dimensional A_{02} curve (given in the ECCS Recommendations, 1976a) as basic buckling-load curves. Both local and torsional buckling states are taken into account by reducing the yield stress for larger b/t ratios according to the following formulas:

Hot-rolled Equal-leg Angles

$$\left(\frac{b}{t}\right)_{\text{limit}} = 0{\cdot}567\sqrt{\frac{E}{\sigma_y}} = \frac{260}{\sqrt{\sigma_y}} \tag{38}$$

For $b/t < (b/t)_{\text{limit}}$,

$$\sigma_{\text{y effective}} = \sigma_y \tag{39}$$

For $(b/t)_{\text{limit}} < b/t < 4/3\,(b/t)_{\text{limit}}$,

$$\sigma_{\text{y effective}} = \sigma_y\left[2 - \frac{(b/t)}{(b/t)_{\text{limit}}}\right] \tag{40}$$

For $b/t > 4/3\,(b/t)_{\text{limit}}$,

$$\sigma_{\text{y effective}} = \frac{\pi^2 E}{(5{\cdot}1\,b/t)^2} = \frac{80\,000}{(b/t)^2} \tag{41}$$

Due to the lack of heel, cold-formed angles are torsionally less stiff than hot-rolled angles. On the other hand, cold working produces strain hardening and increases the yield stress at the corners. These two influences lead to smaller buckling loads for cold-formed angles at low slenderness and b/t ratios. This is reflected in the following specifications:

Equal-leg Cold-formed Angles

$$\left(\frac{b}{t}\right)_{\text{limit}} = 0{\cdot}503\sqrt{\frac{E}{\sigma_y}} = \frac{231}{\sqrt{\sigma_y}} \tag{42}$$

For $(b/t) < (b/t)_{\text{limit}}$,

$$\sigma_{\text{y effective}} = \sigma_y \tag{43}$$

For $(b/t)_{\text{limit}} < b/t < 3/2\,(b/t)_{\text{limit}}$,

$$\sigma_{\text{y effective}} = \sigma_y\left[\frac{5}{3} - \frac{2}{3}\frac{(b/t)}{(b/t)_{\text{limit}}}\right] \tag{44}$$

For $b/t > 3/2\,(b/t)_{\text{limit}}$,

$$\sigma_{\text{y effective}} = \frac{\pi^2 E}{(5{\cdot}1\,b/t)^2} = \frac{80\,000}{(b/t)^2} \tag{45}$$

For legs and chords braced as in Fig. 7, the method of computation of slenderness ratios is as shown in the same figure.

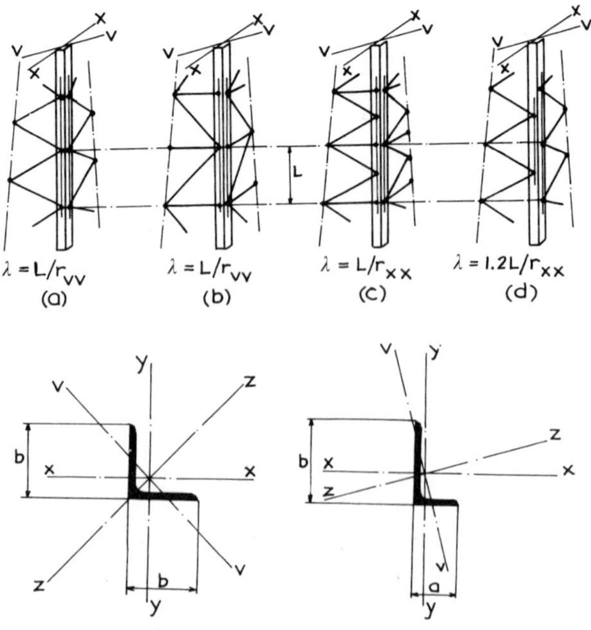

FIG. 7. Critical slenderness ratios for leg angles of latticed towers for different bracing arrangements.

Compound Angles
In order to take into account the possible additional deformations due to shear, the slenderness ratio λ of a compound angle shown in Fig. 8 is computed from

$$\lambda^2 = \lambda_0^2 + \lambda_1^2 \tag{46}$$

in which λ_0 = slenderness ratio of the compound member assuming composite action; and λ_1 = slenderness ratio of one component angle C/r_{vv} (see Fig. 8 for definition of C, and Fig. 7 for definition of the vv-axis). It is good practice to have $\lambda_0 > \lambda_1$ and λ_1 approximately 40 to 50.

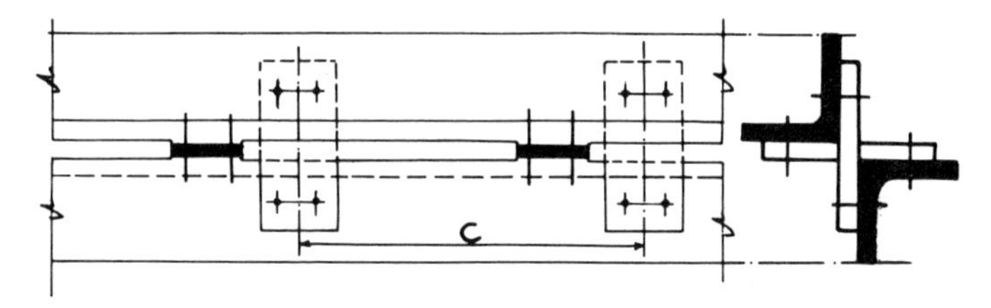

FIG. 8. Arrangement of stitch plates and stitch bolts in compound angles.

The design of intermediate connections is based on a shear force of 1 to 1·5% of the total buckling load.

Buckling Curves for Web Members
Because web members are eccentrically loaded, the effect of eccentricity is taken into account by modifying the non-dimensional slenderness ratios. This procedure is the same as that followed in the ASCE Manual No. 52 (1971).

For $\lambda/\lambda_y \leqslant \sqrt{2}$:

(a) *Eccentricity at one end only:* Buckling about axis v–v,

$$\left(\frac{\lambda}{\lambda_y}\right)_{\text{modified}} = 0{\cdot}25 + 0{\cdot}8232\left(\frac{\lambda}{\lambda_y}\right) \tag{47}$$

Buckling about axis x–x or y–y (Fig. 7),

$$\left(\frac{\lambda}{\lambda_y}\right)_{\text{modified}} = 0{\cdot}35 + 0{\cdot}7525\left(\frac{\lambda}{\lambda_y}\right) \tag{48}$$

(b) *Eccentricities at both ends:* Buckling about axis v–v,

$$\left(\frac{\lambda}{\lambda_y}\right)_{\text{modified}} = 0{\cdot}35 + 0{\cdot}7525\left(\frac{\lambda}{\lambda_y}\right) \tag{48}$$

Buckling about axis x–x or y–y,

$$\left(\frac{\lambda}{\lambda_y}\right)_{\text{modified}} = 0{\cdot}50 + 0{\cdot}6464\left(\frac{\lambda}{\lambda_y}\right) \tag{49}$$

These modified slenderness ratios are based on two or more bolts at each end for single angles and three or more bolts in line at each end for compound angles back to back. If there are only two bolts in line for compound angles back to back, eqns. (48) and (49) are to be modified as follows for buckling about the x–x axis. For eccentricity at one end only,

$$\left(\frac{\lambda}{\lambda_y}\right)_{\text{modified}} = 0{\cdot}35 + 0{\cdot}82\left(\frac{\lambda}{\lambda_y}\right) \tag{50}$$

For eccentricities at both ends,

$$\left(\frac{\lambda}{\lambda_y}\right)_{\text{modified}} = 0{\cdot}50 + 0{\cdot}82\left(\frac{\lambda}{\lambda_y}\right) \tag{51}$$

For $\lambda/\lambda_y > \sqrt{2}$, end restraint has a greater effect than eccentricity; this is accounted for by modifying the non-dimensional slenderness ratios as follows:

(a) *Restraint at one end only:* Buckling about axis v–v, x–x or y–y,

$$\left(\frac{\lambda}{\lambda_y}\right)_{\text{modified}} = 0{\cdot}35 + 0{\cdot}7525\left(\frac{\lambda}{\lambda_y}\right) \tag{48}$$

(b) *Restraint at both ends:* Buckling about axis v–v, x–x or y–y,

$$\left(\frac{\lambda}{\lambda_y}\right)_{\text{modified}} = 0{\cdot}50 + 0{\cdot}6464\left(\frac{\lambda}{\lambda_y}\right) \tag{49}$$

Calculation of λ Values for Web Members

The ECCS Stability Manual also gives the unsupported lengths and the corresponding radii of gyration required for calculating the slenderness ratios of various configurations of web bracing. For compound angles used as web members, it is good practice to limit the spacing of stitch bolts such that the maximum slenderness ratio of component angles is 90, or 0·75 times the slenderness ratio of the compound angle.

6.5 DESIGN AIDS

The exact solution of flexural buckling of eccentrically loaded columns for an ideal elastic–plastic material was presented by Ježek (1934). Subsequently, Ježek (1935, 1936) gave an approximate simple method which yielded satisfactory results; an approximate method of designing eccentrically loaded columns was also developed by Bleich (1952). The use of an equivalent radius of gyration for torsional–flexural buckling does simplify the computation of the corresponding buckling load; to aid designers, equivalent slenderness ratios of angle sections, rolled according to Czechoslovak standards, are given by Mrazik and Sadovsky (1971) in tables for both equal-leg and unequal-leg angles for simply supported end condition.

A simple method to obtain exact interaction relationships of general sections composed of rectangular elements which meet each other at right angles was presented by Chen and Atsuta (1972, 1974). The method is applicable to unequal angles, equal angles, compound angles, etc. Typical interaction curves are shown in Figs. 9 and 10. These are based on the following equations:

$$\left.\begin{aligned} \frac{P_{\mathrm{cr}}}{P_{\mathrm{yield}}} &= 1 - 2\sum A_i/A_0 \\ \frac{M_x}{M_{\mathrm{p}x}} &= 2\sum Q_{xi}/Z_x \\ \text{and}\qquad \frac{M_y}{M_{\mathrm{p}y}} &= -2\sum Q_{yi}/Z_y \end{aligned}\right\} \tag{50}$$

in which A_i = area of ith element above the neutral axis; A_0 = area of the entire cross-section; Q_{xi}, Q_{yi} = static moments of area above the neutral axis of ith element; Z_x, Z_y = plastic section moduli, P_{cr} = axial force; P_{yield} = axial force at yielding; and $M_{\mathrm{p}x}$, $M_{\mathrm{p}y}$ = plastic moments about the x and y axes. The formulation and the method are found to be extremely powerful and efficient for computer programming. Figure 9 shows the interaction curve of a 102 × 76 × 9·5 mm angle section. The largest oval loop represents the interaction curve without any axial force ($P_{\mathrm{cr}}/P_{\mathrm{yield}} = 0$); as the axial force increases, the loop becomes smaller; since the section has no axis of symmetry, the interaction curves also have no symmetry. Figure 10 shows the interaction curves for a compound angle section consisting of two equal angles 152 × 152 × 12·7 mm connected back to back; here the interaction

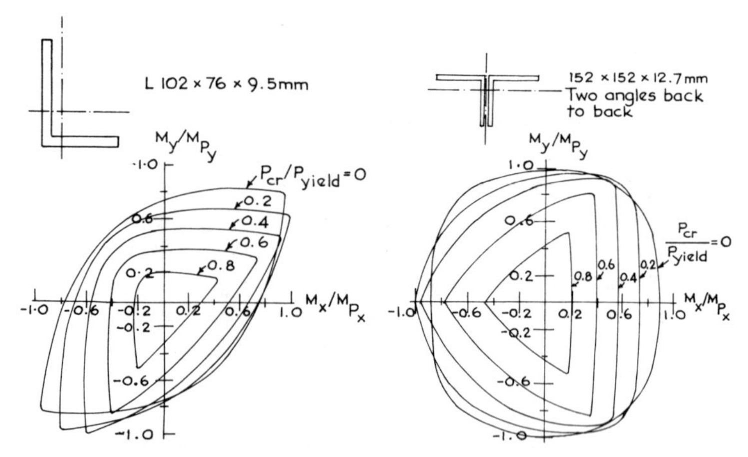

FIG. 9. Interaction curves of an angle section.

FIG. 10. Interaction curves of a double-angle section.

curves have one axis of symmetry. In order to investigate the effect of cross-section sizes on the interaction curves, two extreme sizes for angle section 229 × 102 × 25·4 mm and 152 × 152 × 25·4 mm were investigated; their interaction curves are shown in Fig. 11. These curves were found to be close to the interaction curves for the

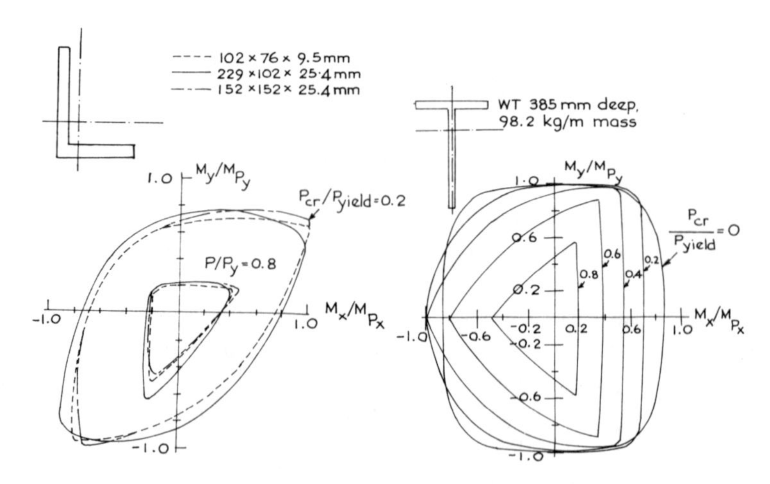

FIG. 11. Comparison of interaction curves for various angle sections.

FIG. 12. Interaction curves of a structural tee section.

102 × 76 × 9·5 mm angle section. Variations of the interaction curves due to different thicknesses of the same shape were also investigated and were found to have an insignificant effect. Based on these results, Chen and Atsuta (1974) recommended that Fig. 9 be taken to represent the interaction curve for all single angles. The same procedure was followed for compound angles and similar conclusions were arrived at; therefore, Chen and Atsuta (1974) recommended that Fig. 10 be taken to represent the interaction curves for all compound angles consisting of two angles connected back to back. As expected, the interaction curves for a T-section, shown in Fig. 12, are almost identical to the interaction curves for a double-angle section shown in Fig. 10.

The following are some of the design tables available:

1. *Handbook of Steel Construction* published by the Canadian Institute of Steel Construction (1980) gives factored axial compressive resistances for double-angle struts on pages 4–122 to 4–141; these are based on clause 13.3 of CAN3–S16.1–M78 of the Canadian Standards Association (1978).
2. *Manual of Steel Construction* published by the American Institute of Steel Construction (1980) gives the allowable concentric loads on double-angle columns on pages 3–49 to 3–75; these are based on Section 1.5.1.3 of the Specifications of the American Institute of Steel Construction (1978).
3. *Handbook of Structural Stability* edited by the Column Research Committee of Japan (1971) has several tables and graphs to aid the designer in routine design calculations, two of which are shown in Figs. 13 and 14. Figure 13 is used for the design of concentrically loaded single and compound angles and T-sections which fail by flexural buckling. The table in Fig. 13 gives the approximate value of section number 'k' ($k = A^2/I$, where A is the area of the cross-section and I is the moment of inertia) for the chosen cross-section; the influence of the wall thickness has been ignored in the tabulated 'k' values which correspond to average values of commonly used width-thickness ratios. The coefficient μ is the coefficient of end restraint and is obtained from Fig. 13(b). The term 'q' can then be computed and the corresponding 'w' for steel columns is obtained from the upper curve in Fig. 13(a). (The lower curve, viz., the square parabola $q = w^2$, gives values of 'w' required when the column buckles in the elastic range according to the Euler formula.) The required cross-sectional area

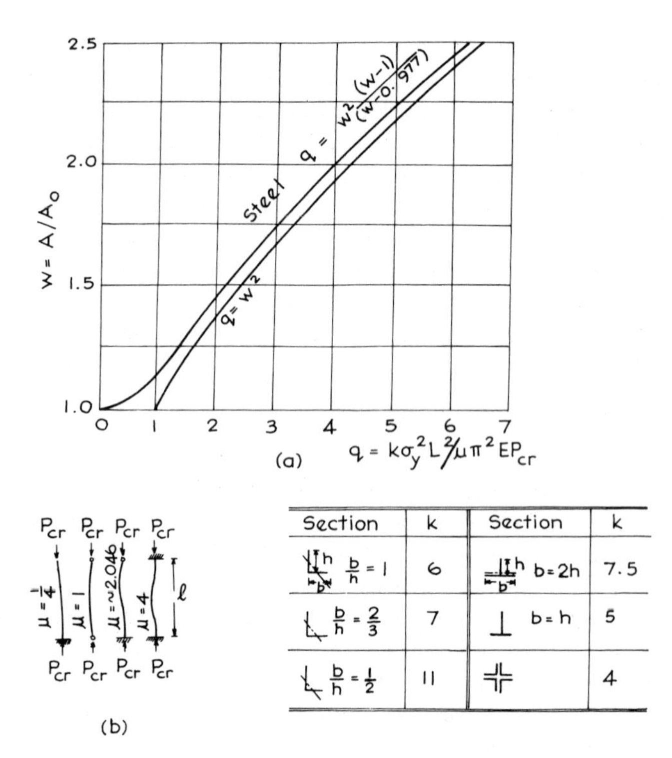

Section	k	Section	k
$b/h = 1$	6	$b = 2h$	7.5
$b/h = 2/3$	7	$b = h$	5
$b/h = 1/2$	11		4

FIG. 13. Design of concentrically loaded single and compound angles and tee sections for flexural buckling.

of the column is then $A = (w)(A_0) = (w)(P_{cr}/\sigma_y)$. Similarly, Fig. 14 is used to compute the torsional–flexural buckling loads of double-angle sections (connected back to back) and T-sections. From Fig. 14(a), $\lambda_{eq}/\lambda_{yy}$ is obtained from dimensions and end conditions of the member; λ_{eq} is computed after finding the value of λ_{yy} from Fig. 14(b) (λ_{yy} is the slenderness ratio with respect to the symmetric y–y axis). The torsional–flexural buckling load is then calculated as $P_{cr} = \pi^2 EA/\lambda_{eq}^2$.

6.6 CONCLUDING REMARKS

Review of the work done on the buckling of single and compound angle members reveals scope for further research. For example, while the

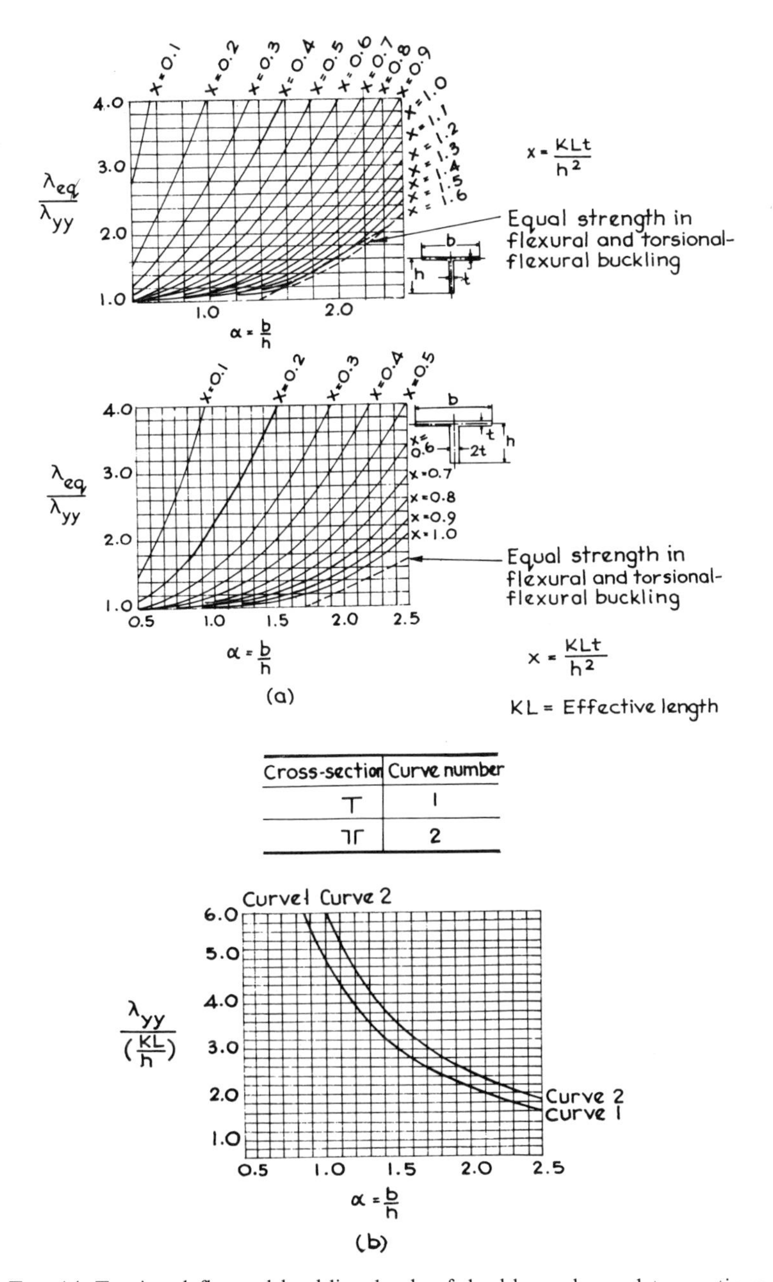

Cross-section	Curve number
T	1
⅂Γ	2

FIG. 14. Torsional–flexural buckling loads of double angles and tee sections.

design of angle members in transmission towers is based on empirical formulae derived from full-scale tower tests, analytical work is required to account for: the interaction between plate and torsional–flexural buckling; interaction between the designed member and its adjoining members; effect of spatial deformation of the tower; and the restraint provided by the end connections. Because of the smaller torsional stiffness and greater yield stress in corners of cold-formed angles versus hot-rolled angles, separate design formulae should be established in North America for cold-formed construction used in transmission towers. The effects of the number of interconnectors in compound angles, and gap width between angles connected back to back require further study; and the recent findings, contrary to some specifications, on the increased load-carrying capacity of double-angle members with positive load eccentricity should be verified. For partial restraint of compound angle members in transmission towers, the ECCS Manual requires at least three bolts in line while North American practice requires only two bolts in line; this discrepancy in the two design practices should be examined. The buckling of leg members of latticed towers with staggered bracing (Fig. 7(d)) needs further work since the ECCS Stability Manual gives the critical slenderness ratio as $(1{\cdot}2L/r_{xx})$ versus the value of (L/r_{xx}) given in the ASCE Manual No. 52. Finally, for the design of an earthquake-resistant bracing system, the load-carrying and energy-absorption capacities of angle members under alternate repeated loading should be assessed.

REFERENCES

American Institute of Steel Construction (1978). *Specification for the Design, Fabrication and Erection of Structural Steel for Buildings*, New York, N.Y.

American Institute of Steel Construction (1980). *Manual of Steel Construction*, Eighth Edition, Chicago, Illinois.

American Iron and Steel Institute (1980). *Specification for the Design of Cold-formed Steel Structural Members*, Washington, D.C.

ASCE Manual No. 52, (1971). *Guide for Design of Steel Transmission Towers*, American Society of Civil Engineers, New York.

Bleich, F. and Bleich, H. (1936). Bending, Torsion and Buckling of Bars Composed of Thin Walls, *Prelim. Pub., 2nd Cong. International Association for Bridge and Structural Engineering*, English edition, p. 871, Berlin.

Bleich, F. (1952). *Buckling Strength of Metal Structures*, McGraw-Hill Book Company, Inc., New York.

BRITISH STANDARDS INSTITUTION (1970). *The Use of Structural Steel in Building*, BS449: Part 1 London, England.
CANADIAN INSTITUTE OF STEEL CONSTRUCTION (1977). *Limit States Design Steel Manual*, First Edition, Willowdale, Ontario.
CANADIAN INSTITUTE OF STEEL CONSTRUCTION (1980). *Handbook of Steel Construction*, Third Edition, Willowdale, Ontario.
CANADIAN STANDARDS ASSOCIATION (1978). *Steel Structures for Buildings—Limit States Design*, CAN3–S16.1–M78, Rexdale, Ontario.
CARPENA, A., CAUZILLO, B. A., and NICOLINI, P. (1976). Modern Technical and Constructional Solutions for the New Italian Power Lines, CIGRÉ, 1976, paper no. 22-13.
CHAJES, A. and WINTER, G. (1965). Torsional–Flexural Buckling of Thin-Walled Members, *Journal of the Structural Division, ASCE*, **91**(ST4), 103–124, Paper No. 4442.
CHAJES, A., FANG, P. J., and WINTER, G. (1966). *Torsional–Flexural Buckling, Elastic and Inelastic, of Cold-formed Thin-walled Columns*, Cornell Engineering Research Bulletin 66-1.
CHEN, SHAO-FAN (1980). Lateral Torsional Buckling of T-Section Steel Beam Columns, Paper Presented at SSRC Annual Technical Session in New York City, April 1980.
CHEN, W. F. and ATSUTA, T. (1972). Interaction Equations for Biaxially Loaded Sections, *Journal of the Structural Division, ASCE*, **98**(ST5) 1035–52.
CHEN, W. F. and ATSUTA, T. (1974). Interaction Curves for Steel Sections Under Axial Load and Biaxial Bending, *Engineering Journal*, **57**(3/4), 49–56.
CIGRÉ (1975). *Buckling Tests on Crossed Diagonals in Lattice Towers*, CIGRÉ SC22-WG08, Electra, No. 38, Paris, 89–99.
COLUMN RESEARCH COMMITTEE OF JAPAN, Editor (1971). *Handbook of Structural Stability*, Corona Publishing Company Ltd, Tokyo.
CULVER, C. G. (1966). Exact Solution of the Biaxial Bending Equations, *Journal of the Structural Division, ASCE*, **92**(ST2), Proc. Paper 4772, 63–83.
DABROWSKI, R. (1961). Dünnwandige Stäbe unter zweiachsig aussermittigem Druck, *Der Stahlbau*, December 1961.
DIN Specification 4114 (1952). *Stabilitätsfälle (Knickung, Kippung, Beulung)* English translation by T. V. Galambos and J. Jones, Column Research Council, June 1957.
ECCS (EUROPEAN CONVENTION FOR CONSTRUCTIONAL STEEL WORK) (1976a). *European Recommendations for Steel Construction.*
ECCS (EUROPEAN CONVENTION FOR CONSTRUCTIONAL STEEL WORK) (1976b). *Second International Colloquium on Stability*, Tokyo, September 9, 1976; Liege, April 13–15, 1977, Washington, May 17–19, 1977, Budapest, October 1977; In Cooperation with International Association for Bridge and Structural Engineering (IABSE), Structural Stability Research Council (SSRC), Column Research Committee (CRC) of Japan. Introductory Report, Second Edition, Sub-chapter 3.1.5, Angles (pp. 98–103), Sub-chapter 9.2, Angles in Lattice Transmission Towers (pp. 263–274).
EL DARWISH, I. A. and JOHNSTON, B. G. (1965). Torsion of Structural Shapes, *Journal of the Structural Division, ASCE*, **91**(ST1), Proc. Paper 4228, 203–227.
EULER, L. (1759). *Sur la force de colonnes*, Mémoires de l'Académie de Berlin.

GOODIER, J. N. (1941). *The Buckling of Compressed Bars by Torsion and Flexure*, Cornell Univ. Eng. Expt. Sta. Bull. No. 27.

GOODIER, J. N. (1942). *Flexural–Torsional Buckling of Bars of Open Section*, Cornell Univ. Eng. Expt. Sta. Bull. No. 28.

HAAIJER, G., CARSKADDAN, P. S., and GRUBB, M. A. (1981). Eccentric Load Test of Angle Column Simulated with MSC/NASTRAN Finite Element Program; Paper presented at the Annual Meeting of Structural Stability Research Council, Chicago, Illinois, April 7–8.

ISHIDA, A. (1968). Experimental Study on Column Carrying Capacity of 'SHY Steel' angles, *Yawata Technical Report No. 265*, December 1968, pp. 8564–82 and pp. 8761–63, Yawata Iron & Steel Co. Ltd, Tekko Building, Tokyo.

JAIN, A. K., GOEL, S. C., and HANSON, R. D. (1980). Hysteretic Cycles of Axially Loaded Steel Members, *Journal of the Structural Division, ASCE*, **106**(ST8), 1777–95, Proc. Paper No. 15607.

JEŽEK, K. (1934). Die Tragfähigkeit des exzentrisch beanspruchten und des querbelasteten Druckstabes aus einem ideal plastischen Material, *Sitzungsberichte der Akademie der Wissenschaften in Wien*, Abt. IIa, Vol. 143.

JEŽEK, K. (1935). Näherungsberechnung der Tragkraft exzentrisch gedrückter Stahlstäbe, *Der Stahlbau*, **8**(12), 89–96.

JEŽEK, K. (1936). Die Tragfähigkeit axial gedrückter und auf Biegung beanspruchter Stahlstäbe, *Der Stahlbau*, **9**(2), 12–14, **9**(3), 22–24, **9**(5), 39–40.

KAPPUS, R. (1937). Drillknicken zentrisch gedrückter Stäbe mit offenem Profil im elastischen Bereich, *Luftfahrtforschung*, **14**(9), 444–57. Translated in NACA Tech. Mem. 851, March 1938.

KENNEDY, J. B. and MURTY, MADUGULA, K. S. (1972). Buckling of Steel Angle and Tee Struts, *Journal of the Structural Division, ASCE*, **98**(ST11), Proc. Paper 9348, 2507–2522.

KENNEDY, J. B. and SINCLAIR, G. R. (1969). Ultimate Capacity of Single Bolted Angle Connections, *Journal of the Structural Division, ASCE*, **95**(ST8), Proc. Paper 6721, 1645–60.

KLÖPPEL, K. and RAMM, W. (1972). Zur Stabilitätsuntersuchung von mehrteiligen Gitterstäben, *Der Stahlbau*, No. 1, 14–21.

KOLLBRUNNER, C. F. (1935). Das Ausbeulen des auf Druck Beanspruchten Freistehenden Winkels, *Mitteilungen*, *4*, Institut für Baustatik, Eidgenössische Technische Hochschule, Zürich.

KOLLBRUNNER, C. F. (1946). Das Ausbeulen der auf einseitigen, gleichmässig verteilten Druck Beanspruchten Platten im elastischen und plastischen Bereich, *Mitteilungen*, *17*, Institut für Baustatik, Eidgenössische Technische Hochschule, Zürich.

LORIN, M. and CUILLE, J. P. (1977). An Experimental Study of the Influence of the Connections of the Transmission Tower Web-Members on their Buckling Resistance, *Second International Colloquium on Stability, Preliminary Report*, Liege, Belgium, pp. 447–56.

LUNDQUIST, E. E. and FLIGG, C. M. (1937). *A Theory for Primary Failure of Straight Centrally Loaded Columns*, NACA Tech. Rept. 582.

MACKEY, S. and WILLIAMSON, N. W. (1953). *Report on Experimental Investigation of Two Mild Steel Lattice Girders*, British Constructional Steelwork

Association, Artillery House, Westminster, SW1, Publication No. 7, pp. 19–36.

MARSH, C. (1969). Single Angle Members in Tension and Compression, *Journal of the Structural Division, ASCE*, **95**(ST5), 1043–49.

MARSHALL, W. T., NELSON, H. M., and SMITH, I. A. (1963). Experiments on Single-Angle Aluminum Alloy Struts, *Symposium on Aluminum in Structural Engineering*, Kensington Palace Hotel, London, W8, June 11 & 12, Session One, Paper 3.

MASSONNET, Ch. and PLUMIER, A. (1981). Essais de Flambement Sur Cornieres de Pylones, effectués à la demande de L'UNERG, March 31, 1981, Liege, Belgium.

MRAZIK, A. and SADOVSKY, Z. (1971). The Buckling of Steel Angle Bars (in slovak), *Stavebnicky Casopis*, **19**(7), 504–519.

OSTENFELD, A. (1931). Politecknisk Laereanstalts Laboratorium for Bygningsstatik, Meddelelse No. 5, Kopenhagen.

PEKÖZ, T. B. and CELEBI, N. (1969). *Torsional Flexural Buckling of Thin-Walled Sections Under Eccentric Load*, Cornell Engineering Research Bulletin 69–1.

PEKÖZ, T. B. and WINTER, G. (1969). Torsional–Flexural Buckling of Thin-Walled Sections Under Eccentric Load, *Journal of the Structural Division, ASCE*, **95**(ST5), Proc. Paper 6571, 941–63.

PRAWEL, S. P., Jr. and LEE, G. C. (1964). Biaxial Flexure of Columns by Analog Computers, *Journal of the Engineering Mechanics Division, ASCE*, **90**(EM1), Proc. Paper 3805, 83–111.

SHORT, J. (1977a). The Buckling of Compound Members Consisting of Two Angles Stitch-bolted Together, *Second International Colloquium on Stability of Steel Structures, Liege, Preliminary Report*, pp. 137–142.

SHORT, J. (1977b). The Buckling of Compound Angles with Varying Gap Between Angles, *Second International Colloquium on Stability of Steel Structures, Liege, Final Report*, pp. 75–77.

SHORT, J. (1977c). The Buckling of Single Angles About the XX and VV Axes, *Second International Colloquium on Stability of Steel Structures, Liege, Final Report*, pp. 271–2.

SHORT, J. and MORSE, J. (1979). The Variation Between Predicted and Actual Performance of Transmission Towers under Test Conditions, Paper Presented at the *1979 IEE Conference on Progress in Cables and Overhead Lines for 220 kV and Above*, pp. 125–130.

STRUCTURAL STABILITY RESEARCH COUNCIL, (1976). Johnston, B. G., editor. *Guide to Stability Design Criteria for Metal Structures*, Third Edition, John Wiley & Sons, New York.

TEMPLE, M. C. and SCHEPERS, J. A. (1980). The Interconnection of Starred Angle Compression Members, *Proceedings, Annual Conference of Canadian Society for Civil Engineering*, May 29 and 30, 1980, Winnipeg, Manitoba, pp. S/22:1–8.

THOMAS, E. W. (1941). Torsional Instability of Thin Angle Section Struts, *Structural Engineer*, **19**(5), 73–82.

THÜRLIMANN, B. (1953). *Deformations of and Stresses in Initially Twisted and Eccentrically Loaded Columns of Thin-Walled Open Cross Section*, Brown Univ. Report No. E696-3, Providence, R. I.

TIMOSHENKO, S. P. (1945). Theory of Bending, Torsion and Buckling of Thin-Walled Members of Open Cross-Section, *Jour. Franklin Inst.*, Philadelphia, Pa., **239**(3), 201–19; **239**(4), 249–68; **239**(5), 343–61.

TIMOSHENKO, S. P. and GERE, J. M. (1961). *Theory of Elastic Stability*, Second Edition, McGraw-Hill Book Company.

TRAHAIR, N. S., USAMI, T., and GALAMBOS, T. V. (1969). *Eccentrically Loaded Single Angle Columns*, Research Report No. 11, Washington University, Department of Civil and Environmental Engineering, August 1969. Also, Supplemental Research Report No. 11A by Galambos, January 1970.

USAMI, T. and FUKUMOTO, Y. (1972). Compressive Strength and Design of Bracing Members with Angle or Tee Section, *Proceedings of Japan Society of Civil Engineers*, **201**, 43–50 (in Japanese). Abstract in English in *Transactions of JSCE*, **4**, 28–9.

USAMI, T. and GALAMBOS, T. V. (1971a). Eccentrically Loaded Single Angle Columns, *IABSE*, Zurich, **31-II**, 153–84.

USAMI, T. and GALAMBOS, T. V. (1971b). On the Strength of Restrained Single-Angle Columns under Biaxial Bending, *Proc. Japan Society of Civil Engineers*, **191**, 31–44 (in Japanese). Abstract in English in *Transactions of JSCE*, **3**, Part 2, 118–19.

VLASOV, V. Z. (1940). *Thin-Walled Elastic Beams*, Moscow. Translated from Russian by the Israel Program for Scientific Translations, 1961.

WAGNER, H. (1929). *Verdrehung und Knickung von offenen Profilen*, 25th Anniversary Publication, Technische Hochschule Danzig, 1904–1929. Translated in NACA Tech. Mem. No. 807, October 1936.

WAGNER, H. and PRETSCHNER, W. (1934). Verdrehung und Knickung von offenen Profilen, *Luftfahrtforschung*, **11**(6), 174–180. Translated in NACA Tech. Memorandum No. 784, Jan. 1936.

WAKABAYASHI, M., NAKAMURA, T., and YOSHIDA, N. (1980). Experimental Studies on the Elastic-Plastic Behaviour of Braced Frames under Repeated Horizontal Loading; Part 2. Experiments of braces composed of steel circular tubes, angle-shapes, flat bars or round bars, *Bulletin of the Disaster Prevention Research Institute*, Kyoto University, **29**, Part 3, 99–127.

WAKABAYASHI, M. and NONAKA, T. (1965). On the Buckling Strength of Angles in Transmission Towers, *Bulletin of the Disaster Prevention Research Institute*, Kyoto University, **15**, Part 2, 1–18.

YOKOO, Y., WAKABAYASHI, M. and NONAKA, T. (1968). An Experimental Study on Buckling of Angles, *Yawata Technical Report No. 265*, December 1968, pp. 8543–63 and pp. 8759–60, Yawata Iron & Steel Co. Ltd, Tekko Building, Tokyo, Japan.

Chapter 7

CENTRALLY COMPRESSED BUILT-UP STRUCTURES

W. UHLMANN
Department of Construction Engineering
Technische Hochschule Darmstadt, West Germany

and

W. RAMM
Fachgebiet Massivbau und Baukonstruktion,
University of Kaiserslautern, West Germany

SUMMARY

This chapter begins with a general explanation of the calculation procedure for built-up struts by modelling them as full members in which bending and shear stiffness are smeared continuously along their lengths. After a brief review of the historical development of methods of analysing built-up struts, the two main steps of the current design method are dealt with. First, the strut is treated as a full member having reduced shear stiffness. After establishing local equilibrium the second step is to design the members (such as chords and web elements) according to the relevant standards. Methods of taking into account the flexibility and lack of fit at the joints as well as eccentricities of the axial forces acting at the connections are also dealt with. Finally, some topics for future research are discussed.

7.1 INTRODUCTION

This chapter deals with centrally compressed built-up members subjected to externally applied axially compressive forces acting at the centroidal axis. It is pre-supposed that the cross-section of the built-up structure is double-symmetrical so that the longitudinal axes of the centres of gravity and shear coincide at every position along the length of the member. In the following, therefore, only built-up structures which act as columns are considered, even though it is possible to use these members in other structural forms.

The treatment is confined to built-up struts which consist of single members interconnected in such a way that a discontinuous structure having a very high degree of statical redundancy results. It seems reasonable to ignore these discontinuities by smearing the single members along the length of the strut and to approximate the structure as an ordinary column with continuous cross-section. In the following, the structure regarded in such a manner will be called 'full member'. On the other hand a strut with flanges continuously interconnected by the web will be referred to as 'solid member'. The implications involved in the simplification into the 'full' member will be discussed in detail.

It is well known that columns must have adequate stiffness to resist buckling of various types. As a rule, instabilities are caused by strong-axis buckling, weak-axis buckling, and lateral–torsional buckling. In the case of a built-up column, in general, torsional instability can be disregarded since the torsional rigidity of these struts is normally sufficient to prevent it. Thus we have only to deal with buckling caused by bending moments which occurs on account of second-order effects.

These bending moments depend upon lateral deflections and therefore will not be uniform so that shear forces also occur. In general, the latter are small. In the case of normal columns the shear rigidity is large enough to permit the deflections caused by shear forces to be neglected. On the other hand, built-up struts when treated as full members are significantly weaker against shear deflections compared with normal columns. This will be discussed in detail in sub-section 7.3.3.

Bending and shear stiffness of the full member are furnished by different parts of the structure. The chords (which correspond to the flanges of an ordinary column) have to supply the required bending stiffness, whereas resisting the shear is assigned to the web members. This is different from an ordinary column in which the web has to bear a part of the bending moment, too. In this sense, the built-up strut can be compared with a member having a sandwich cross-section. The difference is, however, that the latter is assumed to be continuously connected between web and flanges so that the danger of instability for the single members of the structure is reduced.

There are two types of built-up columns, based on the formation of the web members. In the case of 'laced struts', the web members consist of single struts so that the whole structure behaves similarly to a grid. In the second type, called 'struts with batten-plates', the chords are connected by batten-plates at discrete positions along the length of the strut so that bending moments can be interchanged between the chords. In this case, the structural behaviour resembles that of a Vierendeel system. Figure 1 shows the basic types of built-up columns.

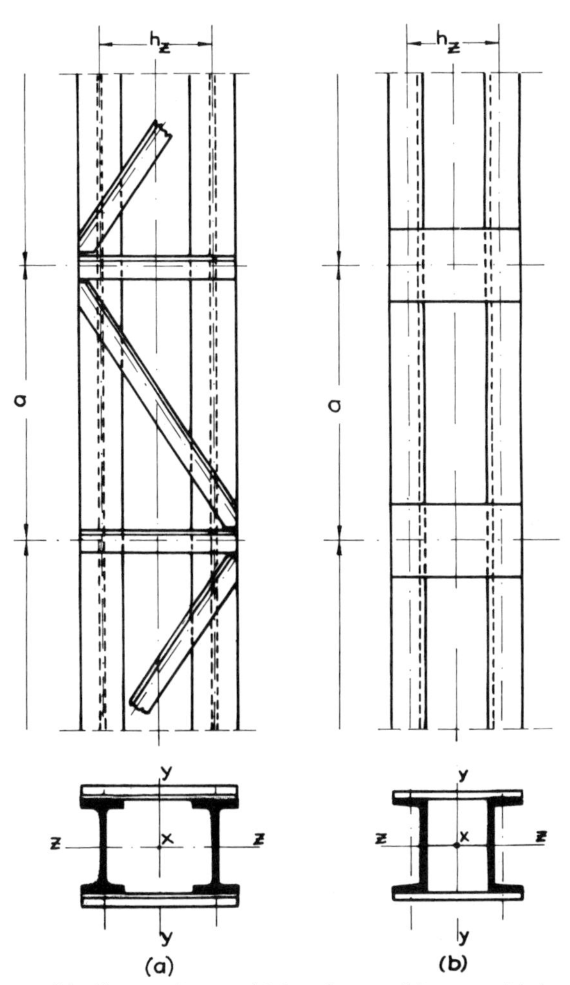

FIG. 1. Types of built-up columns. (a) laced strut; (b) strut with batten plates.

The stiffness of a truss against deflections in its own plane is well known. However, trusses, having comparatively small depths are uneconomical. In general, laced struts are preferred in cases where the distance between the chords is relatively large, whereas struts with batten-plates are used for smaller depths.

Both types of built-up struts modelled as full members display a great deal of similarity so that it is possible to treat them largely in the same way for calculation purposes.

In all cases the instability of the chord members has to be taken into account. In the case of laced struts, web members can also fail by instability produced by exceeding their critical axial compressive force. The batten-plates, however, act as simple short-spanned beams which, generally speaking, will not fail by instability of any kind assuming that connection problems at the joints are disregarded. The method of computing built-up struts as full members has been known for a long time. Before discussing the prevailing calculation procedures, it seems useful to look at some milestones of their development in the past.

The first known method to calculate built-up struts was to assign half of the external load equally to both chords. This was done until two accidents made it necessary to revise this design practice. In 1907, the Quebec Bridge collapsed; two years later, the gas holder in Hamburg broke down. In both cases collapse was triggered by failure of a built-up column (Foerster, 1911).

As early as 1908, Krohn pointed out that the deflection of the full member during buckling causes unequal distribution of the external load to the chords. This was also verified experimentally by Rudeloff (1914) in connection with the disaster of Hamburg; he concluded that the bearing capacity of the built-up strut is exhausted if the proportional limit is reached at the most stressed fibre, the position of which was supposed to be at the mid-span of that chord bearing the bigger part of the load.

The merit of developing, for the first time, a consistent theory concerning the buckling behaviour of built-up struts should be attributed to Engesser (1891). Starting from the classical Euler equation for solid columns, he took into account additional deflections of the chords and batten-plates caused by shear forces. Later Engesser (1909) refined this calculation method, establishing an equivalent slenderness ratio λ_{vi} for built-up struts and obtaining the expression:

$$\lambda_{vi}=\sqrt{\lambda_y^2+\lambda_1^2}$$

Here, the term λ_1 takes the shear flexibility of the strut into account, whereas λ_y is the slenderness ratio referring to buckling about the non-substantial axis. In 1952, this formula was incorporated into the German buckling standards DIN 4114.

Mann (1909) made an attempt to analyse the discontinuous system of a centrally compressed strut with batten-plates. He evaluated a buckling condition, confining his investigations to the elastic region.

Müller-Breslau (1913) improved the calculations made by Mann. He was the first to point out that an unintentional eccentricity of the external load related to buckling about the non-substantial axis should be taken into account.

The most sophisticated theoretical investigation of the centrally compressed strut with batten-plates was performed by Chwalla (1933). His calculations extended to the region of inelastic behaviour. The unintentional imperfections were approximately taken into account by transferring the drop of the ultimate load determined in solid columns theoretically, (due to increasing load eccentricities) on to built-up struts.

Starting from the fact that ideally straight centrally compressed columns scarcely exist, Klöppel and Uhlmann (1965) carried out an extensive test programme on eccentrically loaded struts with batten-plates. The amount of eccentricity, the slenderness ratio λ_y, the chord section length, a, between the joints, and the depth of the batten-plates were varied. Parallel to this, theoretical investigations of the discontinuous system were performed making full reference to inelastic behaviour, so that the spread of plastic zones into the chords was taken into account.

In 1968, Klöppel and Ramm carried out similar experimental and theoretical investigations on laced struts varying the pattern of lacing as well.

One of the main purposes of the above investigations was to find a scientific basis for improving the design rules for built-up struts in the German standards DIN 4114.

Meanwhile, in Europe efforts were initiated to standardise all specifications existing in various countries. Starting from available knowledge and experience, design specifications for built-up struts consistent with the new trend had to be drafted. Since calculations based on the discontinuous structure are too tedious for practical design purposes, the concept of the built-up strut treated as a full member with reduced shear stiffness was taken as a basis.

The standards for the analysis and design of these structures had to be consistent with those of solid columns which had just been established at

European level. The above concept has been accepted by the ECCS (European Convention for Constructional Steelwork) and is discussed in detail in the following.

7.2 PRELIMINARY REMARKS CONCERNING THE PROPOSED METHOD OF CALCULATION

This chapter is confined to built-up struts containing chords with cross-sections having at least one axis of symmetry parallel to one of the co-ordinate axes of the full member (Fig. 2), since only for these types of struts can existing experimental results be used for comparisons with

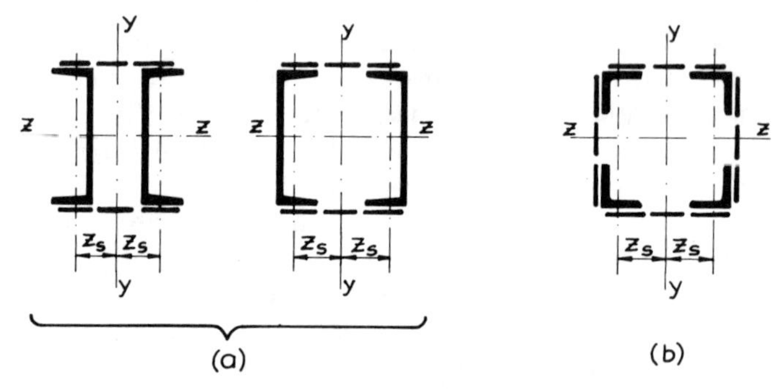

FIG. 2. Cross-sections of built-up columns. (a) strut with one non-substantial axis; (b) laced strut with two non-substantial axes.

theoretical calculations. Included in the above are a large number of built-up struts with batten-plates occurring in practice, with the exception of a group of struts containing chords consisting of single angles. The latter are sometimes used in trusses to carry wind loads. Here no experimental data are available so that in this case German standards, for example, refer partially to prior design rules based on considerations involved with bifurcation loads.

This chapter deals with built-up struts having web members in not more than two planes perpendicular to the non-substantial axes. (Two axes of that kind occur in some laced struts having four chords which, in principle, are included.) It is also assumed that the built-up struts have chords solely in two planes parallel to the non-substantial axes.

The above two restrictions are usually met in many practical cases but in principle it is not difficult to extend the following design concepts to built-up members which fall beyond these limitations. A recent contribution of the authors (1981) deals with this new design concept including detailed formulae and numerical examples.

The proposed design concept takes into account the behaviour of the full member including its susceptibility to shear deformations and the behaviour of the chords and of the web members. All these criteria have an influence on the bearing capacity of the strut. They are not independent of each other so that an involved stability problem results. To take this into account in a consistent manner it seems reasonable to provide a continuous check according to second-order theory. This check can be done simply by splitting it into two steps.

First, the full member is analysed under factored loads according to second-order theory. Here a deviation from the ideal geometrical form must be taken into account. For the present, we confine ourselves to simply-supported built-up struts. So the shape of the initial imperfection can be taken to be sinusoidal according to the ECCS-curves.

This is in good agreement with Timoshenko and Gere (1961) who, referring to built-up struts, stated that when the detrimental effect of shear forces in the corresponding full member is considered, an initial curvature is more unfavourable than an initial constant eccentricity in the application of the load. The analysis of the full member can be carried out in the elastic range for both types of built-up struts. Only in the case of struts with batten-plates are certain easily applied corrections necessary with respect to plastic behaviour of the full member. This part of the calculation results in the knowledge of the stress resultants of the full member: bending moments, axial, and shear forces according to second-order theory.

In the second step the stress resultants of the individual members (chords and web members) can be evaluated by means of local equilibrium conditions. These members are to be checked according to the principles for designing solid struts. As a rule, only axial forces exist, causing either tension or compression. In the first case an ordinary check of tensile strength is adequate, whereas in the second case, check or design of the compression members must be performed according to the ECCS-curves. In addition to this, the connections between chords and web members have to be checked using appropriate design criteria.

7.3 CALCULATION OF THE FULL MEMBER

7.3.1 General Assumptions—Choice of Imperfections

The well-known assumptions concerning second-order theory are taken to be valid. Residual stresses are not taken into account and their effect is allowed for by means of increased initial imperfections. In a similar way, other influences which can have a detrimental effect on the bearing capacity (for example caused by unintentional deviations from the ideal form) are accounted for.

The ECCS-curves are based theoretically on a sinusoidal initial imperfection with a mid-span amplitude $f_0 = l/1000$ (in which l is the column length). We cannot, obviously, choose a smaller value in the case of built-up struts.

In many cases, solid columns consist of hot-rolled steel in which residual stresses caused by non-uniform temperature distributions during manufacture are present. Their influence on the ultimate capacity depends also upon the shape of the cross-section and the direction of loading and is incorporated into the ECCS-curves.

In the case of built-up struts, however, additional residual stresses occur due to the assembling of the strut. If its joints are welded, a further source of residual stresses appears. It is obvious that it is impossible to consider these individual effects in the theoretical calculations; this was therefore done in an empirical manner.

In the experimental investigations carried out at Darmstadt by Klöppel and the authors (1965, 1968), tests were performed systematically with the loads applied eccentrically. In addition to this, an extra sinusoidal out-of-straightness with different mid-span amplitudes f_0 was included in order to match the experimental ultimate loads with the theoretical analysis using the concept described here. As a result of this investigation, the best agreement with the experimental values could be obtained setting f_0 equal to $l/500$ for both types of built-up struts. In the case of laced struts, however, the scatter between experimental values and theoretical predictions obtained in this way is considerable. It varies from 16% on the unsafe side in the most unfavourable case to 38% on the safe side.

To interpret these results, it must be mentioned that the theoretical calculations made in this context were based upon the true yield stresses of the test specimens obtained by tension tests. Since these stresses generally exceed the values prescribed in design standards, an additional safety margin exists in most cases.

Furthermore, laced struts which showed the lowest bearing capacity in the tests, had comparatively short distances between the chords which led to a very close lacing with numerous welded joints. Such an arrangement is not typical for laced struts. Tests with the more typical laced struts, having a wide-meshed lacing pattern, gave much better test results. Considering all facts, it seems reasonable to ignore the results on the unsafe side obtained in exceptional cases in order to avoid uneconomical constructions in most cases.

7.3.2 Bending Stiffness of the Full Member—Effect of Plasticity

As far as laced struts are concerned, only axial forces are present in the web members if we disregard local eccentricities at the joints. Because the whole structure is similar to that of a truss, it seems reasonable to neglect bending moments in the chords, too, even though they can be regarded as continuous beams. As a rule, the distance between the chords is comparatively large so that the full member is stocky and the contribution of these bending moments to the bending capacity of the full member is negligible compared with that due to the axial forces of the chords acting at a large lever arm. Hence the distribution of the axial stress in the chords which are regarded as a part of the full member, is almost uniform. It is therefore proposed (in the case of laced struts) to omit the portion of the moment of inertia of the chords when establishing the bending stiffness of the full member.

Contrary to the laced struts, the spacing between the chords of a strut with batten-plates can be rather small so that the contribution of the chord bending stiffness will be more important. Hence it is uneconomical to neglect it.

On the other hand, if we admit bending stresses in the chords before reaching their ultimate bearing capacity, we assign plastic shape factors exceeding unity to them so that elastic–plastic intermediate states occur which are to be taken into account in the calculation process of the full member.

The real behaviour of the full member being in an elastic–plastic intermediate state will therefore lie between two bounds, both of which can be calculated in the elastic range. The lower bound is obtained by omitting the moments of inertia I_f of the chords as in the case of laced struts, whereas in the upper bound calculations these portions are fully taken into account. Hence it seems reasonable to regard only a reduced term $\eta I_f (0 < \eta < 1)$ when calculating the moment of inertia I_f of the full member. Using this value, the calculation can be carried out in the elastic range. η depends upon special parameters, predominantly upon the slenderness ratio λ_y of the full member referred to the non-substantial axis as well as upon the applied mid-span amplitude of the extra sinusoidal out-of-straightness which is to be chosen appropriately to match experimental results. Here plastic behaviour of the full member is to be accounted for.

The bearing capacity of the strut can be reduced in two different ways:

Firstly, the full member can fail as a solid strut in the elastic–plastic range if equilibrium between the external load and internal stress resultants at the most unfavourable position (in this case at mid-span) is no longer possible. In the elastic–plastic range, equilibrium must be established by trial and error so that divergence of that process indicates exceeding of ultimate theoretical load. This can occur before the chords have reached their bearing capacity. Secondly, the bending stiffness of the full member will be reduced when attaining the plastic region. On account of this, deflections and chord axial compressive forces depending on these will be increased, so that the critical design load in the chords is reached at a lower external load level compared with the case when plastic behaviour of the full member is neglected.

To check this, calculations were performed in three different modes (Uhlmann, 1977):

Mode 1: The above possibilities were fully taken into account setting $\eta = 1$.
Mode 2: Elastic calculation with $\eta = 1$ (upper bound).
Mode 3: Elastic calculation, setting $\eta = 0$ (lower bound).

The slenderness ratio λ_y as well as the amount of mid-span amplitude f_0 of the additional sinusoidal out-of-straightness were varied. λ_y is evaluated in the usual manner, setting $\eta = 1$. The results of these investigations are summarised in Fig. 3 (Ramm and Uhlmann, 1982). Here the differences:

$$\Delta V_n = \frac{F_{n,\,\text{theoretical}} - F_{\text{experimental}}}{F_{\text{experimental}}} \cdot 100$$

are plotted versus a value ΔC defined as $\Delta C = 1/f_0$, so that the complete applied mid-span amplitude f resulting from rectangular eccentricity e as well as from sinusoidal extra out-of-straightness amounts to:

$$f = e + f_0$$

Here n refers to the three modes discussed above, hence varying from 1

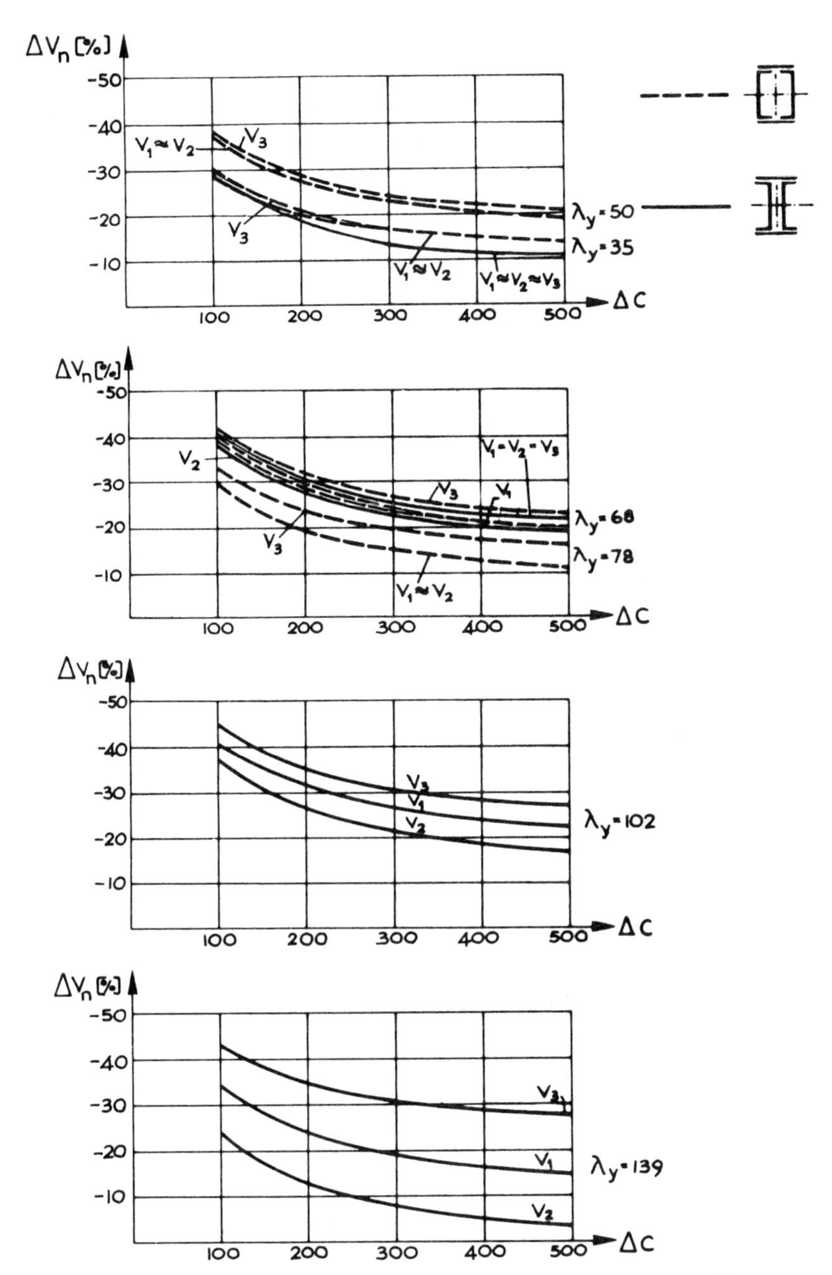

FIG. 3. ΔC versus $\Delta V_n = F_{n,\text{theoretical}} - F_{\text{experimental}}/F_{\text{experimental}}\ 100$.

to 3. In order to enable comparisons with test results, only discrete values of λ_y could be used.

Choosing $\Delta C = 500$ is sufficient to achieve safe results, since ΔV_n is negative in all cases. Furthermore, calculations of the theoretical ultimate load according to Modes 1 to 3 mentioned above show a dependence upon the slenderness ratio λ_y. For low values of λ_y, the differences between the results corresponding to V_1, V_2, V_3 are negligible, whereas these become more pronounced with increasing slenderness ratio λ_y. Compared with the relatively sophisticated calculation Mode V_1, the bearing capacity would be underestimated if calculations according to Mode 3 were carried out, especially in the case of slender struts. On the other hand, the differences between results corresponding to V_1 and V_2 are small. Only for the most slender struts (for which test results are available for][-arrangement of the chords only), the results corresponding to Modes 1 and 2 differ significantly.

Therefore it would be uneconomical to check and design according to Mode 3 for all types of built-up struts, as it is proposed for laced struts. It seems appropriate to consider the above-mentioned differences between Modes 1 and 2 in order to provide a calculation method by Mode 2 for struts with batten-plates after incorporating certain modifications depending on the slenderness ratio λ_y.

Details of these investigations can be found in publications of Uhlmann (1977) as well as Ramm and Uhlmann (1982) wherein the influence of 'test eccentricities e' and the number n of panels have also been taken into account.

It turned out that only in a few cases would the application of Mode 2 lead to slightly unsafe results. This tendency becomes larger with increasing e, λ_y and smaller chord lengths, maintaining the strut length l constant. In view of these studies the following recommendations establishing the coefficient η could be made, leading to theoretical ultimate loads lying on the safe side (Fig. 4).

$$\begin{aligned} 0 \leqslant \lambda_y &\leqslant 75 & \eta &= 1 \\ 75 < \lambda_y &< 150 & \eta &= 2 - \lambda_y/75 \\ \lambda_y &\geqslant 150 & \eta &= 0 \end{aligned} \tag{1}$$

This result is somewhat surprising. It is well known that a solid member, eccentrically loaded by an external compressive force, is most sensitive to the influence of plasticity in the range of small values of λ_y, whereas in the loading histories of slender columns the transition

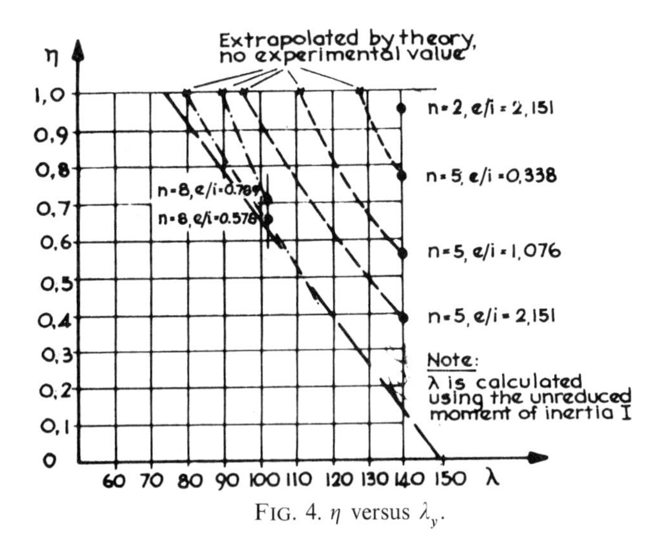

FIG. 4. η versus λ_y.

between the elastic region and the ultimate load level is very small so that the column behaves elastically in principle. Obviously the built-up strut behaves oppositely to this.

The explanation of this deviation from normal behaviour (which is also valid in the case of laced struts) lies mainly in the discrete structural system used. As discussed above, in the case of stocky struts, the external moment is equilibrated predominantly by axial forces in the chords so that the full member does not display any significant plastic deformations before exceeding the critical design load. This is not true for solid members with flanges continuously interconnected by the web, hence they can reach higher load levels in the elasto-plastic range. Slender built-up struts exhibit a rather complex interaction between the influences of the full member and its single parts on the overall behaviour. Both components are of similar dimensions in this case so that plastic deformations can occur at comparatively low load levels.

For this discussion, the bending stiffness of a built-up strut is defined by:

$$EJ_y^* = E \sum_{i=1}^{k} (A_{f,i} \cdot z_{s,i}^2 + \eta \cdot I_{y,f,i}) \tag{2}$$

where A_f = area of a chord member

$I_{y,f}$ = moment of inertia of a chord member related to the axis parallel to the non-substantial axis of the full member

z_s is as defined in Fig. 2

The sum pertains to all chord members lying in planes parallel to the non-substantial axis.

$\eta = 0$ (laced struts)
$\eta =$ to be chosen according to eqn. (1) (struts with batten-plates)
$k =$ number of chords

In the standard case (Fig. 2(a)) k is equal to 2.

7.3.3 Shear Stiffness of the Full Member

It is assumed that the joints between chords and web members can be regarded as rigid. If flexibilities of any kind are to be taken into account, the shear rigidity will be reduced (see Section 5).

The shear rigidity GA_w is defined by the relationship between the shear force Q of the full member and the angle γ produced by Q:

$$GA_w = Q/\gamma \tag{3}$$

Its evaluation exhibits some peculiarities for both types of built-up struts which are discussed below.

(a) Laced Struts

In this case only the axial deformations of the web members are to be taken into account. Elongations or shortenings of the chords form part of the bending deflections of the full member, which have been dealt with in the previous section.

Since the chord members are regarded as simply supported at the joints, the effect of secondary chord bending on the shear rigidity of the full members is negligible compared with the axial stiffness of the web members.

The shear stiffness of laced struts depends on the different types of lacings (Fig. 5). For illustration, the formula corresponding to the type shown in Fig. 5(a) will be explained. In this case the shear force Q is supported completely by the diagonal bars of the lacing, whereas the transverse bars have only to shorten the effective buckling length of the chord sections. p denotes the number of parallel planes with lacings. Considering a panel of the laced strut under the action of Q (Fig. 6) the axial force N_d in one diagonal bar is:

$$N_d = \frac{d}{h_z} \cdot \frac{Q}{p} \tag{4}$$

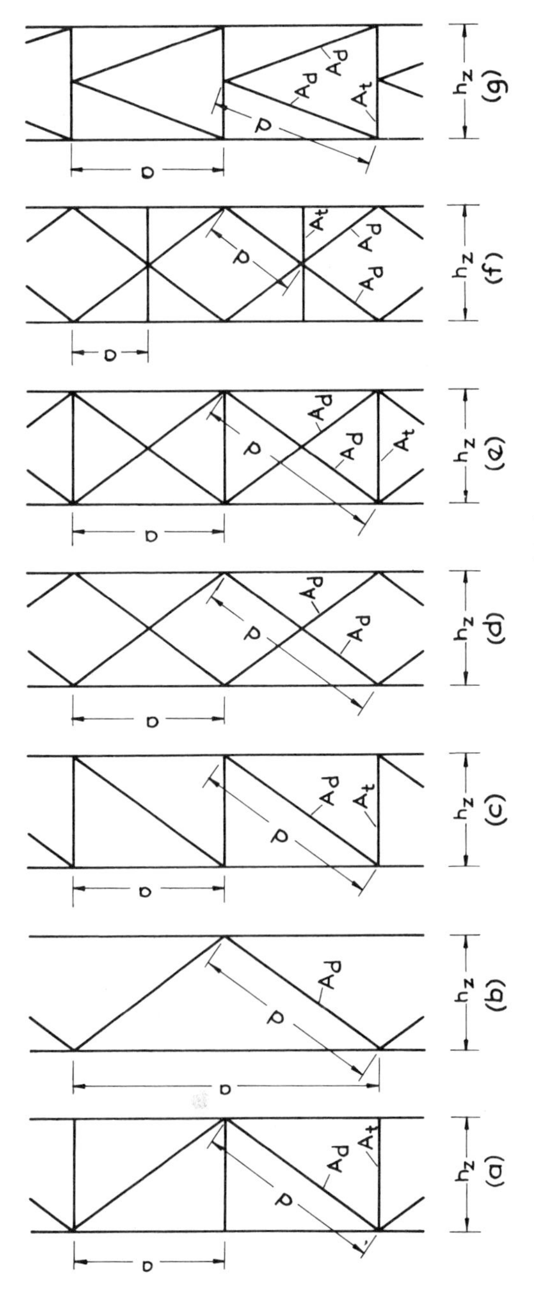

FIG. 5. Different types of lacing.

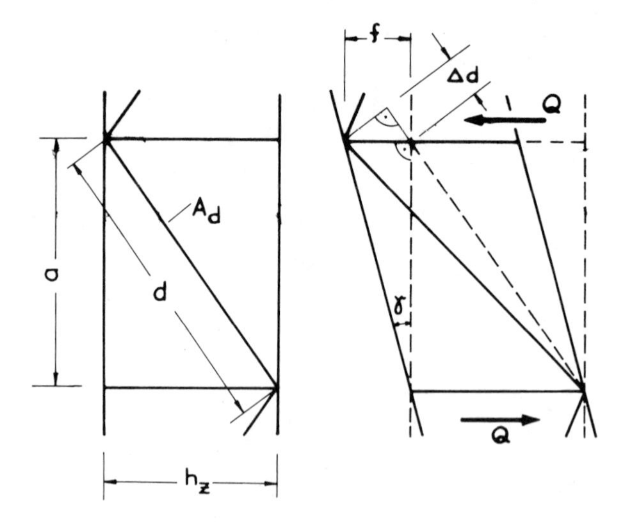

FIG. 6. Deformation of a panel of a laced strut under shear force Q.

The elongation Δd of this bar is given by:

$$\Delta d = \frac{N_{\mathrm{d}} \cdot d}{EA_{\mathrm{d}}} = \frac{d^2}{h_z} \cdot \frac{Q}{p \cdot EA_{\mathrm{d}}} \tag{5}$$

The lateral deflection f can easily be found as:

$$f = \frac{d}{h_z} \cdot \Delta d = \frac{d^3}{h_z^2} \cdot \frac{Q}{p \cdot EA_{\mathrm{d}}} \tag{6}$$

This leads to reciprocal value of the shear stiffness (i.e. shear flexibility):

$$\frac{1}{GA_{\mathrm{w}}} = \frac{\gamma}{Q} = \frac{f}{aQ} = \frac{d^3}{pah_z^2 \cdot EA_{\mathrm{d}}} \tag{7}$$

It is evident that the same formula can be used for the lacing of Fig. 5(b) if a is replaced by $a/2$:

$$\frac{1}{GA_{\mathrm{w}}} = \frac{2d^3}{pah_z^2 \cdot EA_{\mathrm{d}}} \tag{8}$$

The formulae of the shear stiffness for the other types of lacings can be easily found in a similar way as shown before. In the case of Fig. 5(c) the transverse bars are stressed by the shear force, too. Therefore their elongations must be taken into account when calculating the shear stiffness:

$$\frac{1}{GA_{\mathrm{w}}} = \frac{1}{p}\left\{\frac{d^3}{ah_z^2 \cdot EA_{\mathrm{d}}} + \frac{h_z}{a \cdot EA_{\mathrm{t}}}\right\} \tag{9}$$

Figure 5(d) shows a lacing consisting of crossing diagonal bars, which must be designed for the action of tensile and compressive forces. The corresponding shear stiffness is twice that of the lacing of Fig. 5(c):

$$\frac{1}{GA_{\mathrm{w}}} = \frac{d^3}{2pah_z^2 \cdot EA_{\mathrm{d}}} \tag{10}$$

This form of lacing needs at least one transverse bar in one of the rhombuses if the effective buckling length of the chords is to be shortened to the length a specified in Fig. 5(d). In most cases such transverse bars are present at the end of the column.

If the crossing diagonal bars of the lacing of Fig. 5(e) are designed for tensile and compressive forces, the above formula can still be used. If the diagonal bars are not designed with respect to buckling, the same formula as for the lacing of Fig. 5(c) must be used.

The next type of lacing (Fig. 5(f)) is similar to that of Fig. 5(d). Only a has to be replaced by $2a$ and d by $2d$:

$$\frac{1}{GA_{\mathrm{w}}} = \frac{2d^3}{pah_z^2 \cdot EA_{\mathrm{d}}} \tag{11}$$

The last form of lacing (Fig. 5(g)) has the following shear stiffness:

$$\frac{1}{GA_{\mathrm{w}}} = \frac{1}{p}\left\{\frac{2d^3}{ah_z^2 \cdot EA_{\mathrm{d}}} + \frac{h_z}{4a\,EA_{\mathrm{t}}}\right\} \tag{12}$$

(b) Struts with Batten-Plates

It was shown by Klöppel and Uhlmann (1965) that the shear flexibility of a panel which is a part of a strut with batten-plates (evaluated by first-order theory) is:

$$\frac{1}{GA_{\mathrm{w}}} = \frac{a^2}{24EI_{\mathrm{y,f}}}$$

in which a is the length of a chord member. Taking into account second-order theory, the multiplier 24 in the denominator of the above formula must be replaced by $2\pi^2$ so that we get:

$$\frac{1}{GA_w} = \frac{a^2}{2\pi^2 EI_{y,f}} \tag{13}$$

Equation (13) was obtained by comparing the bifurcation load of a single panel with that of the full member taking into account deformations due to shear forces. The formula corresponds to the load level at which the shear rigidity of a single panel is cancelled by effects of second-order theory, produced by axial forces in the chords (Uhlmann, 1977).

In deriving eqn. (13) the batten-plates are treated as completely rigid. Indeed, their deflections caused by shear forces and bending moments are insignificant, compared with those of the chords.

In general this is valid so long as the relationship between depth and span length is comparatively large. In exceptional cases, however, it might be necessary to account for the influence of deflections of the batten-plates upon the shear rigidity. In these cases, additional terms appear in eqn. (13):

$$\frac{1}{GA_w} = \frac{1}{E}\left\{\frac{a^2}{2\pi^2 I_{y,f}} + \frac{ah_z}{24 I_b} + \frac{Ea}{GA_{w,b} \cdot 2h_z}\right\} \tag{14}$$

where h_z –distance between the centroidal axes of the chords
I_b –moment of inertia of a single batten-plate
$GA_{w,b}$ –shear rigidity of a single batten-plate

Equation (14) refers to the standard case in which two parallel batten-plates are present at every joint.

7.4 EVALUATION OF STRESS RESULTANTS OF THE FULL MEMBER

The starting point for the calculation of the stress resultants of the full member is the well-known differential equation of a column having an initial imperfection of sinusoidal shape with mid-span amplitude f_0, taking into account second-order theory and deformations due to shear forces (Fig. 7):

$$\frac{d^2z}{dx^2} + \alpha z = -\frac{1}{1-F/(GA_w)} \cdot \frac{F}{F_{cr}} \cdot \frac{\pi^2}{l^2} f^0 \sin\frac{\pi x}{l} \tag{15}$$

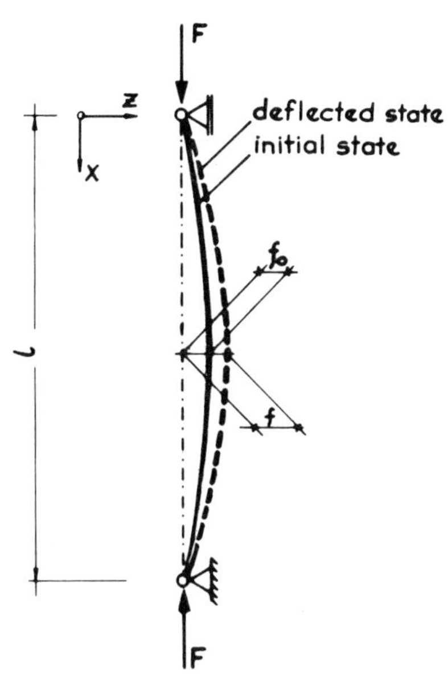

FIG. 7. Strut with unintentional sinusoidal out-of-straightness having midspan amplitude f_0.

where $\alpha^2 = \dfrac{F}{EI^*_y} \dfrac{1}{1 - F/(GA_w)}$

F = factored external compressive load

$$F_{cr} = \frac{E\bar{I}_y \pi^2}{l^2} \frac{1}{1 + (\pi^2 E\bar{I}_y)/(l^2 GA_w)} \tag{16}$$

F_{cr} = bifurcation load of a column having reduced shear rigidity

$\bar{I}_y = I_y{}^*$, according to eqn. (2) setting $\eta = 1$

Equation (15) is to be solved using the boundary conditions $z(0) = z(l) = 0$, referred to the standard case.

Hence the bending moment at mid-span is:

$$\max M_y = F \,.\, \max z = \frac{Ff_0}{1 - F/F_{cr}} \tag{17}$$

The shear force Q is most critical near the supports of the strut.

$$\max Q = Q(0) = \frac{\pi}{l} \cdot \max M_y = \frac{\pi}{l} \frac{Ff_0}{1 - F/F_{cr}} \tag{18}$$

7.5 CHECK OR DESIGN OF SINGLE MEMBERS

7.5.1 Chord Members

Starting from eqn. (17), the axial force N_f in the most stressed chord member situated at mid-span is:

$$\max N_f = \frac{F}{2} + \frac{\max M_y}{h_z} = F\left(\frac{1}{2} + \frac{f_0}{h_z}\frac{1}{1-F/F_{cr}}\right) \tag{19}$$

Chord design is satisfactory if:

$$\max N_f \leqslant N_{cr,f} \tag{20}$$

The right-hand side of (20) refers to the critical load of simply-supported chord members, established by appropriate design standard (ECCS-curves). Here reference is to be made to the cross-section of the chord member. As a rule, the critical length of the member can be set equal to a, neglecting conservatively the influence of chord continuity at the joints. Only in the case of laced struts having two non-substantial axes some peculiarities exist (Section 7.9).

7.5.2 Web Members of Laced Struts

Using the maximum shear force of the full member, obtained by eqn. (18), the maximum axial forces of the web members can be found easily by establishing equilibrium conditions. In the cases of Figs. 5(a), (b), (c), and (g) the maximum axial force N_d of the diagonal bars has the following value:

$$N_d = \frac{1}{p}\cdot\frac{d}{h_z}\cdot \max Q = \frac{1}{p}\cdot\frac{d}{h_z}\cdot\frac{\pi}{l}\cdot\frac{F.f_0}{1-F/F_{cr}} \tag{21}$$

In the other cases shown in Figs. 5(d), (e), and (f) the force N_d has half the value. The maximum axial force N_t in the transverse bars of the lacing of Fig. 5(c) amounts to:

$$N_t = \frac{1}{p}.\max Q \tag{22}$$

N_t of the lacing in Fig. 5(g) reaches half the value.

It has to be checked that all these forces of the web members remain below the critical buckling load of a simply-supported solid column. Here the length of the web members can be used as effective buckling length. In addition to this it is also necessary to design the joints.

7.5.3 Batten-Plates

The experimental investigations (Klöppel and Uhlmann, 1965) incorporating variation of the depth of the batten-plates reveal that these parts of the strut can be considered as less critical. It is sufficient to design the batten-plates and connections according to the rules for simple beams subjected to bending only.

7.5.4 Failure of the Strut with Batten-Plates by Developing a Panel Mechanism

This possibility of failure is dealt with in detail by Uhlmann (1977). In most of the critical cases, this check is not relevant for strut design.

Hence, the essential features of this method are summarised here. The shear force of the full member is equilibrated by shear forces acting at the chords. When these reach a certain magnitude, plastic hinges develop at the ends of the chords. In the simply-supported case, the largest shear force occurs at the ends of the strut. Equilibrium yields to the condition:

$$\sum_{1}^{4} |M_{\mathrm{pl},N}| > \max Q \cdot a \tag{23}$$

in which the sum pertains to the chord end moments of the critical panel. The index N means that the influence of axial forces has to be accounted for when calculating the fully plastic moment M_{pl}. The right-hand side of (23) refers to eqn. (18).

If single-symmetrical cross-section of the chords (for example, channels) are chosen, the interaction between M_{pl} and N is ambiguous depending on the sign of M_{pl}. In this case it is satisfactory to use the average value of all $M_{\mathrm{pl},N}$ occurring at the left-hand side of (23). An example of the above is presented by Ramm and Uhlmann (1982).

7.6 ADDITIONAL DEFORMATIONS OF THE JOINTS BETWEEN CHORDS AND WEB MEMBERS

If the connections between web members and chords are not rigid as assumed before, the occurrence of additional deformations in the joints can reduce the bearing capacity of a built-up column considerably. Such effects therefore must be taken into account.

Two different kinds of deformations must be distinguished: first elastic deformations of the mountings can be found which are increasing proportionally with the loading. Second, there may exist a certain lack of fit at the joints independent of the load level.

In the case of additional elastic deformations of the mountings, the shear stiffness of the full member will be reduced. It is possible to take account of this effect while establishing the shear stiffness. This will be illustrated for laced struts, taking the system of lacing shown in Fig. 6. The relationship between the elastic deformation Δ_{j} within the connections of the diagonal bars measured in the direction of these bars and the axial forces N_{d} may be taken as:

$$\Delta_{\mathrm{j}} = k_{\mathrm{j}} \cdot N_{\mathrm{d}}$$

where k_{j} is the spring constant. Hence the total relative displacement of the nodal points at the chords at the end of a diagonal bar is:

$$\Delta D = \Delta d + 2\Delta_{\mathrm{j}} = N_{\mathrm{d}}\left(\frac{d}{EA_{\mathrm{d}}} + 2k_{\mathrm{j}}\right)$$

This leads to the following reciprocal value of shear stiffness:

$$\frac{1}{GA_{\mathrm{w}}} = \frac{2}{p} \cdot \frac{d^2}{ah_{\mathrm{z}}^{\ 2}}\left(\frac{d}{EA_{\mathrm{d}}} + 2k_{\mathrm{j}}\right) \tag{24}$$

A similar treatment is possible if there is an additional elastic rotation Δ_{ϕ} in the connections between batten-plates and chords. The corresponding relationship is:

$$\Delta_{\phi} = k_{\phi} \cdot M_{\mathrm{b}}$$

where k_{ϕ} = spring constant in rotation

$M_{\mathrm{b}} = \dfrac{a}{2}$ = bending moment between batten-plates and chords under the action of a shear force $Q = 1$ in the full member.

The reduction of shear stiffness can be done by adding the following term in eqn. (13):

$$\Delta\left(\frac{1}{GA_{\mathrm{w}}}\right) = \frac{a}{4} \cdot k_{\phi}$$

A lack of fit at the joints between chords and web members increases the initial imperfections. The example of Fig. 6 may be taken once more to explain this. The connections between the diagonal bars and the chords may admit local relative displacements of size δ, which lead to the following inclination γ:

$$\gamma = 2 \cdot \frac{d}{ah_z} \cdot \delta$$

According to this, in the worst case an initial deflection of the whole strut as shown in Fig. 8 can occur, having a maximum ordinate of:

$$f_1 = \gamma \cdot \frac{l}{2} = \frac{dl}{ah_z} \cdot \delta$$

This displacement has to be taken into account in addition to the initial imperfection. Simplifying and remaining on the safe side, this can be done by using $f_0 + f_1$ instead of f_0.

In the case of struts with batten-plates the same treatment is possible. The angle γ can be taken directly from the lack of fit at the joints.

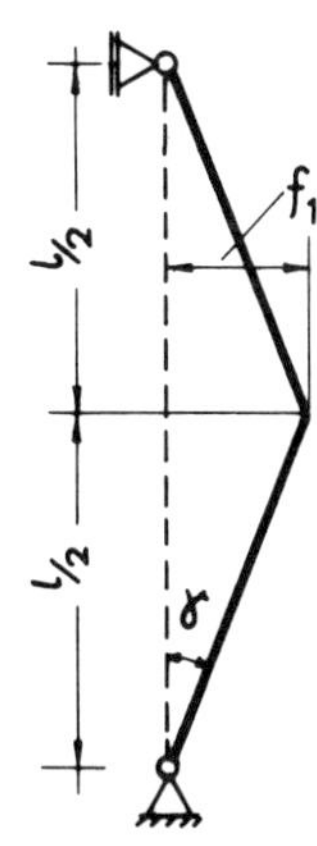

FIG. 8. Possible displacement of built-up struts due to backlash in the connections of the members.

7.7 ECCENTRICITIES AT THE CONNECTIONS BETWEEN LACING AND CHORDS

With laced struts the lacing often is connected directly to the chords without intermediate connection plates. As a consequence and because of the narrow space at the joints it is impossible to ensure that the centre lines of all members run through one point. The resulting eccentricities cause additional bending moments in the chords and consequently additional deformations of the whole strut. These are not negligible in the stability check of the built-up column unless the eccentricities are small or the bending stiffness of the chords is comparatively high.

With regard to the full member, the eccentricities in the joints will increase the shear deformations mentioned earlier. Therefore this effect can be taken into account by a suitable reduction of the shear stiffness. How to do this may be demonstrated by the example of Fig. 9(a). The laced strut has a system of centre lines and eccentricities s at the joints as shown in Fig. 9(b). The shear force Q of the full member must be supported by the chords over the distance of $2s$ within the reach of the connections. This produces bending moments and corresponding shear forces in the chords. Due to antisymmetry, the bending moments of the chords become zero at the ends of the chord sections as shown in Fig. 9(c).

Equilibrium gives:

$$Q_1 - Q_2 = Q$$
$$s \cdot Q_1 = (a/2 - s) \cdot Q_2$$

and by this:

$$Q_1 = Q \cdot (a - 2s)/(a - 4s)$$
$$Q_2 = Q \cdot 2s/(a - 4s)$$

The maximum bending moment of the chords is:

$$\max M_f = Q_1 \cdot s = Q \cdot s \cdot (a - 2s)/(a - 4s)$$

The horizontal component of the axial force N_d of the diagonal bar exceeds Q and the inclination of this bar is smaller due to eccentricities s:

$$N_d = (Q_1 + Q_2) \cdot d/h_z = Qad/\{h_z(a - 4s)\}$$

Analogous to sub-section 7.3.3(a), the reciprocal value of the shear stiffness can be easily established. It consists of two terms: the first one is related to the elongation of the diagonal bars, whereas the second one includes the deflection of the chords.

$$\frac{1}{GA_w} = \frac{2ad^3}{(a-4s)^2 h_z^{\,2}} \cdot \frac{1}{EA_d} + \left(\frac{a-2s}{a-4s}\right)^2 \cdot \frac{2s^2}{3EJ_{y,f}}$$

For other types of lacing the shear stiffness can be easily found in a similar way.

It must be mentioned that the influence of the eccentricities has also to

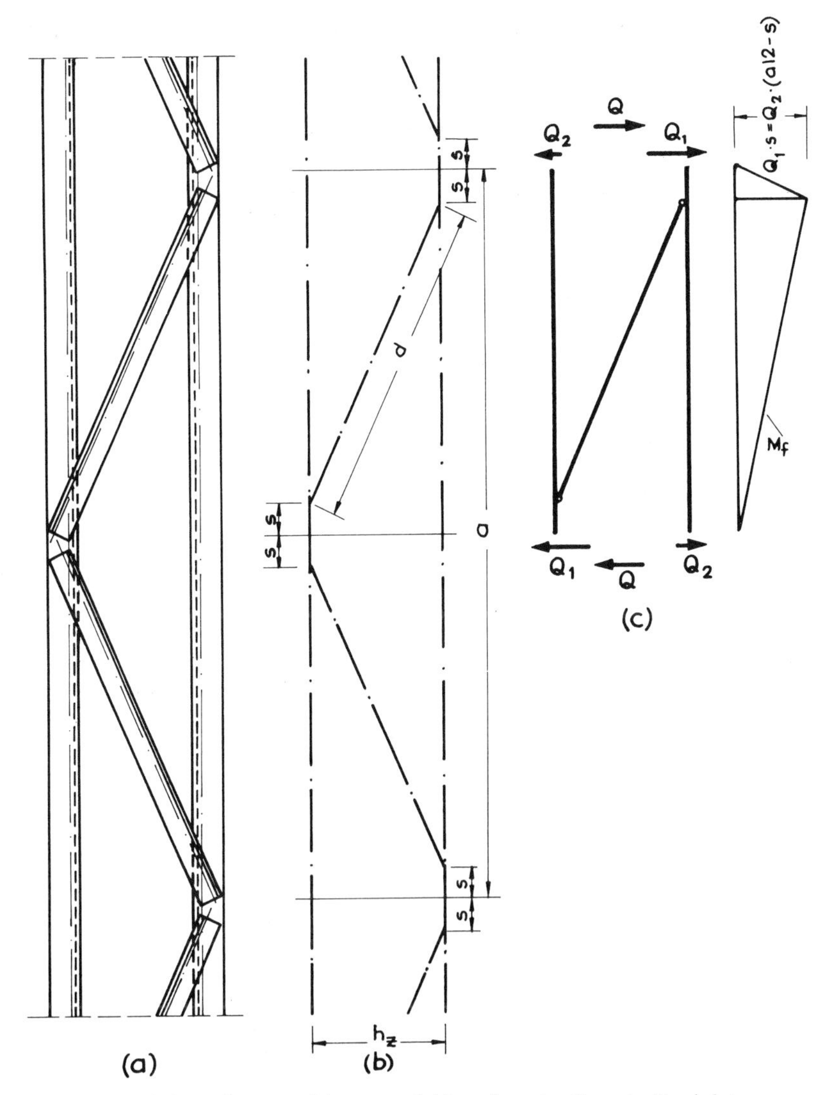

FIG. 9. Laced strut with eccentricities of centre lines in the joints.

be taken into account when checking the single members such as chord-sections and web members. As mentioned before, the axial force of the diagonal bars is large and additional bending moments of the chords will reduce their bearing capacity.

7.8 REDUCED EFFECTIVENESS OF FIXED ENDS IN THE CASE OF BUILT-UP COLUMNS

It has to be investigated whether the boundary conditions are different from those corresponding to the solid column if built-up columns are designed taking into account the shear deformation of the full member. v may denote the lateral deflection as shown in Fig. 10. The boundary

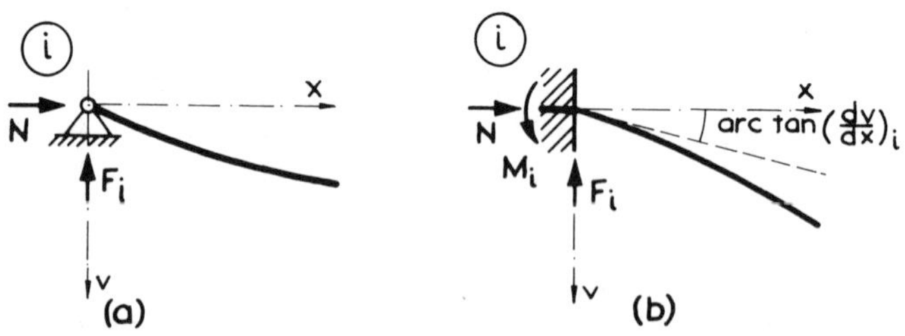

FIG. 10. Deflections near the supports when shear deformations are taken into account.

condition $v_i=0$ remains valid for hinged and fixed ends. Contrary to this, the boundary condition $(dv/dx)_i=0$ is only correct for fixed ends if shear deformations are negligible. Otherwise there will be an inclination of the centre line at the fixed end due to shear deformation (Fig. 10(b)):

$$(dv/dx)_i = Q_i/GA_w$$

Following second-order theory the shear force Q_i at the fixed end contains a component of the axial force N:

$$Q_i = F_i + N \cdot (dv/dx)_i$$

This finally gives the boundary condition of a fixed end:

$$(dv/dx)_i = F_i/(GA_w - N)$$

Hence the effectiveness of fixed ends is reduced in the case of built-up columns, if shear deformation cannot be neglected (Amstutz, 1941).

7.9 SPECIAL BUCKLING LENGTHS FOR ANGLES WHICH ARE CHORDS OF FOUR-PIECED LACED STRUTS WITH TWO NON-SUBSTANTIAL AXES

Whereas the effective buckling lengths of the chords normally equal the distances between the joints of lacing, there are some cases of four-pieced laced struts with special bearing conditions of the chords. To check the chord members in these cases approximately in the same way as described in sub-section 7.5.1, the corresponding effective lengths are required.

With these four-pieced struts the chords usually consist of simple angles. The commonly used types of lacing are shown in Fig. 11. With

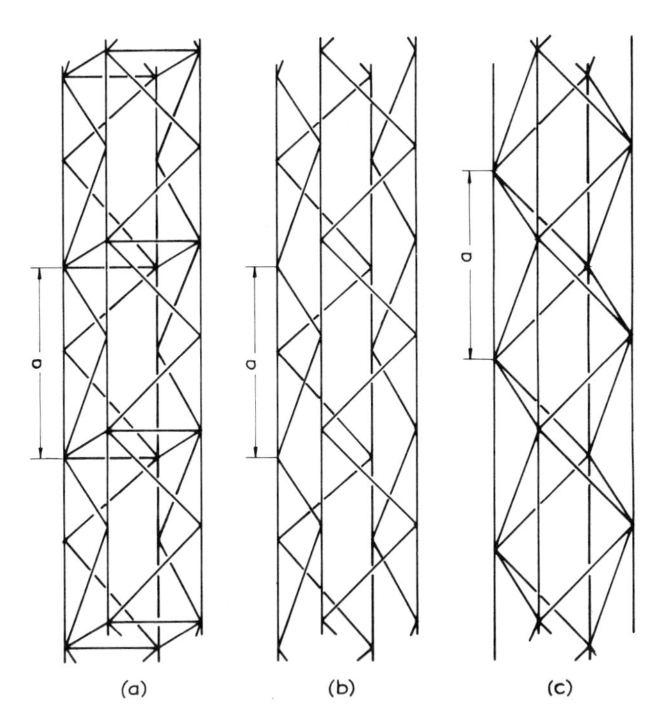

FIG. 11. Types of four-pieced laced struts.

type (a) the effective length of the chords is, undoubtedly, equal to the distance a between the joints. The case of type (b), however, is of some difficulty, since the angles representing the chords are supported by the lacing alternately in two planes perpendicular to each other. The effective length belonging to this special case works out to $0{\cdot}76a$, whereas $0{\cdot}64a$ has to be used for the type of Fig. 11(c).

The two effective lengths were established by earlier research workers, analysing the bifurcation load of such chords. Their results were proved by experiments carried out by Klöppel and Ramm (1972).

7.10 ANALYSIS OF BUILT-UP STRUTS UNDER ECCENTRIC AND TRANSVERSE LOADS

The design concepts outlined above can be easily extended to cases in which the external load acts eccentrically or transverse loads are present. The full member having reduced shear stiffness has to be analysed as a beam-column according to second-order theory. In the appendix of the new German standards DIN 18 800, Part 2, formulae for single members with and without reduced shear rigidity ready for use in different cases of loading are presented, by which the corresponding calculations can be carried out. In general, laced struts are more suitable to resist such loading because of their larger bending and shear rigidity compared with that of a strut with batten-plates.

The applicability of the above concepts to built-up struts having higher yield stresses than Grade 43 steel is discussed by Uhlmann (1977). He concluded that one will be on the safe side under comparable circumstances with respect to residual stresses and plastic behaviour of the full member.

7.11 SOME PROBLEMS FOR FUTURE RESEARCH

Some types of lacing lead to additional stress resultants which are not included in the method described above. The existence of such stress resultants is a consequence of high statical indeterminacy. These forces and moments are not necessary for global equilibrium. Therefore they are usually neglected when checking the stability of the struts. Whether this is possible or not in principle with respect to safety should be a subject for further research.

To explain this problem in more detail, two systems of lacing which behave quite differently in this respect are compared in Fig. 12. For simplicity, it may be assumed that there is only an axial force acting centrally in the full member. Each of the chords has to bear half of it and therefore all sections of the chords will be shortened equally.

In the case of Fig. 12(a) this causes an extension of the distance between the chords only. Such behaviour is not possible in the case of Fig. 12(b) because of the transverse bars. In consequence additional deflections of chords occur combined with local bending moments of the chords and additional axial forces in the lacing. The local bending moment of the chords will probably be reduced owing to the plasticity of material, if the

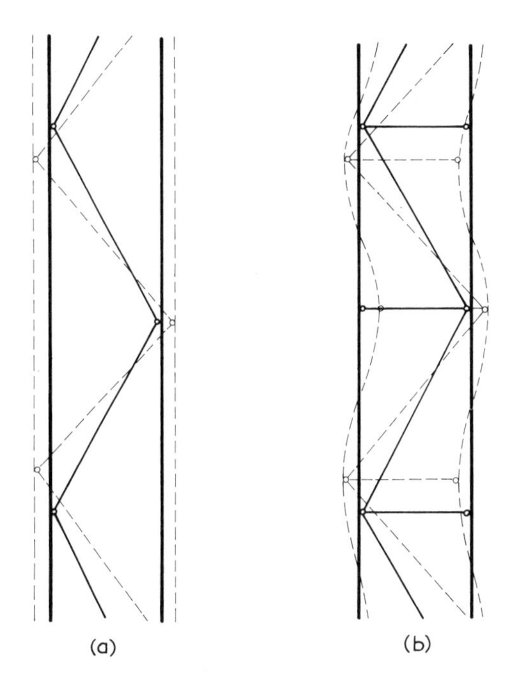

FIG. 12. Local deformations of different lacings.

limit of elasticity is exceeded. Whether there exists a favourable behaviour if the ultimate bearing capacity of the diagonal bars with respect to buckling is reached first by the additional axial forces is not yet clear.

As already mentioned in Section 7.1, experimental data as well as satisfactory theoretical investigations are not available for the important group of two-pieced struts with batten-plates having single angles as chords. The current research programme of the first writer is aimed at finding out if these struts can be modelled in a similar way.

7.12 CONCLUDING REMARKS

The discontinuous statical system of a built-up strut has a very high degree of statical redundancy. It has been shown that the calculation can be simplified significantly by treating it as a full member having constant bending and shear rigidity along its length. With respect to material consumption a built-up strut is much more economical than a single member with continuously interconnected web and flanges having the same stiffness properties. This is caused by the very appropriate arrangement of its single parts which follow essentially the flux of internal forces. On the other hand, the costs of manufacturing a built-up strut in general are higher.

Aesthetic criteria can also be of significance, so that the question of whether to use a built-up member or not, must be decided in each case. Use of a built-up strut as constructional element will in some cases doubtless be a powerful alternative compared with conventional solutions.

REFERENCES

AMSTUTZ, E. (1941). Die Knicklast gegliederter Stäbe, *Schweizerische Bauzeitung* **118** (9), 97.

CHWALLA, E. (1933). Genaue Theorie der Knickung von Rahmenstäben, *HDI–Mitteilungen des Hauptvereins Deutscher Ingenieure in der Tschecho-slowakischen Republik*, Brünn.

ENGEßER, F. (1891). Die Knickfestigkeit gerader Stäbe, *Zentralblatt der Bauverwaltung* **11** 483.

ENGEßER, F. (1909). Über die Knickfestigkeit von Rahmenstäben, *Zentralblatt der Bauverwaltung* **29**, 136.

FOERSTER, M. (1911). Die Gründe des Einsturzes des großen Gasbehälters am großen Graasbrook zu Hamburg, *Der Eisenbau* **2**, 178.

KLÖPPEL, K. and RAMM, W. (1968). Versuche und Berechnung zur Bestimmung der Traglast mehrteiliger Gitterstäbe unter außermittiger Belastung, *Der Stahlbau* **6**, 164, 236.

KLÖPPEL, K. and RAMM, W. (1979). Zur Stabilitätsuntersuchung von mehrteiligen Gitterstäben, *Der Stahlbau* **41**, 14 (see here for further references).

KLÖPPEL, K. and UHLMANN, W. (1965). Versuchsmäßige und rechnerische Bestimmung der Traglasten mehrteiliger Rahmenstäbe unter Verwendung elektronischer Rechenautomaten, *Der Stahlbau* **6**, 161, 199, 231.

KROHN, R. (1908). Beitrag zur Untersuchung der Knickfestigkeit gegliederter Stäbe, *Zentralblatt der Bauverwaltung* **28**, 559.

MANN, L. (1909). *Statische Berechnung steifer Vierecknetze*. Thesis, Berlin.

MÜLLER–BRESLAU, H. (1913). *Die neueren Methoden der Festigkeitslehre und der Statik der Baukonstruktionen*, p. 380, Kröner, Leipzig.

RAMM, W. and UHLMANN, W. (1982). Zur Anpaßung des Stabilitätsnachweises für mehrteilige Druckstäbe an das europäische Nachweiskonzept, *Der Stahlbau* **50**, 161–72.

RUDELOFF, M. (1914). Untersuchung von Knickstäben auf Knickfestigkeit, *Lichterfelder Versuche*, Berlin.

TIMOSHENKO, S. P. and GERE, J. M. (1961). *Theory of Elastic Stability*, McGraw Hill Book Co., New York.

UHLMANN, W. (1977). Some Problems Concerning Design Recommendations for Centrally Compressed Built-up Members, *2nd Conference on Stability of Steel Structures, Liège 1977, Preliminary Report*, p. 121.

Chapter 8

BATTENED COLUMNS—RECENT DEVELOPMENTS

D. M. PORTER
University College, Cardiff, UK

SUMMARY

The existing design rules of 'built-up' steel columns generally and of battened columns in particular, were devised before powerful computation and sophisticated testing facilities were available. Also there is now a better understanding of the inelastic failure process of steel columns. A re-appraisal of these design rules may prove beneficial and this chapter describes, briefly, theoretical and experimental work which could assist such a re-appraisal.

A simple method for the determination of the elastic critical loads which takes account of the discontinuous nature of battened columns and a plastic collapse theory for these columns are briefly outlined. An experimental programme in which these theories are confirmed is then summarised.

It is concluded that the restrictive design rules relating to the maximum batten interval and the slenderness of the main members between battens for battened columns are unnecessary as the failure load for all columns can be accurately and safely assessed by using the 'effective' slenderness in existing column curves.

NOMENCLATURE

$\bar{A}$	Reduced area of column for shear resistance
A	Cross-sectional area of column
A_f	Cross-sectional area of main member
E	Modulus of elasticity

F Transverse shear force on column
G Shear modulus
I Second moment of area of overall column
I_f Second moment of area of main member
K Effective length factor
L Length of column
M_p Full plastic moment of resistance
M_{pr} Reduced plastic moment of resistance
N Number of main members
P Vertical load on column
P_y Squash load of column
P_{cr} Critical load of column
a, b, c Matrix coefficients
a_1 Constant in design formula (15)
d Distance of main member from column centroid
e Eccentricity of loading
k EI/l
l Interval distance
m Moments at ends of main members
n, o Stability functions
r Radius of gyration of overall column
α I/I_f
β $M_p/P_y \cdot r$, shape factor overall column section
β_f Shape factor of main member section
γ l/L
δ Vertical end displacement of column
∂_p Vertical end displacement due to plastic deformation
∂_e Vertical end displacement due to elastic deformation
η_1 $0{\cdot}001\ a_1(\lambda - \lambda_0)$
η_0 $1{\cdot}2\ e/r$
θ Rotation
λ Slenderness of overall column
λ_f Slenderness of main member
λ_b Effective slenderness of battened column
λ_0 $\sqrt{\pi^2 E/\sigma_y}$
λ' λ/λ_0
λ'_f λ_f/λ_0
λ'_b λ_b/λ_0
σ Design failure stress of column

σ_{cr} Critical stress
σ_{y} Yield stress

8.1 INTRODUCTION

In the early part of the century it was common practice to 'build-up' the required cross-sectional area of steel compression members from a number of smaller sections but since then with the increasing availability of larger steel sections and the ever-increasing fabrication costs the proportion of 'built-up' compression members in steelwork construction has fallen considerably. The restrictive design rules of most steel codes in regard to 'built-up' compression members introduced in the aftermath of the Quebec Bridge failure (1907) was another important factor. However, there is a continuing use of 'built-up' members where stiffness with lightness is required as in transmission line towers and support works for concrete construction. In addition, in developing countries, where fabrication costs are relatively cheaper and large sections are unavailable, 'built-up' members are an economic and possibly the only solution.

Most of the research work on 'built-up' compression members dates back to the time of the widespread use of this form of construction. For instance, the latest reported major study in the UK was due to Ng (1948) and the clauses referring to battened columns in the existing code BS:449 were based on his findings.

The disrepute of 'built-up' compression members probably arises from the belief that the 'built-up' member should have the capacity of and behave in every respect in an identical manner to the corresponding solid member and the designers were disappointed to discover that this was not possible. Even the existing design rules appear to have been formulated in the attempt to ensure that 'built-up' members will have the load-bearing properties of the corresponding solid member and by so doing have constrained their use.

With the advent of powerful computation and sophisticated servo-controlled testing facilities unavailable to our predecessors, the time is now propitious for further theoretical and experimental studies into the behaviour of 'built-up' compression members. The first stages of such studies have been completed in the Department of Civil and Structural Engineering, University College, Cardiff. This work is confined to battened and related types of columns, as shown in Fig. 1, and will be briefly described and the findings discussed in the following sections.

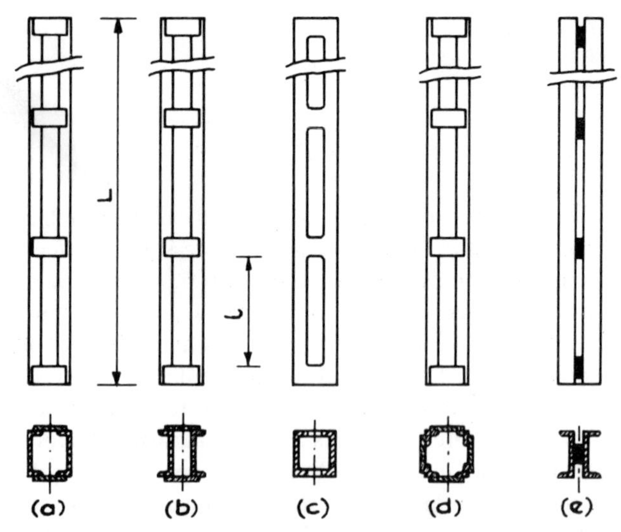

FIG. 1. Battened Columns: (a), (b), and (d) Battened plated, (c) Perforated tube, (e) Stitch welded or bolted.

8.2 ELASTIC CRITICAL BEHAVIOUR

8.2.1 Critical Loads—General

In general the failure of solid steel columns occurs at loads below the respective elastic critical loads as a result of the elasto-plastic bending initiated by the imperfections inherent in all practical columns. This is also the case with battened columns but, nevertheless, the study of their elastic buckling behaviour is still necessary because, as will be shown later, the elastic critical load is an important parameter in the determination of allowable stress and an understanding of the effect of the variation of the geometry of the battened column will help the designer to deploy the material in the column to the optimum advantage.

The critical loads (P_{cr}) which are referred to, are those about the axis perpendicular to the plane of the battens determined by small deflection theory and correspond to the Euler load of a solid column.

8.2.2 Modified Continuum Equation

The essential feature producing the difference in the elastic behaviour of the battened column compared with the corresponding solid column is the shear flexibility of the 'openings' between battens. In practical columns the battens are very stiff and being, normally, welded to the vertical members can be considered as rigid connectors. Thus the shear flexibility will be due only to the lateral displacement of the vertical members and will have the first-order value of $l^2/(24EI_f)$, where l = batten interval and I_f = second moment of area of the vertical members. The Euler formula for a uniform column can be modified to take account of the shear flexibility of the section as follows:

$$P_{cr} = \frac{\pi^2 EI/L^2}{1 + (\pi^2 EI/L^2)/\bar{A}G} \tag{1}$$

where $1/\bar{A}G$ is the shear flexibility.

Increasing the equivalent shear flexibility of the battened strut from $l^2/(24EI_f)$ to $l^2/(2\pi^2 EI_f)$ empirically, allows for the destabilising effect of the axial loads in the vertical members, and using this value, eqn. (1) will now give approximate critical loads for the battened columns. This modified formula can be further reduced to the following simple form to give λ_b, the effective slenderness of the battened column:

$$\lambda_b = \sqrt{\lambda_f^2 + \lambda^2} \tag{2}$$

where λ_f = slenderness of vertical members between batten intervals, and
λ = slenderness of the overall column.

Equation (2) which dates back to Engesser (1909) is the basis in many steel codes of determining the critical loads and clearly demonstrates the influence of the slenderness of the vertical members between battens in that the effective slenderness of a battened column with uniform batten spacing will always be greater than that of the vertical members between battens.

However, it must be remembered that eqn. (2) is an approximation and whilst, being a continuum formula, it could be expected to give accurate values for columns with a large number of uniformly spaced battens, it has been shown by Porter and Williams (1978) actually to give accurate and safe values over the entire practical range of parameters for uniform columns with normal depth battens and, therefore, for such columns the use of this simple equation is recommended.

8.2.3 Rigidly Connected Cell Method

For non-uniform columns or for columns which are subjected to an axial load varying along the length, the use of continuum formulae is inappropriate and the critical load for these columns can only be determined by a method which takes into account the discontinuous nature of battened columns.

A possible solution for these conditions would be to consider the battened column as a rigidly jointed frame and make use of a computer frame program which will obtain elastic buckling loads, providing the program used takes account of the changes in length of the members and, preferably, also the finite depth of the members and connections.

When recourse to such a frame program is not appropriate, a simpler method has been devised by Porter and Williams (1978). In this method, which can be readily programmed for very small, even pocket-size, computers, the battened column is considered as an assembly of rigidly connected cells as shown in Fig. 2. In these cells all the vertical members

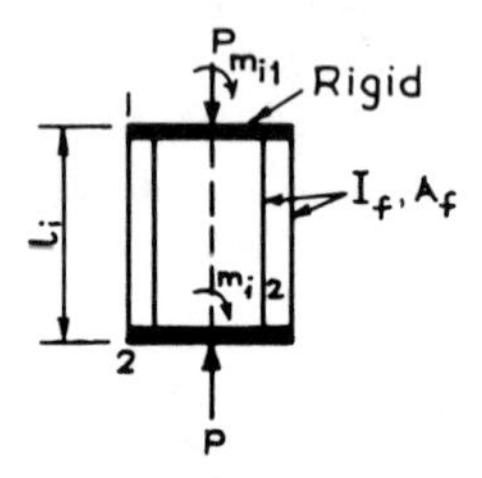

FIG. 2. Rigidly connected cell $I = N(I_f + A_f \Sigma d^2)$.

are identical and subjected to the same axial loads. When this cell is given an end rotation θ_i and allowed to deflect laterally without restraint then, in small deflection theory, the deformations and forces which arise can be considered as comprising the two separate effects, viz., the identical bending of the individual cell members and the change of length of the cell members. See Fig. 3.

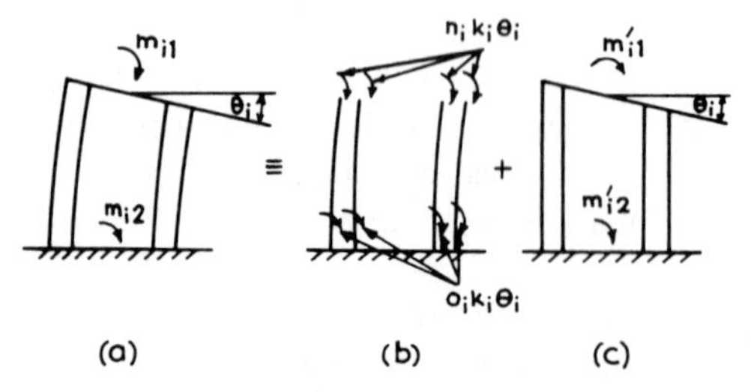

FIG. 3. Moments for rotation θ at end 1, fixed at end 2.

For a typical cell denoted by subscript i, if the moment at rotated joint $= m_{i1}$ and at the fixed end $= m_{i2}$ then:

$$m_{i1} = (nNk)_i \theta_i + m'_{i1}$$

and

$$m_{i2} = (-oNk)_i \theta_i + m'_{i2} \tag{3}$$

Using the expression for I, the overall second moment of area given in Fig. 2, it follows from Fig. 3(c):

$$\begin{aligned} m'_{i1} &= -m'_{i2} \\ &= ENA_{\mathrm{f}} \sum d^2 \frac{\theta_i}{l} \\ &= E(I - NI_{\mathrm{f}})_i \theta_i / l_i \\ &= (\alpha - N)_i k_i \theta_i \end{aligned} \tag{4}$$

where $k_i = (EI/l)_i$, $\alpha = I/I_{\mathrm{f}}$, and n and o are the stability functions given by Livesley and Chandler (1956) for bending with no horizontal restraint and depend on the value of the axial force on the member. For the majority of battened columns N, the number of vertical members $= 2$.

The end moments for the typical i cell when subjected to rotations θ_i and θ_{i+1} at the ends can then be determined in terms of these rotations only.

$$\begin{aligned} m_{i1} &= (nN + \alpha - N)k_i \theta_i - (oN + \alpha - N)k_i \theta_{i+1} \\ m_{i2} &= -(oN + \alpha - N)k_i \theta_i + (nN + \alpha - N)k_i \theta_{i+1} \end{aligned} \tag{5}$$

The cells $(i - 1)$ and i will have some rotation θ_i at the batten junction and thus from the moment joint equilibrium equations:

$$m_{(i-1)2} + m_{i1} = 0 \tag{6}$$

The overall stiffness matrix K corresponding to $(\theta_i, \theta_2 \ldots \theta_n)$ for a column with the top free and base fixed will be assembled in the tri-diagonal form below:

$$K = \begin{vmatrix} a_1 & b_1 & & & 0 \\ b_1 & a_2 & b_2 & & \\ & b_2 & a_3 & b_3 & \\ & & \ddots & \ddots & \\ 0 & & & b_{n-1} & a_n \end{vmatrix} \tag{7}$$

in which a_1, b_1, etc., are coefficients obtained from the equilibrium equations (6).

The lowest buckling load, the elastic critical load, of the above column can be found by obtaining the smallest eigenvalue of the tri-diagonal matrix. A recommended method is given by Williams (1977) and a suitable program utilising this method by Porter and Williams (1978).

For pin-ended battened columns the critical loads will be the same as those of the fixed-base column of half the length. Columns which are partially restrained at the ends can be analysed by using imaginary end-cells of the requisite rotational stiffness.

Another useful feature of the rigidly connected cell method is that the strengthening effect of the depth of the battens can be determined. The batten depth is considered as a solid cell, i.e. consisting of one member with the cross-sectional stiffness of the overall column. Comparison of Fig. 4(b) with Fig. 4(a) highlights the increase in the critical load which is available by taking the depth of batten into consideration.

In the critical failure mode the slope of the deflected column will be maximum at the ends and thus improving the shear resistance at the ends could be an effective method for increasing the critical loads. The beneficial effect of deeper end battens is confirmed by the results given in Fig. 4(c).

8.2.4 Eccentric Loading

Halabia (1981) in his recent study has analysed battened columns with eccentric loads by adapting the cell method to take account of the different axial load produced in the vertical members by the eccentric load. His results are shown in Fig. 5 and it is noted that for very small unintentional eccentricities the reduction of the critical load can be safely ignored because it appears that for these small eccentricities the destabilising due to the increase of axial load in one vertical member is matched by the stiffening effect of the corresponding reduction of the axial load in the other member. However, it is seen that as the eccentricity increases to the order of half the width of the column the reduction in the critical load is becoming significant and, as will be described later in Section 8.4, the eccentricity should be always taken into account to determine the failure load.

8.2.5 End-Restraint

Because of the inherent shear flexibility of the battened column the amount of enhancement in the critical load for the fixed-end condition

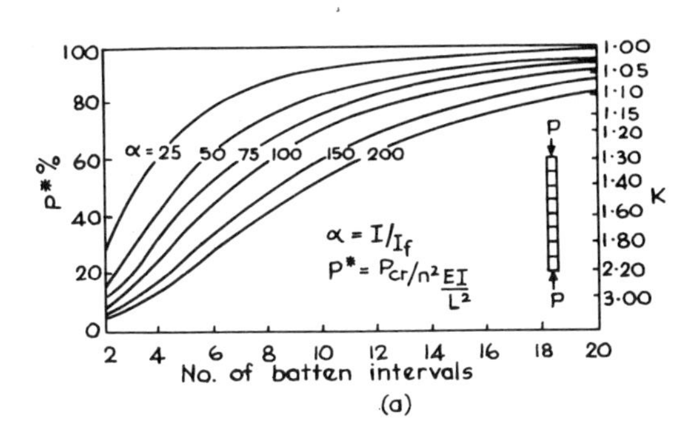

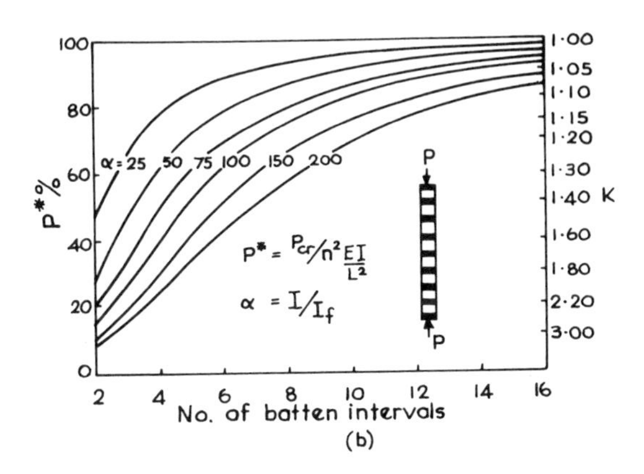

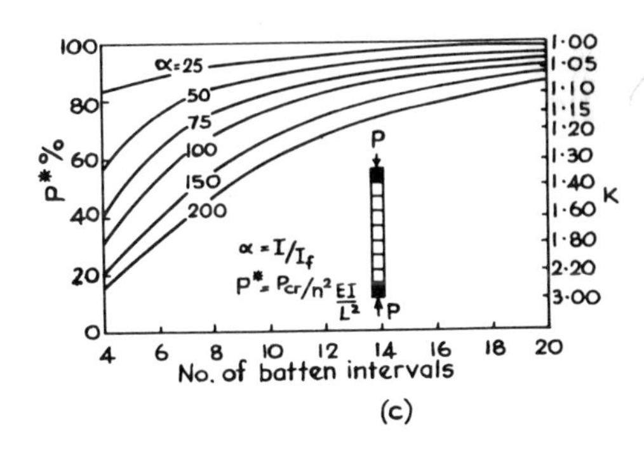

FIG. 4. (a) Critical load factors for pin-ended columns neglecting depth of batten. (b) Critical load factors for pin-ended columns with batten depth/batten interval = 0·2. (c) Critical load factors for pin-ended columns with end batten depth = batten interval.

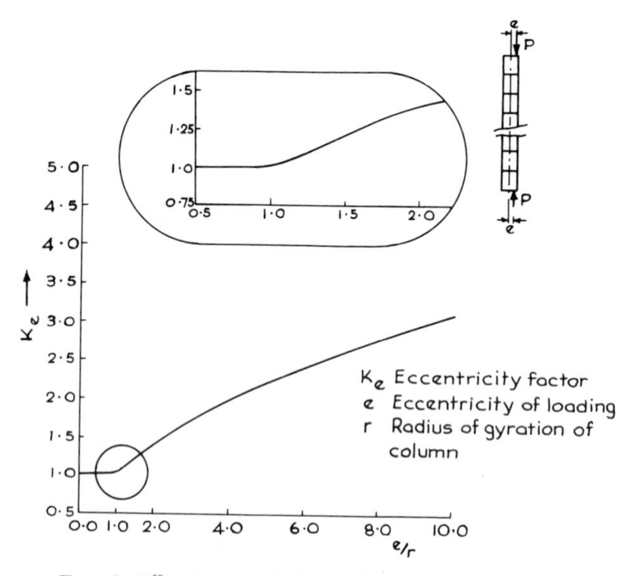

FIG. 5. Effective length factor for eccentric loading.

will always be less than that associated with the corresponding solid column. This is very well illustrated in Fig. 6. Halabia (1981) has analysed fixed-ended battened columns and his results given in Fig. 7(a) show that as the relative shear flexibility of the battened column increases the end restraint effect diminishes rapidly. Results are also available for battened columns with partial fixity at the ends such as provided to a continuous column in a building frame. Figure 7(b) gives factors for that amount of restraint sufficient to give an effective length factor of 0·85 for a solid column.

8.3 FAILURE BEHAVIOUR

8.3.1 Plastic Collapse Theory

The earlier workers whose tests are largely the basis for the existing battened column design rules confined their theoretical treatment to elastic critical behaviour even though, it is now evident, failure, in general, occurs as the result of elasto-plastic action. Indeed, even in the case of solid columns, it was not until 1947 that Shanley (1947) produced a realistic failure model for the steel column. However, it is the plastic

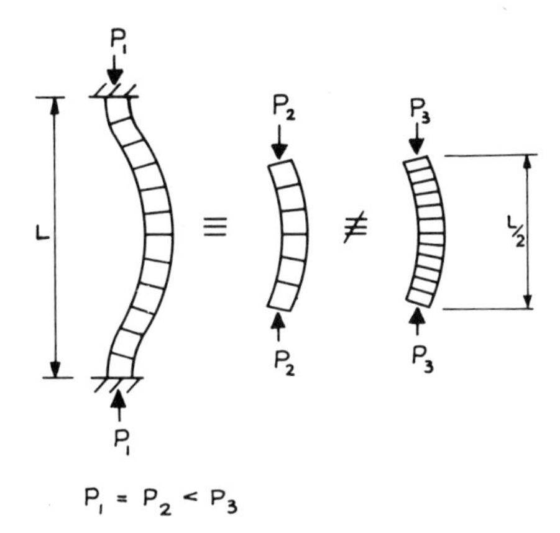

FIG. 6. Effect of fixed end on critical load of battened columns.

collapse mechanism theory of Horne and Merchant (1965) which best correlates with the observed failure behaviour.

For a pin-ended solid steel column, failure occurs when a plastic hinge develops at the mid-height position, under the combined effect of the axial load and the small but inevitable bending of the practical column, see Fig. 8. The plastic mechanism formed will deflect laterally, the load-carrying capacity of the column reducing with the increasing deflection.

As failure by the formation of plastic mechanisms also applies to battened steel columns it is worth while examining the plastic failure process in some detail with the aid of Fig. 8 and Fig. 9.

Figure 9 gives the normalised equilibrium load–displacement characteristics for the perfect column. The use of normalised plots facilitates comparison of the behaviour of columns of differing slenderness and material properties. The normalised slenderness λ' is obtained by dividing the actual slenderness λ by λ_0, the slenderness ratio at which the critical stress $\sigma_{cr} = \sigma_y$, i.e. $\lambda_0 = \sqrt{\pi^2 E/\sigma_y}$.

In Fig. 9, 0a is the normalised displacement characteristic for the axial loading from zero load of the perfect straight column and at this stage the vertical displacement is entirely elastic and is given by $\partial = PL/AE$ or alternatively the displacement normalised against the column length will be:

$$\frac{\partial}{L} = \frac{P}{AE} = \frac{P}{P_y} \cdot \frac{\sigma_y}{E} \tag{8}$$

At the squash load, P_y, a slight disturbance produces the collapse

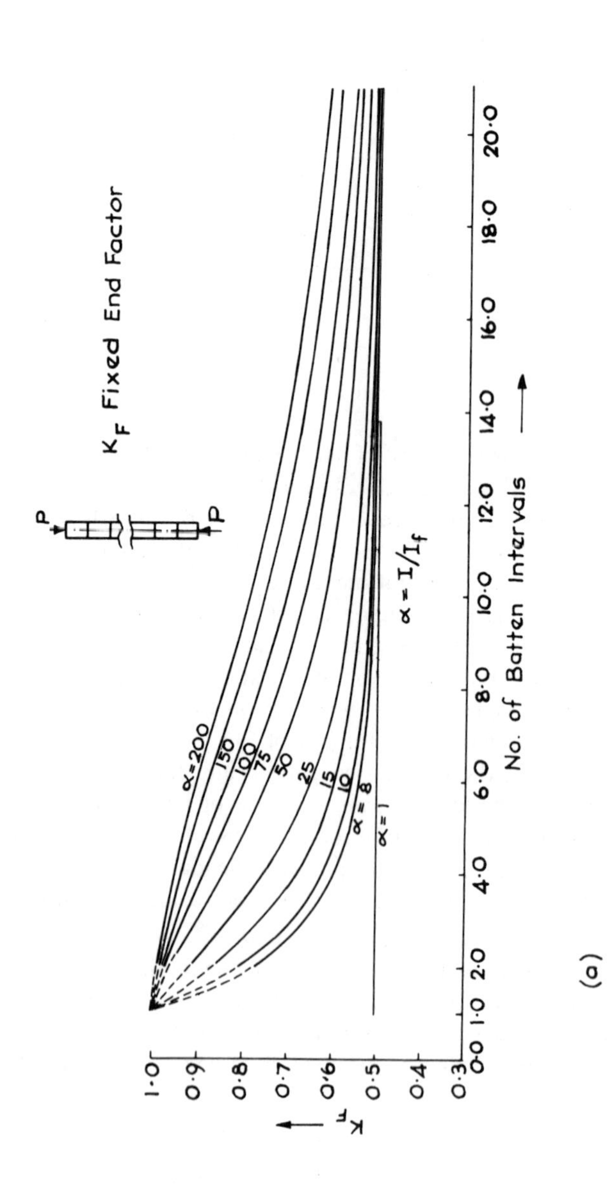

FIG. 7. (a) Effective length factors for fixed-ended columns. (b) Effective length factor for partially restrained ends$\equiv 0{\cdot}85L$ for solid columns.

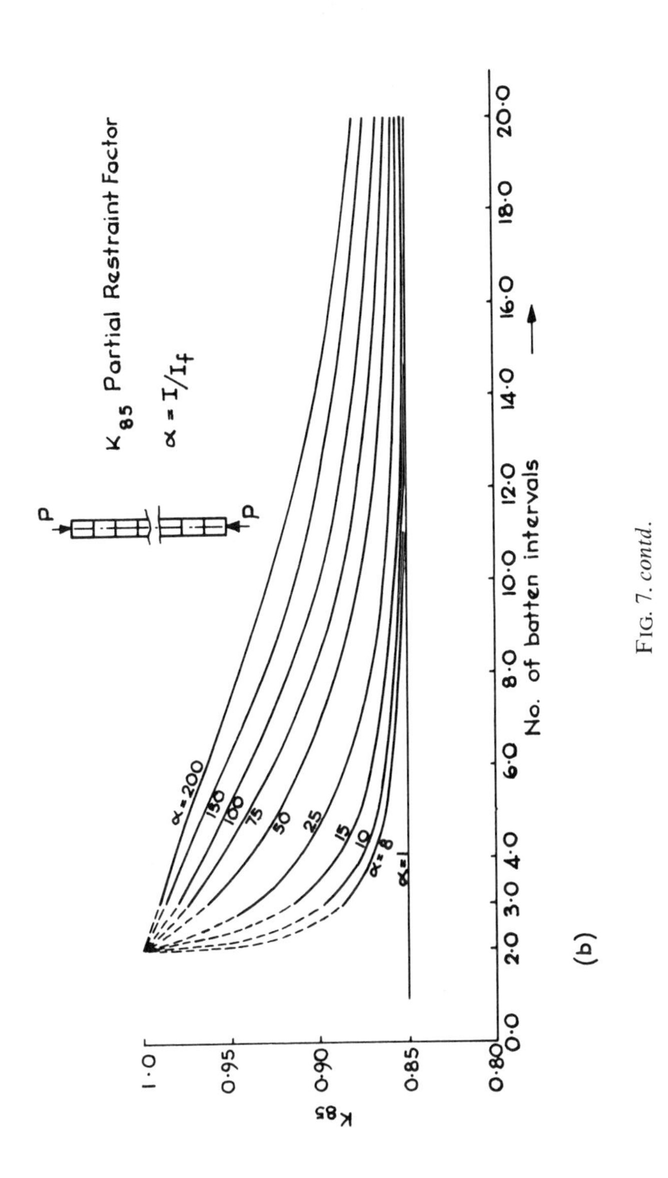

K85 Partial Restraint Factor
α = I/I_f
P
P
K85
1·0
0·95
0·90
0·85
0·80
0·0
2·0
3·0
4·0
6·0
8·0
10·0
12·0
14·0
16·0
18·0
20·0
No. of batten intervals
α=200
150
100
75
50
25
15
10
α=8
α=1
(b)

FIG. 7. *contd.*

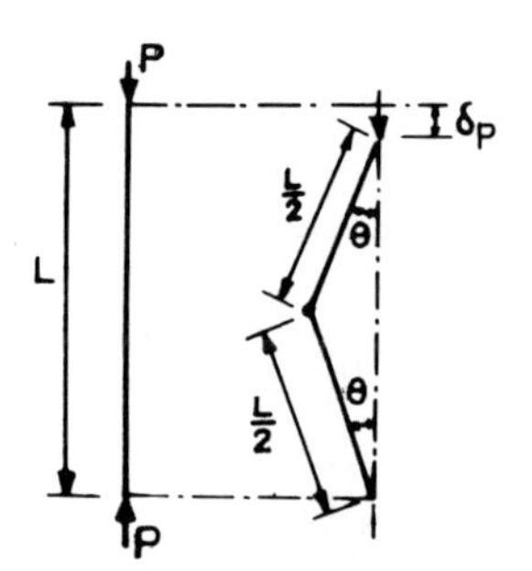

FIG. 8. Central hinge collapse mechanism.

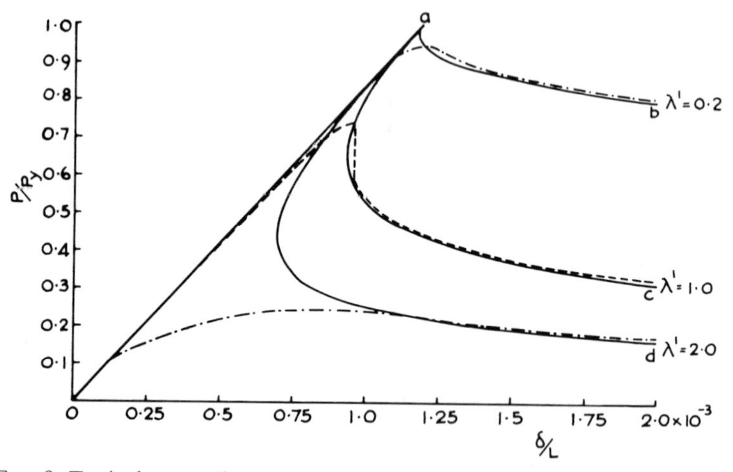

FIG. 9. Typical normalised end-displacement characteristics, solid column.

mechanism and the full lines ab, ac, and ad represent the equilibrium 'off loading' mechanism displacement characteristics for columns of slenderness $\lambda' = 0{\cdot}2$, $\lambda' = 1{\cdot}0$, and $\lambda' = 2{\cdot}0$, respectively.

This equilibrium 'off loading' mechanism characteristic is determined by considering the equilibrium of the deflected column under the vertical load P. From Fig. 8

$$\frac{PL}{2}\theta = M_{pr} \tag{9}$$

where M_{pr} is the reduced plastic moment of resistance of the column section due to the presence of the axial load.

For small lateral deflections, the vertical displacement due to the plastic deformation at the hinge $\partial_p = L \cdot \theta^2/2$. Assuming the interaction relationship

$$M_{pr} = M_p \left[1 - \left(\frac{P}{P_y} \right)^2 \right]$$

putting M_p in terms of P_y as follows, $M_p = \beta r P_y$ where r = radius of gyration and β is a factor depending on the shape of the cross-section (e.g. for a rectangular solid section $\beta = 0{\cdot}866$ but for many practical sections $\beta \to 1{\cdot}0$), and then substituting in eqn. (9), ∂_p becomes:

$$\partial_p = \frac{2\beta^2 r^2}{L} \left[\frac{P_y}{P} - \frac{P}{P_y} \right]^2 \tag{10}$$

In addition to the above displacement there will be the vertical displacement due to the elastic shortening and curvature of the column outside the plastic hinge zone. The vertical displacement due to curvature can be generally ignored compared with the other effects and, hence, the elastic vertical displacement for small lateral deflections is given by:

$$\partial_e = \frac{PL}{AE} = \frac{P}{P_y} \cdot \frac{\sigma_y}{E} \cdot L \tag{11}$$

The total vertical displacement $\partial = \partial_p + \partial_e$ and the expression for the displacement normalised by the length L of the column becomes:

$$\frac{\partial}{L} = \frac{2}{\pi^2} \left(\frac{\beta}{\lambda'} \right)^2 \left[\frac{P_y}{P} - \frac{P}{P_y} \right]^2 \cdot \frac{\sigma_y}{E} + \frac{P}{P_y} \cdot \frac{\sigma_y}{E} \tag{12}$$

In the above displacement expression the first term, representing the effect of plastic deformation, increases as the load decreases but the second term, which represents the elastic recovery, will decrease. From Fig. 9 it is seen that close to the squash load this elastic recovery exceeds the effect of plasticity but at some lower load, depending on the slenderness of the column, the plastic effect will overtake the elastic recovery; the initial concave curve of the 'off loading' mechanism characteristics ending in a vertical slope of Fig. 9 correspond to these load stages.

So far the displacement characteristics have been described using an idealised loading procedure. Consider now the actual behaviour of a practical column loaded in a modern servo-controlled testing machine under 'displacement control'. Under 'displacement control' the column is given a known displacement and the testing machine automatically provides that equilibrium load associated with this given displacement. The actual displacement plot obtained under these test conditions for the column of slenderness $\lambda' = 1{\cdot}0$ is given by the dashed line in Fig. 9. This displacement characteristic departs from that of the perfect column due to initial curvature and residual stress effects and, as partial plasticity occurs, will then bend over until it intersects the mechanism characteristic. This intersection point represents the failure load of the column with the formation of the collapse mechanism. Even with the testing machine held stationary at this displacement the load will fall dramatically until it reaches the lower load associated with this displacement; when the displacement is subsequently increased the load will continue to decrease but at a much reduced rate, following the 'off loading' characteristic indicated by the dashed line. Thus for columns of intermediate slenderness the failure is 'brittle' and this 'brittle' failure feature is responsible for the scatter of failure test results in the intermediate range of slenderness. In this range of slendernesses, failure occurs on the near-vertical part of the displacement characteristic and small changes in imperfections could give rise to relatively large differences in failure loads. For very stocky and for very slender columns the failure intersections occur on the near-horizontal parts of the respective displacement characteristics producing 'ductile' failures, the imperfections having less effect on the value of the failure load. The chain dotted lines in Fig. 9 represent the actual displacement characteristics for these two classes of columns.

8.3.2 Plastic Collapse Mechanisms for Battened Columns

In the case of solid steel columns there is only one possible collapse mechanism but for battened columns there is a series of possible mechanisms and Halabia (1981) has identified the following five possible collapse mechanisms for practical battened columns:

1. Single central hinge (bending mode) Fig. 8
2. Panel mechanism (shear mode) Fig. 10
3. Four hinge short cell (bending mode) Fig. 11
4. Four hinge long cell (bending mode) Fig. 12
5. Three hinge short cell (bending mode) Fig. 13

It should be noted that in practical columns the stiffness of the battens relative to that of the axially loaded vertical members will be such that possible mechanisms which require the formation of plastic hinges in the battens need not be considered.

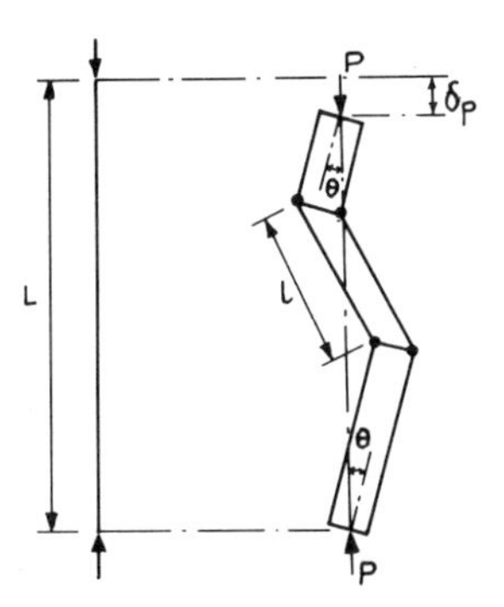

FIG. 10. Panel mechanism.

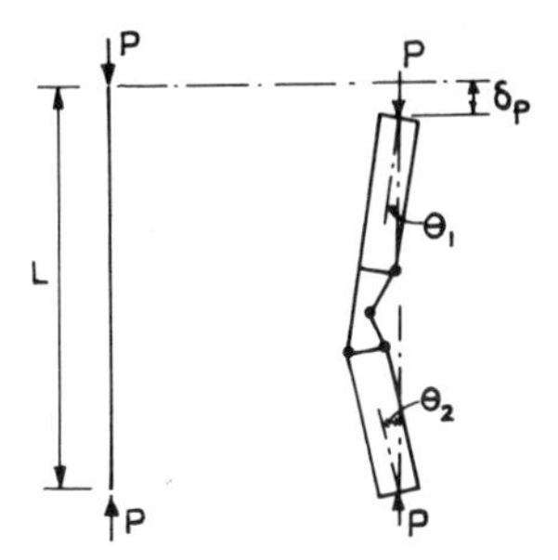

FIG. 11. Four-hinge short cell mechanism.

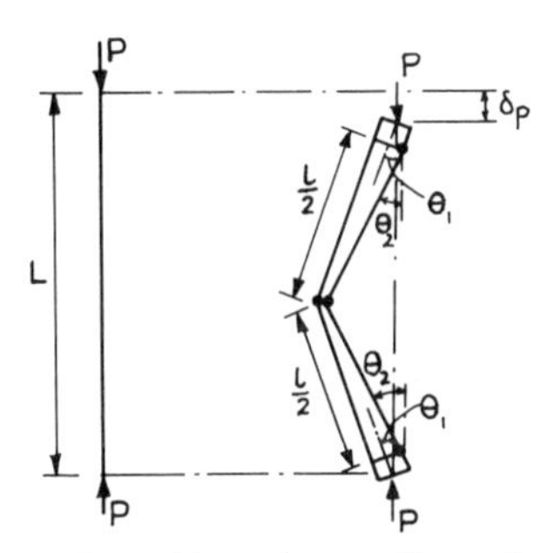

FIG. 12. Four-hinge long cell mechanism.

All the first four mechanisms listed above can form immediately from the straight column but in the case of mechanism No. 5 the column has to be sufficiently bowed either elastically or by elasto-plastic action so that the positions of the three hinges are in line before the mechanism can occur. Whilst it may be possible to predict the failure mechanism by

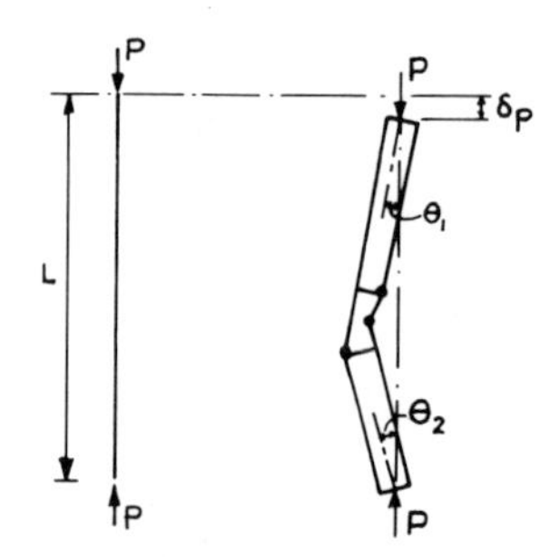

FIG. 13. Three-hinge short cell mechanism.

inspection in respect to battened columns with very stocky cell members (mechanism No. 1) or very slender cell members (mechanism No. 4), in general, prediction of the collapse mechanism will only be possible by examination of the respective 'off loading' displacement characteristics.

The displacement characteristics for all the above collapse mechanisms are derived by considering the equilibrium of the deflected mechanism in the manner previously outlined for the central hinge mechanism and full details are given by Halabia (1981). The displacement characteristic of the central hinge mechanism was obtained in a simple form (eqn. (12)) but for all the other mechanisms, excepting the panel mechanism, the characteristic expressions do not reduce to a simple form and have to be plotted using a computation process. Fortunately, however, the expression for the panel mechanism displacement is found in the same simple closed form as that of the central hinge mechanism and is as follows:

$$\frac{\delta}{L}=\frac{2}{\pi^2}\left(\frac{\beta_f}{\lambda'_f}\right)^2\frac{\gamma}{(1-\gamma)}\left(\frac{P_y}{P}-\frac{P}{P_y}\right)^2\cdot\frac{\sigma_y}{E}+\frac{P}{P_y}\cdot\frac{\sigma_y}{E} \tag{13}$$

where β_f is a factor depending on the section shape of the vertical member and $\gamma = l/L$. The above displacement characteristic depends on λ_f, the slenderness of the cell members, and this is also the case for the other mechanisms which have plastic hinges in the cell members. For any given battened column these will be a 'family' of five displacement characteristics and those characteristics which depend on the slenderness of the cell members will all have a very similar form and will tend to bunch, see Fig. 14. Failure will be by that mechanism associated with the characteristic first encountered by the loading path. Thus for the particular column concerned in Fig. 14, failure will be by a panel mechanism.

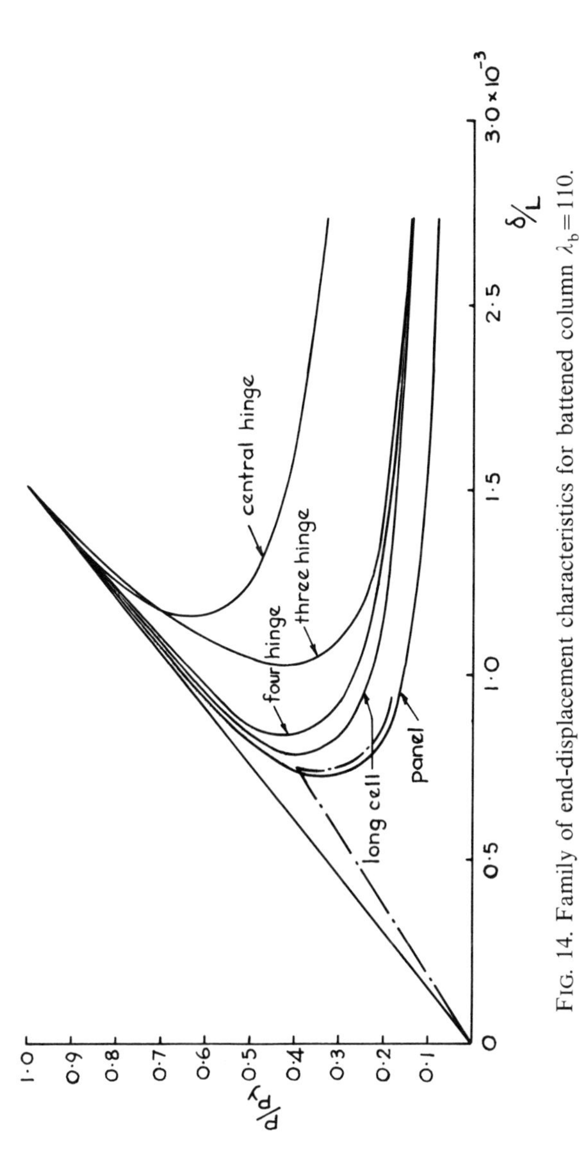

FIG. 14. Family of end-displacement characteristics for battened column $\lambda_b = 110$.

8.3.3 Minimum Value of λ_f/λ to Ensure Solid Column Failure

The expressions for central hinge and panel mechanisms may be compared directly without resort to full plotting and thus it is possible to determine for a column of given overall properties the maximum spacing at which the battens may be placed so that the column would still fail in the central hinge mechanism in preference to a panel mechanism. Let this maximum batten interval be defined by γ^* and at this value the displacement characteristics of the central hinge and panel mechanism coincide. Hence, equating eqns. (12) and (13) it follows that:

$$\frac{\gamma^*}{(1-\gamma^*)} = \left(\frac{\lambda'^2_f}{\lambda'}\right) \cdot \left(\frac{\beta}{\beta_f}\right)^2 \tag{14}$$

The maximum batten interval permitted by BS:449 is $\frac{1}{3}$ of the column length. From eqn. (14) with $\gamma^* = \frac{1}{3}$ the corresponding value of λ_f/λ is approximately 0·7, since for practical columns $\beta \approx \beta_f$. $\lambda_f/\lambda = 0{\cdot}7$ is the maximum value allowed by existing and revised versions of the British Standard Code on Structural Use of Steelwork in Building (1977) and thus columns designed in accordance with these codes should never fail with the formation of a panel mechanism. It would seem then that the purpose of the restrictions on the maximum spacing of battens and ratio of cell slenderness to overall column slenderness is to ensure that battened columns will always fail in the same manner as the equivalent solid column.

8.4 DESIGN FAILURE LOAD OF BATTENED COLUMNS

The designer is primarily interested in the pre-failure serviceability and the value of the failure load of the battened column and to the designer its post-failure behaviour is of concern only in that close agreement of the actual post-failure behaviour with the predicted behaviour would effectively confirm the failure theory.

It has been previously shown in Fig. 9 that it is the form of the 'off loading' displacement characteristic (which in turn is dependent on the slenderness of the column) that essentially determines the failure load of the column. For instance, for stocky columns, ($\lambda' < 0{\cdot}2$) the failure load is close to the squash load, for very slender columns ($\lambda' > 2{\cdot}0$) the failure load approaches the respective critical loads and for columns in the intermediate range of slenderness, ($\lambda' \approx 1{\cdot}0$) the failure load, in general, will fall below the critical load, being influenced by the extent of the initial imperfections.

This feature of column failure behaviour is taken account of in most steel design codes by relating the design failure load to the slenderness by column design curves which have been obtained by 'fitting' the curves to test results. In the interest of economic design the codes often provide different design curves for different classes of section shapes and also for welded columns.

In UK codes the Perry–Robertson formula is the basis of design column curves. This formula has the merit of a continuous form and, also, the theoretical justification of predicting failure on yielding at the outer fibres if a pseudo imperfection factor is employed to fit the test results. In the draft code (1977) the Perry–Robertson formula for the failure stress σ, is given the following form:

For columns $\lambda' > 0{\cdot}2$,

$$\sigma = \frac{\sigma_y + (\eta_1 + 1)\sigma_{cr}}{2} - \sqrt{\left|\frac{\sigma_y + (\eta_1 + 1)\sigma_{cr}}{2}\right|^2 - \sigma_y \sigma_{cr}} \qquad (15)$$

where the imperfection factor $\eta_1 = 0{\cdot}001\ a_1\ (\lambda - 0{\cdot}2\lambda_0)$

For columns $\lambda' \not> 0{\cdot}2$ then $\sigma = \sigma_y$.

The value of constant a_1 to be used in eqn. (15) will vary between 2 and 8 depending on the shape of the section and its susceptibility to imperfections. For 'built-up' columns, a_1 is to be taken as 5·5.

As discussed in sub-section 8.3.3, the 'off loading' displacement characteristics for battened columns will depend on either λ_f, the cell slenderness, or if failure is in the central hinge mechanism, on λ, the overall slenderness. Therefore, failure loads determined by using λ_b, which will be higher than either of the other slendernesses, in a design formula can be expected to be conservative.

Thus, it is proposed that the failure load of battened columns should be determined from the above Perry–Robertson type formula (eqn. (15)), with the value of $a_1 = 5{\cdot}5$ and using λ_b the 'effective slenderness' to determine η_1 and the appropriate critical stress σ_{cr}.

The axial failure load of eccentrically loaded battened columns can also be obtained from the design formula (15) provided that η_1 is increased by the addition of η_0,

$$\eta_0 = 1{\cdot}2\frac{e}{r} \qquad (16)$$

in which e = eccentricity of load and r = radius of gyration of the column as a whole about the appropriate axis, and that the effective slenderness modified for the effect of eccentricity is used.

Although the Perry–Robertson curve has been proposed as a suitable design method largely because of its familiarity in UK practice, without doubt, equivalent curves of other codes would be equally satisfactory.

8.5 EXPERIMENTAL CORRELATION

8.5.1 Recent Experimental Programme

Halabia (1981) has recently concluded a comprehensive experimental study in which he tested 30 model steel columns and 15 large-scale model columns. These columns had a wide range of cell/overall slenderness ratios but the overall slenderness of most of the columns was in the vital range of $\lambda' \approx 1{\cdot}0$.

The majority of the smaller scale model battened columns were made by machining slots in 12 mm × 20 mm rectangular steel bar and were generally 360 mm overall length. Figure 15 gives details of typical model specimens.

Large-scale models up to 2·2 m high were made by welding batten plates to 80 mm × 20 mm × 4 mm channel sections. Figure 16 gives details of typical large-scale specimens.

Much care was taken when machining the small model columns and they can be considered as being largely free from initial stresses. The larger scale welded models were fabricated with care taken to prevent distortion and in many of the specimens the welding strains were carefully monitored, peak compressive strains equivalent to 40% of yield being recorded in the webs and average tensile strains to 20% of yield in the outstands of the main members. In addition, three columns with identical geometry to three welded columns were made by machining appropriate openings in a rectangular steel tube so as to be able to compare the behaviour of welded and unwelded battened columns.

Halabia tested all the columns under ‘displacement control’ and using this technique he was able to obtain accurate failure loads and carefully plot the loading displacement characteristics at all stages, from zero load to failure and then post-failure as the collapse mechanisms deformed. In the large-scale model tests the rotations at the ends of the specimens, shear strains in the battens and axial and bending strains in the vertical members were also recorded.

Figures 18(a) and 18(b) are photographs of the battened column

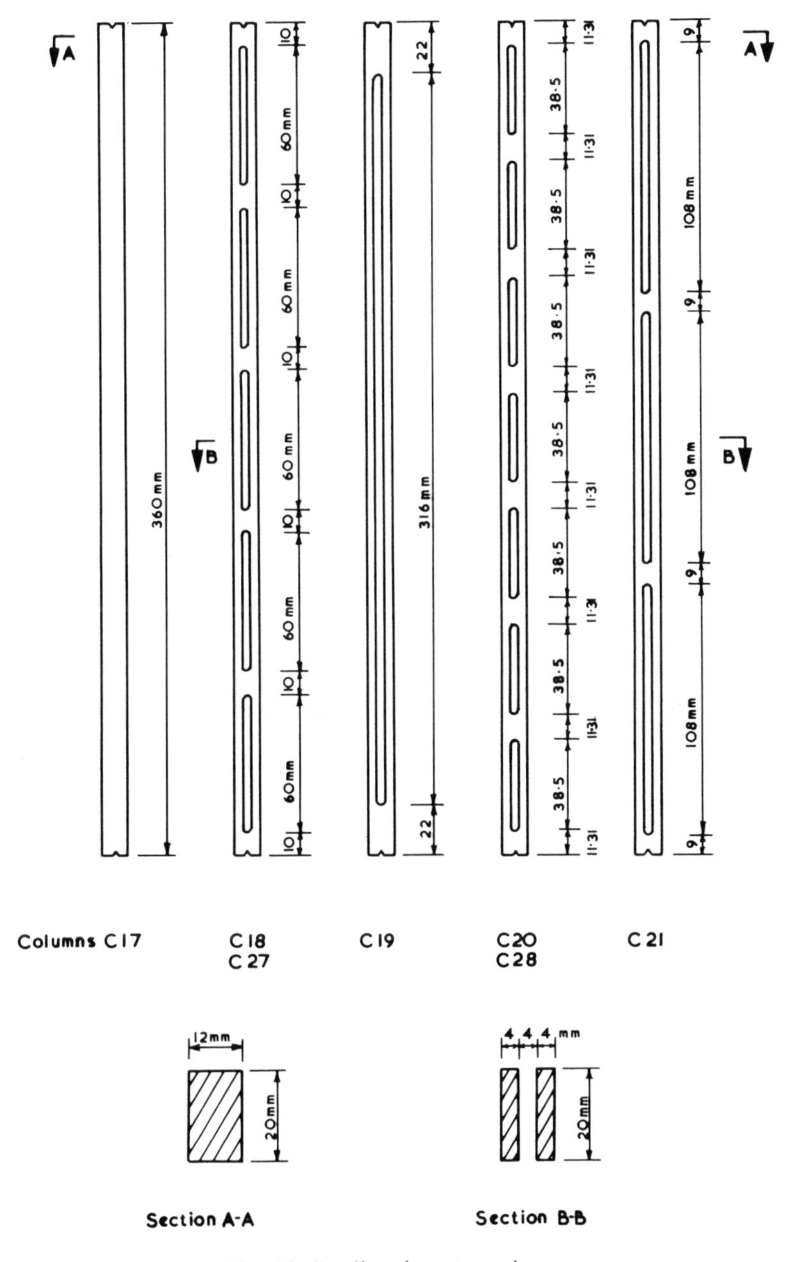

FIG. 15. Small-scale test specimens.

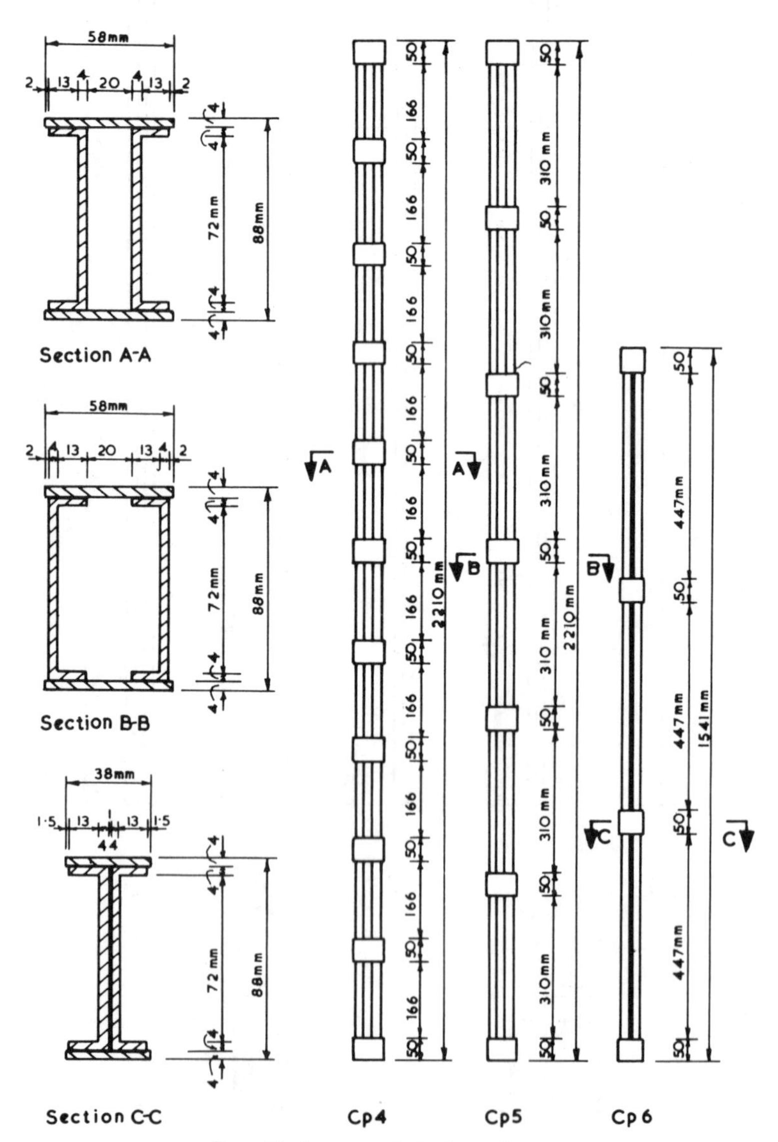

FIG. 16. Large-scale test specimens.

specimens after testing and the form of the collapse mechanisms are readily distinguishable.

8.5.2 Main Results

Halabia's experimental findings can be briefly summarised as follows:

1. All the battened columns failed in the predicted collapse mechanism and the actual load/displacement plots closely followed the theoretical displacement characteristic.
2. For all the columns which covered a wide range of cell/overall column slenderness the failure loads were accurately predicted by the design formula given in Section 8.4. This design formula also catered for welded columns. Although the non-welded columns had a slightly higher failure load than their welded counterparts the difference did not appear to be sufficiently significant to warrant separate design curves for welded and unwelded columns.
3. The shear force due to the vertical loading depended on the initial imperfections and slenderness of the columns, and the usual code value of 2·5% of the failure load was not always conservative, see Fig. 17(a).

8.5.3 Correlation with Previous Work

In addition, Halabia analysed the test results of previous workers and using the proposed design procedure was able to predict the failure loads of the tests carried out by Ng (1947) and Peterman (1931) with accuracy and all the predicted values were conservative.

Ng (1947) also carried out eccentrically loaded tests and all these results also were predicted accurately and conservatively by the proposed design procedure.

8.6 CONCLUSIONS

The failure of steel battened columns is adequately explained by the plastic collapse mechanism theory. However, where the failure load of the battened column is concerned the actual form of collapse mechanism is of secondary importance; the effective slenderness and amount of initial imperfection are the significant factors.

The existing restrictive rules with regard to the design of battened columns have been formulated so that both their elastic behaviour and

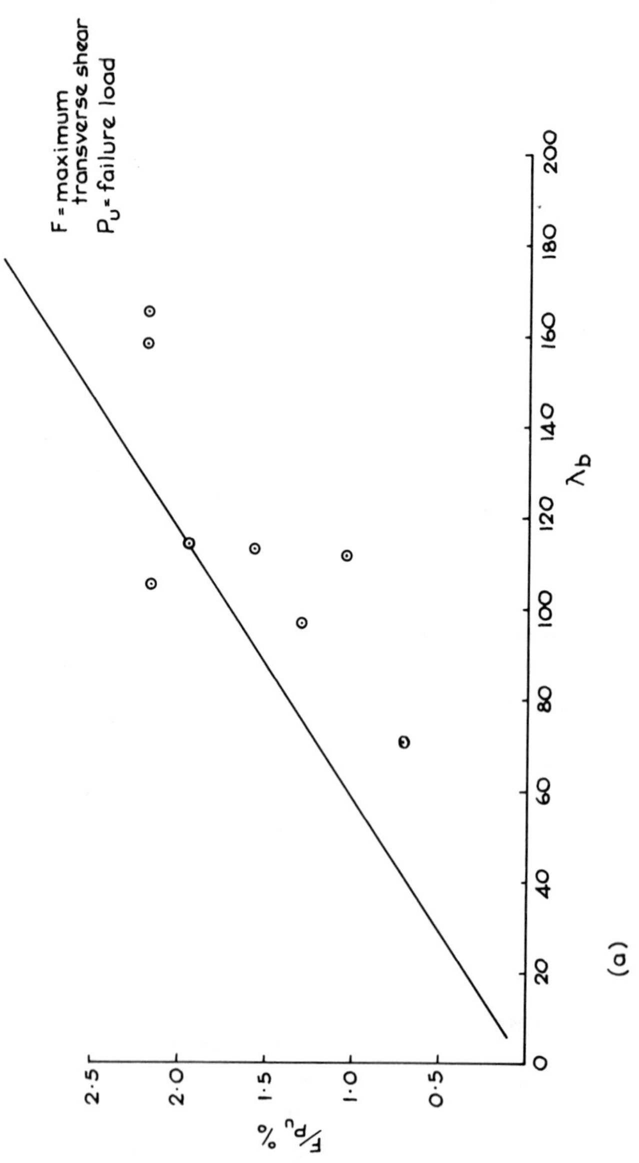

FIG. 17. (a) Variation of maximum transverse shear/failure load with slenderness. (b) Variation of maximum transverse shear/squash load with slenderness.

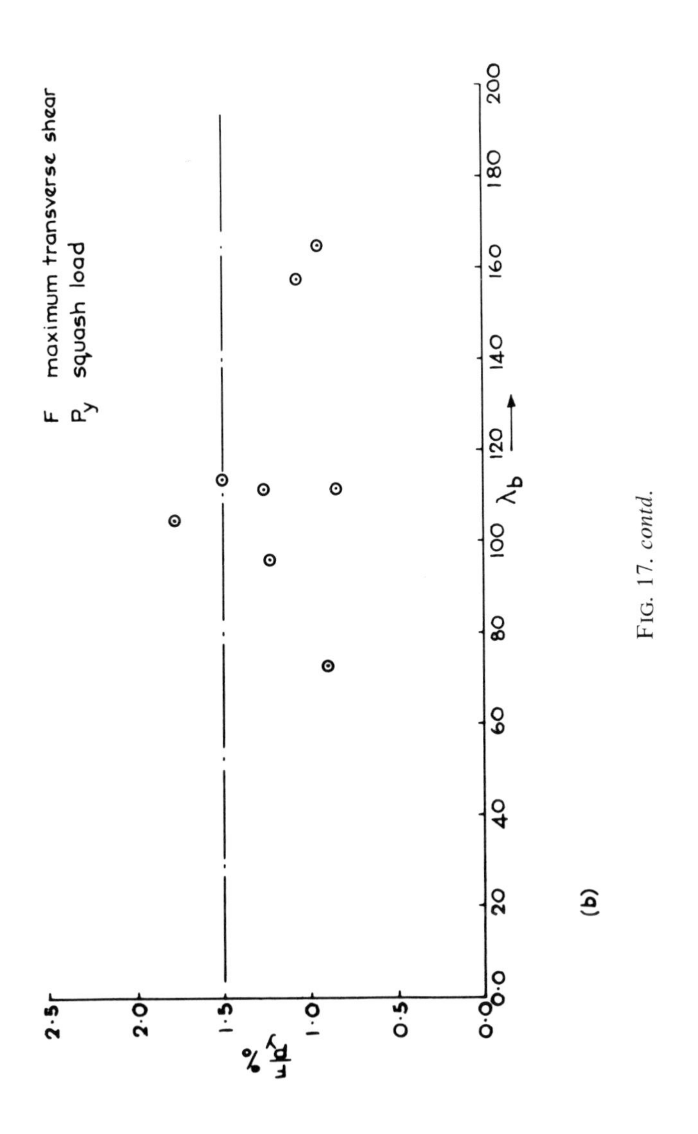
F maximum transverse shear
P_y squash load
$\frac{F}{P_y}$ %
λ_b
(b)

FIG. 17. *contd.*

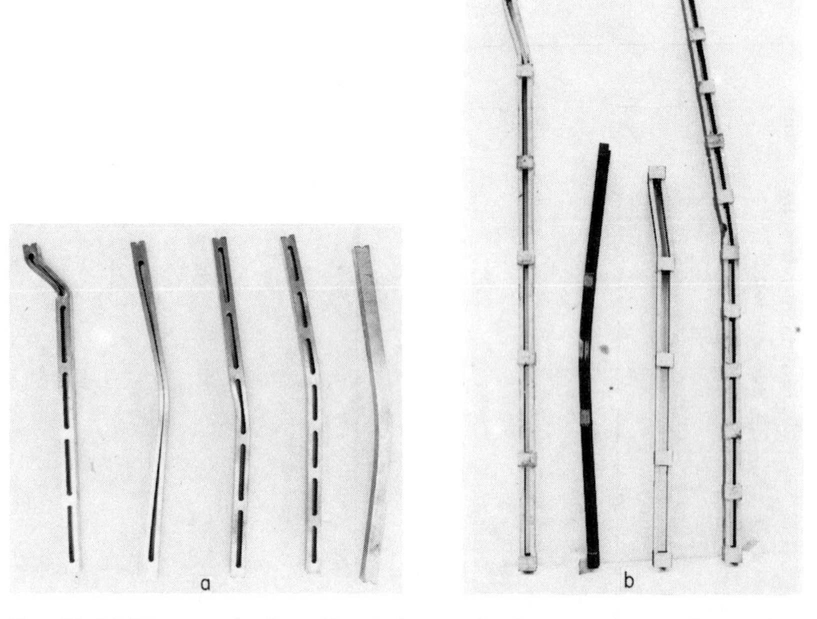

FIG. 18. (a) Photograph of small-scale battened column specimens after testing, clearly illustrating forms of collapse mechanisms. (b) Photograph of large-scale battened column specimens after testing, clearly illustrating forms of collapse mechanisms.

their failure behaviour would be of the same manner as the equivalent solid column. Theoretical and experimental evidence demonstrates that the restrictive rules are unnecessary and that the failure load of any battened column can be accurately and conservatively determined by the design procedure proposed in Section 8.4. The transverse shear force design value of $2\frac{1}{2}\%$ of failure load, the evidence for which appears to date back to tests carried out following the Quebec Bridge failure, could be unconservative in the case of some slender columns and a better design value would be to take 1·5% of squash load (see Fig. 17(b)).

Now that large servo-controlled testing facilities are available, experimental programmes on the behaviour of full-size or near full-size battened or other forms of 'built-up' columns would be desirable to confirm that the results discussed in this chapter apply to columns with sections of practical dimensions.

REFERENCES

BRITISH STANDARDS INSTITUTION (1977). Draft Standard Specification for the Structural Use of Steelwork in Building, Part 1, British Standards Institution, London, Revision of BS:449.

ENGESSER, F. (1909). Über die Knickfestigkeit von Rahmenstäben (on the Buckling Resistance of Battened Columns), *Zentralblatt der Bauverwaltung*, Berlin, **29**, 136.

HALABIA, S. L. (1981). *Stability and behaviour of battened steel struts*, PhD Thesis, University of Wales.

HORNE, M. R. and MERCHANT, W. (1965). *The stability of frames*, Pergamon Press, Oxford.

LIVESLEY, R. K. and CHANDLER, D. B. (1956). *Stability functions for structural frameworks*, Manchester University Press.

NG, W. H. (1947). *The behaviour and design of battened structural members*, PhD Thesis, University of Cambridge.

PETERMANN, A. (1931). Knickversuche mit Rahmenstäben aus st. 48 (Buckling Experiments with Battened Columns of Steel 48), *Der Bauingenieur*, **12**, 509–15.

PORTER, D. M. and WILLIAMS, F. W. (1978). Critical loads of built-up columns and a pocket calculator program, *Proc. Instn Civ. Engrs*, Part 2, **64**, 761–71.

SHANLEY, R. F. (1947). Inelastic column theory, *Journal of Aeronautical Sciences*, **14**(5), 261–7.

WILLIAMS, F. W. (1977). Buckling of multi-storey frames with non-uniform columns using a pocket calculator program, *Computer Structures*, **7**, 631–7.

Chapter 9

ULTIMATE CAPACITY OF COMPRESSION MEMBERS WITH INTERMITTENT LATERAL SUPPORTS

PIERRE DUBAS
Swiss Federal Institute of Technology, Zürich, Switzerland

SUMMARY

The ultimate capacity of continuous compression members can be calculated, with an elasto-plastic second-order analysis, for intermittent rigid or flexible lateral supports. The usual design procedure based on effective lengths given by the bifurcation theory seems to be very conservative for stocky members. This method can be improved by introducing a buckling modulus derived from the suitable column curve.

9.1 INTRODUCTION

In building and bridge construction, constraints may prevent the arrangement of a bracing system in the plane of the compression chords. Old-fashioned through trusses for highway bridges with short spans are a classical example of such structures. The vertical clearance requirements prohibit direct upper bracing: the top chords of the so-called 'pony trusses' are restrained laterally by U-shaped frames including the floor-beams and the truss verticals. Because of railway requirements for flat grades, pony trusses are still necessary for new railway bridges (see Fig. 1), whereas for highway bridges the deck will mostly be at the upper chord level.

For continuous constructions with cross frames (instead of cross

FIG. 1. Double-track railway bridge over the river Landquart, Switzerland. Simply supported span 40·8 m. Opened to traffic 1972.

bracings), the same problem occurs for the bottom chords in the region of intermediate supports (hogging moments). For plate-girder bridges, the analysis of lateral buckling for elastically braced compression flanges can be treated in a similar way.

In roof structures also the compression chords are sometimes supported laterally by frames. Figure 2 shows a storage building with main trusses spanning 72 m and located above the roof plane. The framing arrangement consists of longitudinal girders connected by diagonal tie-beams to the top nodes of the trusses (Prince and Delacoste, 1974).

Mostly the structure lies entirely in the interior of the building and each panel point of the top chord is braced normal to the plane of the truss, the connection to the lateral bracing being constituted by the purlins. For a light-weight roofing, however, the factored upward wind may induce compression in the bottom chord. This loading case frequently requires longitudinal frames composed of the purlins and inclined stays (i.e., an arrangement opposite to that shown in Fig. 2).

The compression members of the structures described above are often designed with an effective length computed from the elastic bifurcation load. If the proportional limit of the chord material is exceeded, the Young's modulus E is sometimes replaced by the tangent modulus or by a kind of buckling modulus.

FIG. 2. Storage hall during erection.

On the following pages reference will first be made to the bifurcation solution for compression members with intermittent lateral supports. The main part of the chapter will, however, be concerned with the ultimate capacity of such members, conforming with the design concept adopted by TC 8 of the ECCS (European Convention for Constructional Steelwork, 1976). An approximate solution using a proper buckling modulus will be compared to the elasto-plastic second-order analysis based on the same assumptions as the ECCS column curves.

Since the case of elastic lateral restraint is the more general one, members with laterally *fixed* intermittent supports will not be treated in an extensive manner.

9.2 BIFURCATION SOLUTION FOR MEMBERS WITH INTERMITTENT LATERAL SUPPORTS

9.2.1 Exact Solution

From a theoretical point of view the determination of the bifurcation load for a compression chord with intermittent supports does not offer major difficulties. Using either the force method (flexibility matrix) or the deformation method (stiffness matrix) a system of homogeneous linear equations is obtained. They can yield non-zero values of the redundant forces or of the deformations only if the determinant of their coefficients vanishes (eigenvalues problem). The fundamental principles of such exact investigations were elaborated more than seventy years ago (for example Zimmermann, 1907; Ostenfeld, 1916; Kriso, 1935 and 1941; see also Johnston, 1976, for an extensive references list).

In practice the difficulties lie in the fact that the true unknown, i.e. the lowest bifurcation load, appears as a variable in trigonometrical functions. The corresponding values of the matrix coefficients, for example the stiffness factors in the second-order deformation method, are available in literature (for example Timoshenko, 1936; Chwalla, 1959). Nevertheless, hand computations are feasible only for systems with a small number of redundant forces or of unknown deformations.

However, the wide use of digital computers in the field of structural engineering and the development of sophisticated procedures for the calculation of eigenvalues have overcome this difficulty and now many FEM programs are available for such eigenvalues problems, based mostly on the work done by (among others) Argyris *et al.* (1964) and Turner *et al.* (1964). The stability criterion is generally written in the following matrix form

$$([K_0]+\lambda.[K_G]).\{w\}=\{0\}$$

with $[K_0]$ = small displacements stiffness matrix
$[K_G]$ = geometrical stiffness matrix
λ = scalar quantity defined as load multiplicator, with the critical values λ_{cr} corresponding to the eigenvalues of the system and with deflected shapes expressed by the related eigen vectors.

In connection with the numerical applications given later, some computer results will be used as examples.

9.2.2 Engesser's Approximate Solution (1884 and 1885)

Engesser's analysis of the top chord buckling in a pony-truss bridge is a typical example of a very simple but nevertheless accurate solution to an intricate problem. Figure 3 shows the simplifying assumptions on which

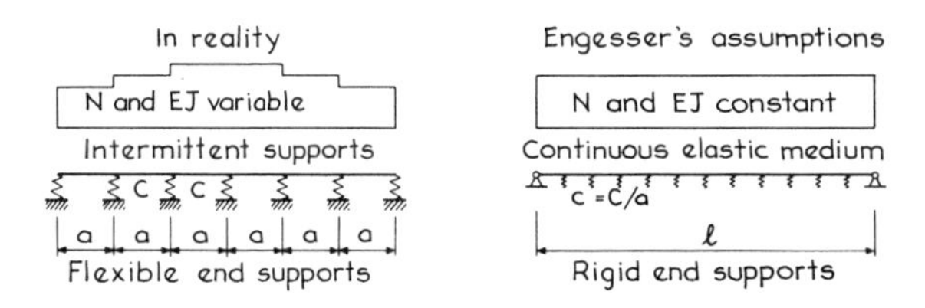

FIG. 3. Simplifying assumptions for Engesser's analysis.

the analysis is based, viz.,

(i) compression chord of uniform cross-section, with a compressive force constant throughout the chord length;
(ii) chord ends pin-connected and rigidly supported;
(iii) equally spaced elastic supports having the same stiffness and assumed 'smeared' over the support spacing, a.

Equilibrium considerations between the elastic bending resistance of the chord,

$$(EJ.w'')''$$

the deviation forces due to the buckling curvature

$$(N.w)''$$

and the reaction of the continuous elastic medium

$$c.w$$

lead, under the above mentioned assumptions of constant values for the chord force N and for the second moment of area J, to the following differential equation

$$EJ.w'''' + N.w'' + c.w = 0.$$

With the boundary conditions $w = w'' = 0$ at both ends, the solution takes the form $w_0.\sin(n\pi x/l)$ and the critical load can be written as

$$N_{cr} = \frac{n^2\pi^2}{l^2}EJ + c\frac{l^2}{n^2\pi^2}$$

i.e. as the sum of the well-known Euler load and of the contribution from the elastic medium.

The determinative number n of half waves corresponds to ${}_{\min}N_{cr}$, i.e.

$$n^2 = \frac{l^2}{\pi^2}\sqrt{\frac{c}{EJ}} \rightarrow {}_{\min}N_{cr} = 2\sqrt{EJ.c} \tag{1}$$

In practice n should be an integral number. Figure 4 shows that the Engesser parabola, corresponding to the above formula (1), leads to safe values in the frame of the simplifying assumptions. The figure demonstrates also clearly the increase of the critical load due to the continuous elastic medium, in comparison with the Euler load N_E of a chord without intermittent supports.

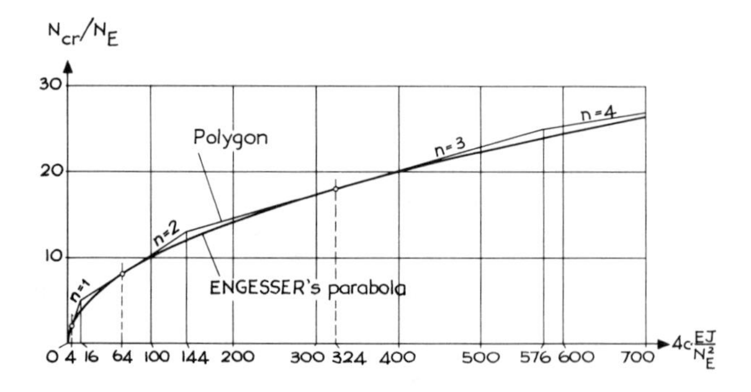

FIG. 4. Eigenvalues as function of the spring stiffness.

For design, it is convenient to introduce an effective length l_K for the chord. By equalising

$${}_{\min}N_{cr} = 2\sqrt{EJ.c} = \frac{\pi^2.EJ}{l_K^2}$$

we obtain

$$l_K = \pi.\sqrt[4]{\frac{EJ}{4c}} = \pi.\sqrt[4]{\frac{EJ.a}{4C}} \tag{2}$$

The effective length does not correspond to the half wavelength l/n of the chord with elastic supports. The factor 2 in the Engesser formula (1) involves

$$l/n = \sqrt{2}.l_K$$

Because of the substitution of a continuous medium for the intermittent supports, the Engesser formula can lead to an effective length which is less than the frame spacing. Physically this cannot occur and the minimal value for l_K is a.

9.2.3 Comparison between Engesser's Analysis and the exact Solution

Comparisons between exact eigenvalues and Engesser's approximate analysis lead to the following conclusions (see also Fig. 11):

(i) The assumption of a uniform cross-section and of a constant compressive chord force involves acceptable errors. In fact, in an actual construction the quantities N and J are correlated.
(ii) The introduction of a continuous medium instead of intermittent supports leads to a reasonable accuracy, provided that the effective length of the continuously supported chord is at least 1·25 times the support spacing (or 1·8a for the half wavelength l/n).
(iii) The flexibility of the end supports may reduce ${}_{\min}N_{\mathrm{cr}}$, especially if the corresponding transverse-frame stiffness is not higher than that of the intermediate supports. The end frames of the bridge shown in Fig. 1, with a polygonal top chord, will automatically be stiffer than the intermediate one, owing to the reduced height of the end posts.

There are other effects which will be mentioned in connection with the elasto-plastic analysis (see sub-section 9.3.4)

The Engesser analysis gives therefore an accurate solution within certain limits and will consequently be used as basis for the approximate method presented in Section 9.4.

9.2.4 Stiffness of a Pony-truss Transverse Frame

As an example of the computation of the spring constant C we consider the transverse frame of a pony-truss bridge as shown in Fig. 5:

$$C = \frac{E}{h_{\mathrm{i}}^3/3J_{\mathrm{V}} + h^2 \cdot b/2J_{\mathrm{B}}}$$

Often the floorbeam is very stiff in comparison with the vertical member of the truss and the second term in the denominator can be neglected. Possibly the reduction in stiffness due to a compressive force in the vertical should be considered (Hrennikoff, 1935, Holt, 1952).

For Warren trusses without verticals, reference should be made to Hilbers, 1977.

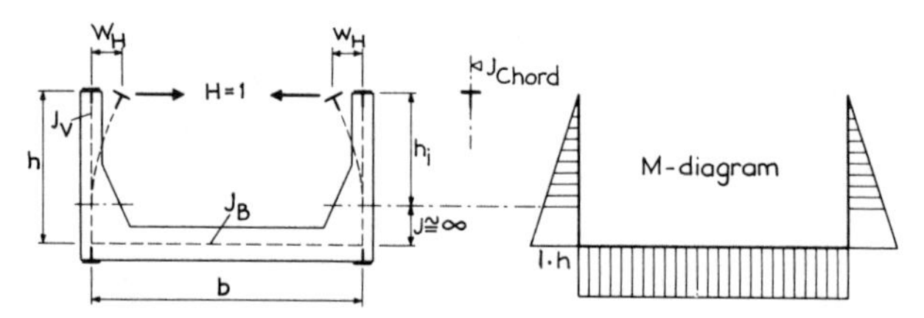

FIG. 5. Typical transverse frame in a pony-truss structure.

Roof structures, as shown in Fig. 2, often possess continuous supporting frames, for which the spring constant C will be computed in a similar way as for a pony truss.

9.3 PARAMETERS AFFECTING THE ULTIMATE CAPACITY OF COMPRESSED MEMBERS WITH INTERMITTENT LATERAL SUPPORTS

9.3.1 Geometrical Imperfections

According to the investigations carried out by the ECCS (1976), the initial out-of-straightness may be assumed to be 1/1000 of the half wavelength. For a continuous chord this length is unknown before the end of the computations, so that a judicious assumption has to be made. Along the chord length, the curvature changes its sign according to the shape of the eigenvector.

9.3.2 Bending of the Floorbeams

Direct loading applied to the floorbeams of a pony-truss bridge involves lateral displacements of the compression chord. For a uniform loading throughout the bridge deck and for floorbeams with equal stiffness, the chord remains straight and its ultimate capacity will not be affected. Live load, however produces lateral displacements varying along the span and thus initial bending in the chord. Schibler (1946), using an elastic second-order analysis, has demonstrated that this effect may be of importance. In the following numerical examples, two kinds of live load will be simulated: a truck with two unequally loaded axles on the one hand, and a standard railway loading on the other.

9.3.3 Residual Stresses

Residual stresses play the main role for the ultimate capacity and consequently furnish the criterion for the selection of an appropriate column curve (ECCS, 1978).

The cross-section shown in Fig. 6 serves as a basis for all calculations in the following examples: it is a slightly idealised wide flange profile HEA 200. Buckling about the weak axis of this rolled I-shape has defined the ECCS column curve c, which is also applicable for all profiles with a height-to-width ratio not higher than 1·2. Therefore it seems appropriate

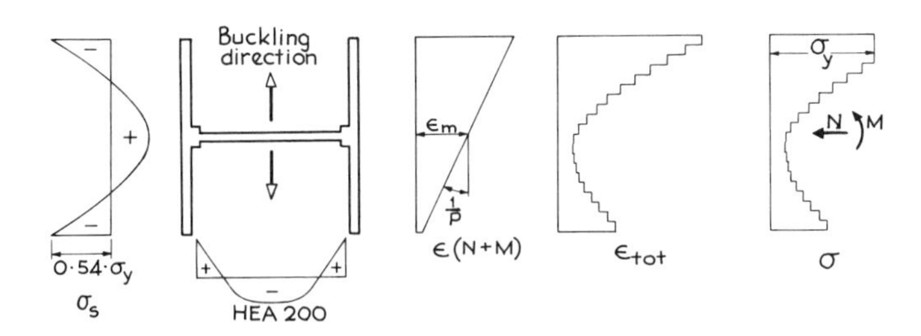

FIG. 6. Chord section selected for the computations. Strain and stress distributions.

to use the weak axis buckling of the HEA 200 for a comparison between a method based on the effective length (see Section 9.4) and 'exact' calculations, although in a 'pony-truss' structure this type of section will generally be used for strong axis buckling.

The residual stress pattern on Fig. 6 corresponds exactly to the distribution introduced for the computation of ECCS curve c. The yield point considered is 235 N/mm^2 appropriate to steel Fe 360, EURONORM 25–72. The stress–strain relationship is idealised as linear elastic till the yield point ($E = 210\,kN/mm^2$) and perfectly plastic at yield (strain hardening neglected).

According to Fig. 6 the interaction curves between the bending moment M and the curvature $1/\rho$, with the normal force N as parameter, will be computed using the following procedure: for an assumed slope $1/\rho$ of the strain distribution $\varepsilon(N+M)$ due to the axial compression and its eccentricity moment, the assumed residual strains $\varepsilon_s = \sigma_s/E$ are added to the values $\varepsilon(M+N)$. The above-mentioned bilinear stress–strain relationship determines now the stress in each fibre and therefore, by elementary summations, the section forces N and M related to the centroid.

Figure 7 represents the resulting interaction diagram for the weak axis bending of the profile HEA 200. For each flange 100 fibres, each 2 mm high, have been adopted, in comparison with 17 for the less important web and 4 for the fillets (admitted σ_s-free). In Fig. 7 the normal force N is kept constant along each curve, despite the varying extent of plasticity,

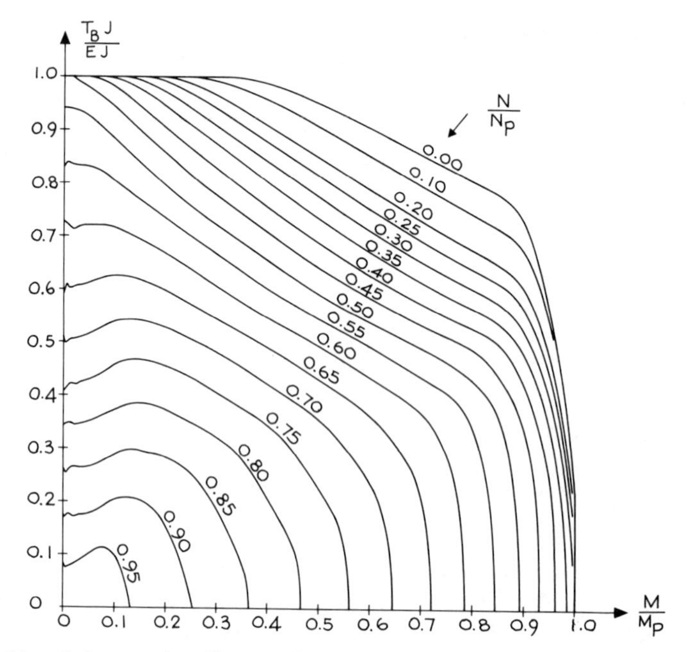

FIG. 7. Interaction diagram for the weak axis bending of HEA 200.

the value ε_m of the centroidal strain being corrected in an appropriate way by using a loop in the computer program. The ordinate in Fig. 7 is the relative bending stiffness $T_B J/EJ$, with $T_B J = \rho.M$. The section forces M and N are both referred to the corresponding maximum value N_p and M_p (plastic moment).

9.3.4 Other Influences

The ultimate capacity may be also affected by additional effects, such as:

(i) Torsional rigidity of the compression chord and of the other truss members. This influence may be important for hollow sections, as shown among others by Baar (1968) for trusses without lateral supports in the plane of the compression chord. Cidect Monograph No. 4 (1981) presents formulae which will make it possible to calculate the lateral fixity coefficient of chords.

(ii) Interaction between lateral and torsional buckling for chords without axis of symmetry in the buckling plane.

(iii) The effect of chord curvature for nonparallel-chord trusses (Fig. 1).

(iv) Secondary stresses occurring in a rigid jointed truss.

For most of these influences spatial calculations are needed, with a corresponding increase of computing time. For this reason, in the following examples only two-dimensional systems have been taken into consideration. However, an example shows how a spatial problem—the lateral buckling of a plate girder with hollow longitudinal stiffeners—can be solved with a planar simulation.

9.3.5 Method of Analysis

The elasto-plastic second-order analysis used here is based on the FEM program BARBU established by Rouvé (1976). The loading will be applied in increments, the equilibrium conditions being satisfied by iterations under corresponding modifications of the stiffness properties, i.e. of the curvature $1/\rho$ and of the centroidal strain ε_m (see Fig. 6). Of course, the interaction between N, M, $1/\rho$, and ε_m is not taken from diagrams analogous to Fig. 7 but computed for each section and each iteration directly from a subroutine included in the program.

The iteration procedure at each increment follows the well-known Newton–Raphson method. An eventual elastic unloading is not considered. More details concerning the computations seem to be interesting only to specialists and are not included here.

9.4 APPROXIMATE METHOD FOR THE ESTIMATION OF THE ULTIMATE CAPACITY

The computer calculations described above are very expensive and therefore not suitable for current design. For this reason an approximate method, using the Engesser formula with a buckling modulus based on the ECCS column curve relating to the chord section, will be presented. In the following numerical examples the predictions of this approximate design procedure will be compared with the results of the elasto-plastic computations.

The buckling modulus T is defined by the following relation for the ultimate buckling stress σ_K

$$\sigma_K = \frac{\pi^2 T}{\lambda_K^2}$$

in analogy to Euler's critical stress

$$\sigma_{cr}=\frac{\pi^2 E}{\lambda_K^2}$$

Therefore we obtain

$$T=E\frac{\sigma_K}{\sigma_{cr}}$$

or, after introduction of the well-known slenderness parameter,

$$\bar{\lambda}_K=\frac{\lambda_K}{\pi}\sqrt{\frac{\sigma_y}{E}}=\sqrt{\frac{\sigma_y}{\sigma_{cr}}}$$

T can be written in the form

$$T=E.\bar{\lambda}_K^2.\sigma_K/\sigma_y$$

Figure 8 demonstrates the iteration procedure needed: the buckling modulus T must first be assumed. For a pony-truss, as an example, the Engesser formula, eqn. (1), with T instead of E yields an approximate value for the ultimate capacity of the chord and therefore for the ultimate stress σ_K. The slenderness parameter $\bar{\lambda}_K$ can now be taken from the

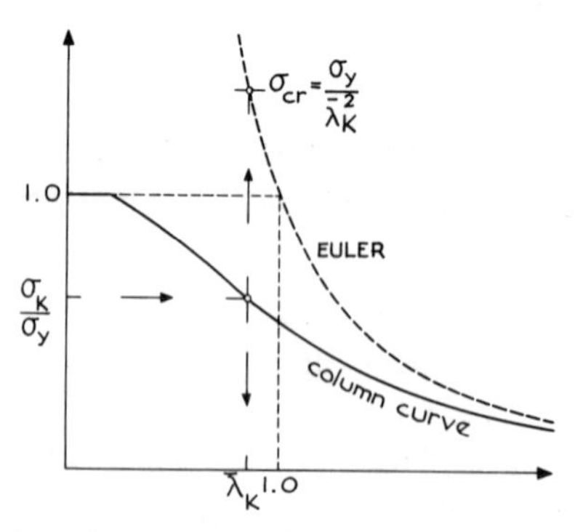

FIG. 8. Procedure for the iterative calculation of the modulus T.

corresponding column curve and the value of T can be adjusted. The convergence is acceptable and each calculation is simple. It is also possible to utilise the Engesser formula, eqn. (2), for the effective length and so to compute first $\bar{\lambda}_K$ and after that σ_K/σ_y from the column curve. At the convergence point, the results will be exactly the same.

Details of application of this procedure for chords with varying forces and sections are given in the corresponding examples.

9.5 ULTIMATE CAPACITY OF THE COMPRESSION CHORD IN A PONY-TRUSS STRUCTURE

9.5.1 Pony-truss Structure selected for the Computations

Figure 9 shows the structural arrangement of the pony-truss structure investigated. The Pratt truss with 12 panels spans 54 m; this seems to be a maximum for a structure without upper bracing. Table 1 contains the chords forces and the corresponding cross-section properties, obtained by affinity from the HEA 200 (see Fig. 6).

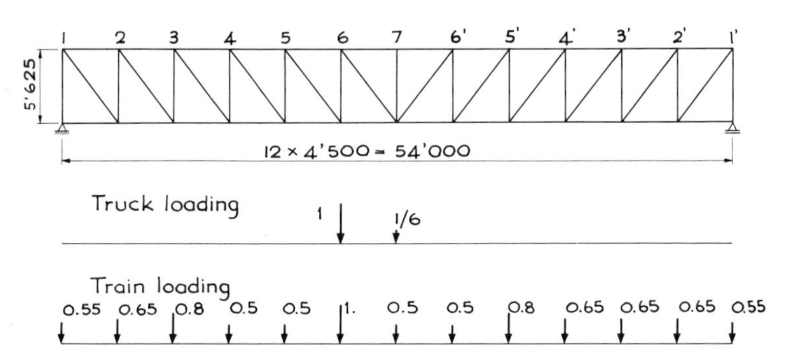

FIG. 9. Truss structure investigated.

TABLE 1

Panel	1–2	2–3	3–4	4–5	5–6	6–7	
Force	594	1 080	1 458	1 728	1 890	1 944	kN
Area A	12·1		13·1	14·8	16·7		10^3 mm^2
Second moment of area, J	67·6		91·2	108·1	121·7		10^6 mm^4

The chord forces vary in a parabolic manner and correspond to a vertical load of 135 kN at each node or to a distributed loading of 30 kN/m. In order to make the comparisons between the different floor loadings easier, the chord axial forces are assumed to be the same for all cases, independently of the distribution of the floorbeam loading schematically shown below the truss system. The two unequal loads near the middle of the structure simulate a heavy truck, whereas the train loading corresponds to the Swiss code for metre-gauge. In both cases the loading arrangement is chosen so that the chord bending is the maximum for the panels in which buckling initiates. The floor loading is applied in increments, with the same multiplicator as for the chord axial forces.

The pattern of the geometrical imperfections should correspond to the number of buckling waves. For the structure examined, independently of the transverse-frame stiffness (see Table 2) the chord collapses with six half-waves, as shown in Fig. 12. The corresponding length is 9 m, and a maximum initial out-of-straightness of ± 9 mm has been introduced.

TABLE 2

Frame type	I	II	III	IV
γ	0·0732	0·1098	0·1830	0·3137

9.5.2 Planar Model

For the computations the chord is considered as straight compressed bar supported by equally spaced springs. The spring constant C corresponds to the stiffness of the transverse frames, as calculated in sub-section 9.2.4 (see also sub-section 9.5.3). In the deformed state, according to Fig. 10, the springs exert reactions perpendicular to the chord, amounting to $C.w$.

At each node the chord axial force is increased by a value ΔN, corresponding to the longitudinal component of the web members. Since

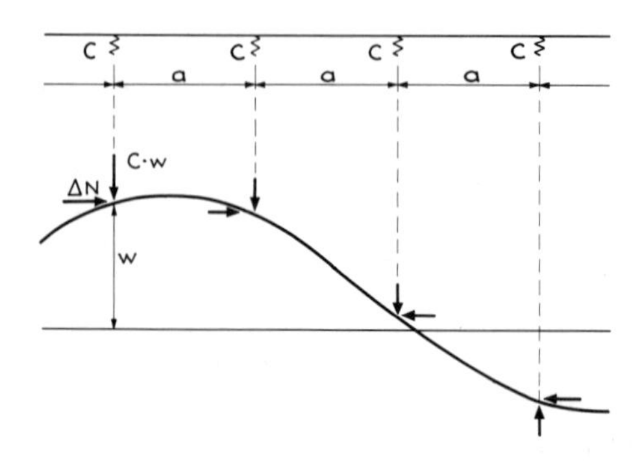

FIG. 10. Planar model and corresponding deformed state.

the bottom chord is laterally braced, it remains straight and therefore the chord loading ΔN acts always parallel to the bottom chord, i.e. also parallel to the undeformed axis of the top chord (conservative loading).

For the computations a linear reduction of 2/3 has been applied, so that the spring spacing is equal to 3 m, instead of 4·5 m as in Fig. 9. In comparison to the section properties stated in Table 1, the area is reduced by a factor $(2/3)^2$ and the second moment of area by $(2/3)^4$; chord 1–2–3 is therefore identical to the wide flange HEA 200 (see Fig. 6). Each panel is subdivided in 9 *FE*.

The influence of the direct floor loading has been simulated by vertical loads applied to the springs. The corresponding elongation of the spring, assumed to be isolated from the chord, must be equal to the lateral deflection at the top of the truss verticals, due to the loading on the floorbeams (also here without connection between the frames and the chord).

For the railway loading, the relative stiffness of the floorbeams and of the truss verticals are similar to those adopted by Schibler (1946) from an actual bridge design. The maximum spring load (node 6, Fig. 9) is then equal to 1/180 of the axial force in the chord member 6–7 (only frame type II, see Table 2, considered here).

For the truck loading, the values are $N_{6-7}/92{\cdot}3$ for frame type I, $N_{6-7}/70$ for II, $N_{6-7}/48{\cdot}6$ for III, and $N_{6-7}/32{\cdot}7$ for IV (see Table 2). At collapse (see Table 3) the elongations of the isolated springs are nearly the same for all types, i.e. the floorbeams are assumed to be identical whereas the stiffness of the truss verticals varies.

9.5.3 Spring Constants Considered for the Computations

The elasto-plastic second-order computations have been performed with four values of the spring constant, i.e. of the frame stiffness, expressed in Table 2 in the following non-dimensional form

$$\gamma = \frac{4a^3C}{\pi^4 EJ}$$

with EJ = lateral bending stiffness of the chord in the elastic range (chord member 5–6–7)

C = elastic transverse-frame stiffness, as defined in sub-section 9.2.4

a = frame spacing

For a stiffness factor $\gamma = 1$, Engesser's equation (2) leads to an effective length $l_K = a$, i.e. to the optimal value in the frame of Engesser's assumptions (see sub-section 9.2.2).

The highest γ-value guarantees, according to the computations, an ultimate capacity of the central panel corresponding nearly to an effective length equal to the frame spacing a. This can be considered in a first approximation as the maximum possible capacity of the chord. In fact, the axial force in panel 5–6 is 3% less than in the adjacent central panels 6–7–6′, whereas the section properties are the same. Member 5–6 therefore slightly restrains the critical chord region and the ultimate capacity can exceed somewhat that of a pin-ended central panel.

Frame type I, on the other hand, is relatively flexible, whereas type II seems to correspond with a practical design for such structures. For this reason the greater part of the computations has been undertaken with this value of γ; the figures given below will also concern frame type II.

For the *end* frames, according to sub-section 9.2.3, the stiffness has been increased to 150%.

9.5.4 Bifurcation Loads and Corresponding Eigenvectors

The bifurcation loads obtained with the computer program BARBU (Rouvé, 1976) are drafted in a non-dimensional form in Fig. 11, $N_{a,cr}$ being the Euler buckling load belonging to an effective length equal to the frame spacing, i.e. $l_K = a$ for the central panel 6–7. In analogy to the indications given in sub-section 9.5.3 this bifurcation load is nearly the

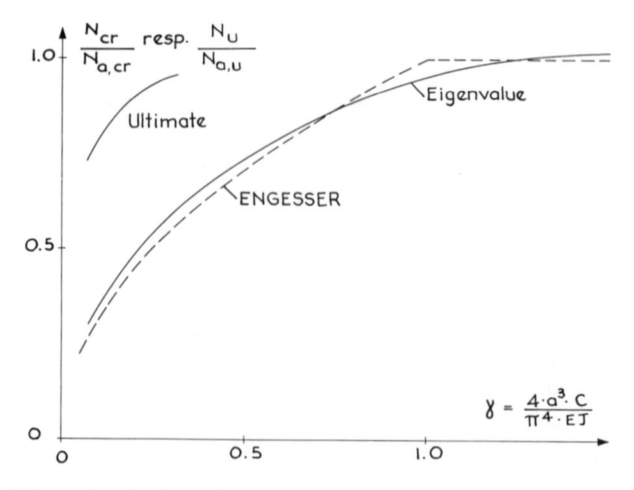

FIG. 11. Bifurcation loads and ultimate loads as function of the stiffness factor γ.

maximum possible value. For computing the eigenvalues, stiffness parameters higher than those in Table 2 have also been considered, up to $\gamma=1{\cdot}5$. The corresponding effective lengths are given by $a{\cdot}\sqrt{N_{a,\mathrm{cr}}/N_{\mathrm{cr}}}$.

The same figure also shows the results given by Engesser's formula, eqn. (1), written for the section properties of chord members 5–6–7. Owing to the variation of the chord forces and of the corresponding chord sections, the accuracy is less good than for a chord with constant forces and sections (Dubas, 1977).

As an example of the eigenvectors, Fig. 12 presents the two patterns corresponding to the first ($N_{\mathrm{cr}}/N_{a,\mathrm{cr}}=0{\cdot}3651$) and to the second ($N_{\mathrm{cr}}/N_{a,\mathrm{cr}}=0{\cdot}3716$) eigenvalues for framing system II. The first eigenvector is symmetrical relating to the bridge centre and comprises five half wavelengths; the second one has six half wavelengths.

The deflection curve resulting from the elasto-plastic computations, on the other hand shows six half waves (Fig. 12, bottom). It is therefore evident that a computation of the ultimate capacity of the chord using an effective length given by the bifurcation theory cannot lead to an accurate solution.

9.5.5 Ultimate Capacity (elasto-plastic computations)

The results of the computations performed as outlined in sub-section 9.3.5 are summarised in Table 3, expressed as load multiplicator relating to chord force 6–7 defined in Table 1.

TABLE 3

Frame stiffness factor γ		0·0732	0·1098	0·1830	0·3137	
Without floor loading	N_{u}	1·26	1·40	1·55	—	$\times N_{6\text{–}7}$
Railway loading	N_{u}	—	1·37	—	—	$\times N_{6\text{–}7}$
Truck loading	N_{u}	1·18	1·30	1·43	1·55	$\times N_{6\text{–}7}$
l_{K} from eigenvalues (9.5.4)		1·817	1·655	1·463	1·292	$\times a$
Ultimate load (ECCS curve c)		1·06	1·16	1·29	1·42	$\times N_{6\text{–}7}$
T-modulus according to 9.4		0·479	0·404	0·310	0·220	$\times E$
Corresponding values l_{K}		1·599	1·385	1·141	(0·916)	$\times a$
Ultimate load (ECCS curve c)		1·20	1·35	1·53	—	$\times N_{6\text{–}7}$

The values corresponding to the truck loading are also included in Fig. 11, where the ultimate capacity N_{u} is related to the corresponding strength $N_{a,\mathrm{u}}$ of the hinged-end column with a length equal to the frame

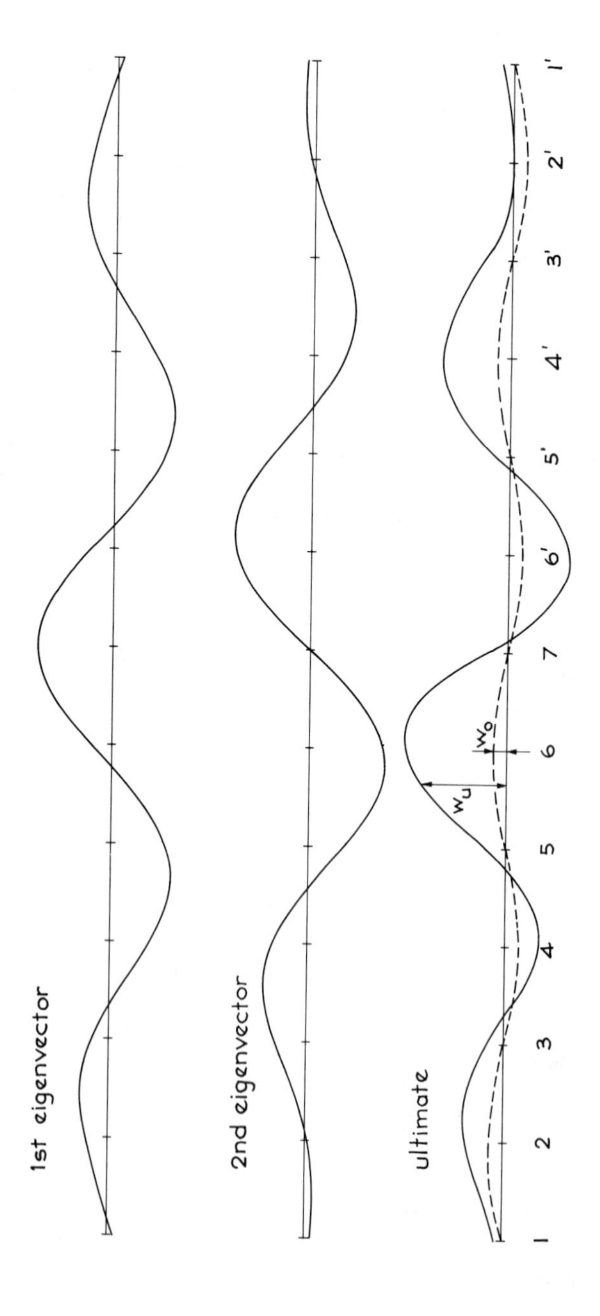

FIG. 12. Eigenvectors and pattern of lateral deflections near collapse.

spacing a and with the section properties of chord member 5–6–7 (see also sub-section 9.5.3)

A comparison with the curve for the eigenvalues clearly shows that the frame stiffness required to reach $N_{a,u}$ is much lower than stipulated from the bifurcation theory. This is due to the decrease of the chord lateral stiffness ρM in the elasto-plastic range, as shown in Fig. 13a for frame type II (see also Fig. 7). This explains why the ultimate load, calculated by introducing into the ECCS column curve c (ECCS 1978) a slenderness factor $\bar{\lambda}_K$ corresponding to the effective length given by the bifurcation theory, is very conservative.

The approximate method presented in Section 9.4 ensures a higher accuracy. In the present case the given values relate also to chord member 6–7 (section properties and reference chord load according to Table 1) and the relation between σ_K/σ_y and $\bar{\lambda}_K$ is fixed from ECCS column curve c (see sub-section 9.3.3). The reduction in the effective length resulting from the introduction of the buckling modulus T instead of E in the Engesser formula eqn. (2) leads to a satisfactory agreement with the 'exact' solution.

To complete the results, Fig. 13a illustrates also the increase of the bending moments in the chord point with the maximum curvature. The form of the M-curves depends to a great extent on the floor loading, whereas according to Table 3 the ultimate capacity is not so much affected.

Figure 13b shows the progression of the deflections at the chord point with the maximum value. Also here, the floor loading exerts a marked influence.

9.5.6 Discussion of the Influence of Floorbeams Loading

The effects of the vertical loads applied directly to the floorbeams of the transverse frames have been examined chiefly for frame type II. Table 3 and Fig. 13 lead to the following conclusions:

(a) The ultimate capacity of the chord is not severely affected by the floorbeams bending. For the railway bridge examined, with a floor loading corresponding to an actual case, the reduction attains about 2% and is therefore not significant. The assumed truck loading, on the other hand, seems to be very heavy in relation to the chord axial force. Indeed the chord force in a highway bridge results mainly from the permanent loads (dead load and floor weight) and also from a distributed live load. For actual bridge

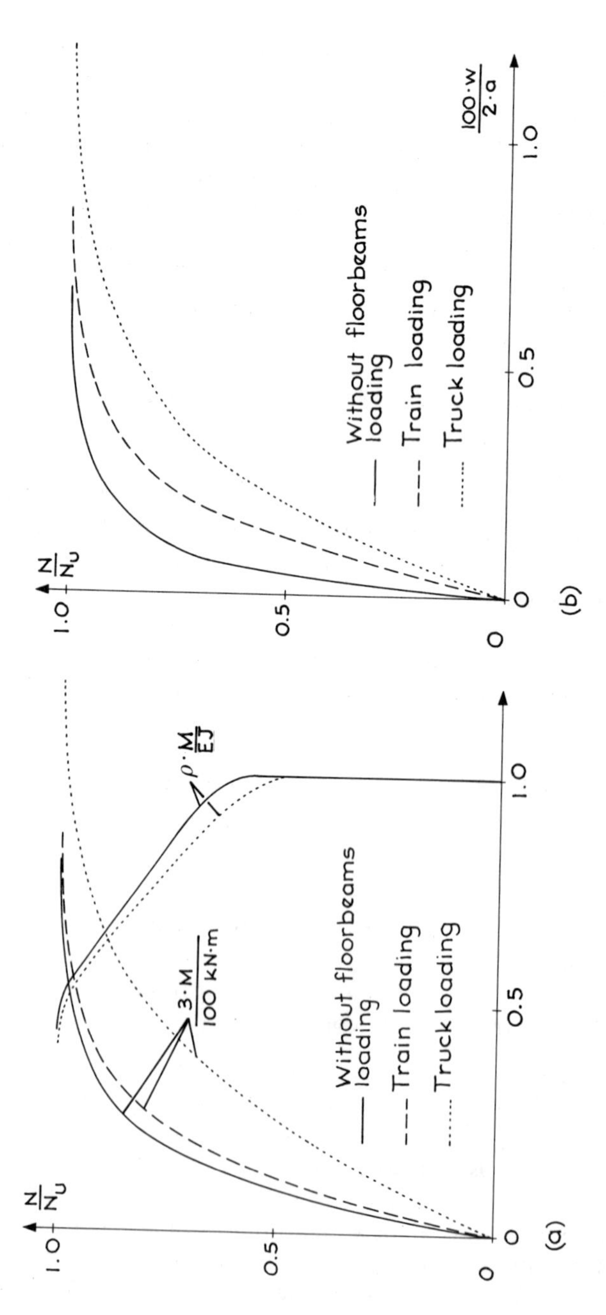

FIG. 13. Variation of bending moments, element stiffness, and deflections.

structures the reduction of the ultimate capacity will certainly be less than indicated by Table 3.

(b) The progression of both bending moments and lateral deflections depends in a great degree on the intensity and the longitudinal distribution of the floor loading. Plasticity in the chord sections will therefore begin earlier for unfavourable floor bending.

9.5.7 Support Reactions

The bifurcation theory, possibly corrected according to sub-section 9.5.5 by the introduction of the *T*-modulus, fixes the frame stiffness required to obtain the desired effective length or the necessary ultimate chord capacity. It furnishes, however, no indication concerning the support reactions: the eigenvectors determine only the pattern of the deflections, without any quantitative value. The spring forces Cw remain therefore indeterminate. Results can only be obtained from second-order computations.

Figure 14, relating to frame type II, shows the progression of the chord

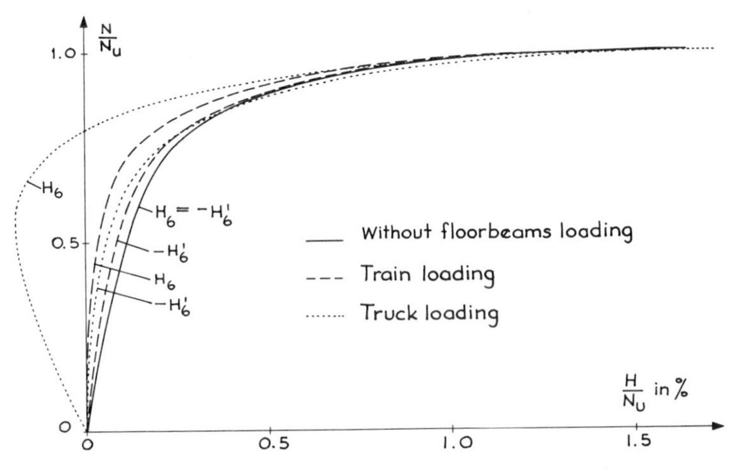

FIG. 14. Progression of the chord reactions for frame type II.

reactions during the loading. The form of the curves depends on the intensity and the distribution of the floorbeam loading. Without such a direct loading, frame 7 lies at a nodal point of the deflections curve and the two adjacent frames 6 and 6′ have to sustain the maximum reactions (opposite in sign). Significant values occur only in the vicinity of the ultimate capacity, a transverse load of 1% of the chord axial force being attained at a level of about 98% of N_u.

For the railway loading the circumstances are similar. By computing the forces exerted by the buckled chord on to the transverse frames, in the planar model of sub-section 9.5.2, the vertical loads are applied directly to the springs and the floorbeam loading should be subtracted from the total spring forces.

The assumed truck loading involves a quite different progression of the lateral forces H: frame 6 with a very high floorbeam loading must at the beginning be sustained by the chord (negative H-values), whereas at the end all transverse frames, including 6, have to stabilise the buckled chord. For frame 6′ without direct floorbeam loading, however, the lateral forces are positive from the beginning. Also here, a value of 1% corresponds to a level of nearly 0·97 N_u.

Of course, the ultimate capacity implies very large lateral deflections and the design of the transverse frames based on the corresponding lateral forces H does not seem to be economical. The value of 1%, already prescribed by some specifications (DIN 4114, 1952; SIA 161/1979) seems adequate, the reduction related to N_u being only 2–3%. For stiffer cross frames, indeed, the ratio H/N_u would be better set at 1·5%.

Special considerations are necessary if plasticity can also occur in the transverse frames. In all computations presented so far, this eventuality has not been accounted for.

9.6 CONTINUOUS COLUMN IN A BRACED STRUCTURE

9.6.1 System Arrangement

The span distribution and the loading arrangement are shown in Fig. 15: only the central part of the column (section HEA 200, weak axis) will be compressed, the lateral spans acting as flexural restraint. This influence is more pronounced in the frame of elasto-plastic computations than for the elastic bifurcation theory, since the lateral spans will still remain elastic and therefore possess a higher stiffness. The half central panel comprises 24 Finite Elements, the lateral spans 16 each. The initial out-of-straightness w_0 is assumed to be 9 mm at mid-span.

9.6.2 Bifurcation Loads and Corresponding Effective Length

The computer program BARBU, already mentioned, gives an eigenvalue $N_{cr} = 948$ kN, corresponding to an effective length of 5·406 m for the

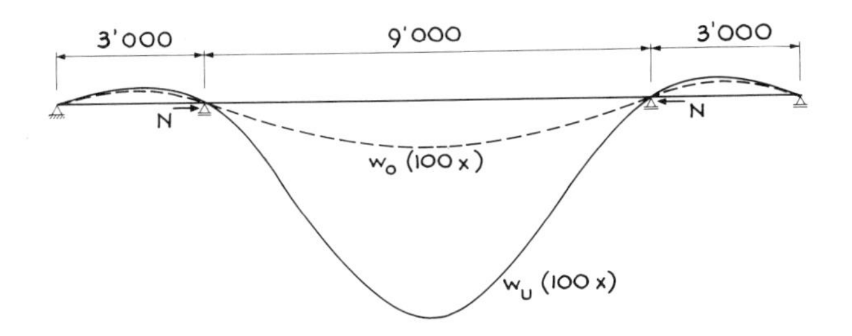

FIG. 15. Column continuous over three spans.

central span, i.e. 0·601*l*. Also, exact solutions can be obtained as described in sub-section 9.2.1 (for the present case, see for example *Stahlbau-Kalender*, 1939 and *Stahlbau-Handbuch*, 1949/50).

9.6.3 Ultimate Capacity

The elasto-plastic second-order computations, based on the assumptions of sub-sections 9.3.1 and 9.3.3, lead to an ultimate capacity

$$N_u = 628 \text{ kN}$$

with the deflection curve w_u shown in Fig. 15.

The corresponding value obtained by using the ECCS column curve c and the above-mentioned effective length is

$$N_u(\text{eigenvalue}) = 580 \text{ kN}$$

On the other hand, the introduction of a buckling modulus $T = 0{\cdot}578E$ for the central span, with the lateral spans remaining elastic, leads to

$$N_u(T) = 627 \text{ kN}$$

This value is equal to the bifurcation load of a continuous column with a bending stiffness EJ in the lateral spans and TJ in the central span. The eigenvalue corresponding to this new stiffness distribution can be determined in the same way as in sub-section 9.6.2. Of course, an iteration procedure is needed, as indicated in Section 9.4.

The usual effective length method is too conservative, since the relative increase of stiffness of the side spans is not included. If the axial force acts at the *ends* of the column, the circumstances are different: for this loading case the effective length is equal to 5597 mm, both with and without the introduction of the T-modulus, and $N_u(\text{eigenvalue}) = N_u(T) = 557$ kN in comparison to $N_u(\text{elasto-plastic}) = 568$ kN.

9.7 LATERAL–TORSIONAL BUCKLING OF A PLATE GIRDER WITH HOLLOW STIFFENERS

9.7.1 Problem to be Solved and Simplifying Assumptions

Figure 16 shows a continuous composite bridge during erection. In the 30-m spans only two intermediate cross bracings are provided. The corresponding spacing of 10 m involves the problem of the lateral stability for the bottom flanges in the region of intermediate supports (hogging moments).

FIG. 16. Composite highway bridge during erection.

The connection between the girders and the concrete deck should be assumed as hinge line since the stud connectors are not designed for tension forces. Preliminary investigations have proved that the influence of the deformation of the transverse stiffeners is negligible so that the model shown in Fig. 17 can be adopted for the control of the lateral–torsional stability of the compression flange.

The figure contains also the response of the longitudinal hollow stiffeners to the lateral displacements w of the flange and to the corresponding twisting angle $\theta = w/h$. Owing to the distance a between stiffener and hinge line, the stabilising effect belonging to the bending alone results from $w(a/h)^2$. Moreover, the variation $\Delta\theta$ of the twisting angle between two transverse stiffeners, spaced Δx, involves a torsional

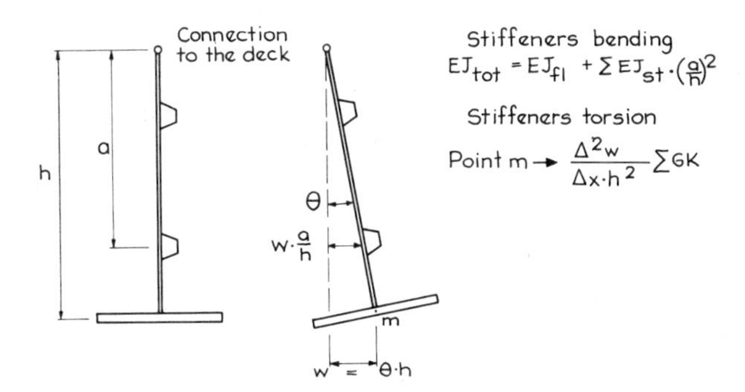

FIG. 17. Assumptions for the lateral buckling analysis.

moment $\Sigma GK{\cdot}\Delta\theta/\Delta x$ or $\Sigma GK{\cdot}\Delta w/(\Delta x{\cdot}h)$ and a stabilising reaction on the flange corresponding to the difference of the torsional moments, right and left of the transverse stiffener, as indicated in Fig. 17.

9.7.2 Planar Model for the Elasto-plastic Computations

In the elasto-plastic computations the influences mentioned in subsections 9.3.1 and 9.3.3 have been included, i.e. geometrical imperfections with a bow of 10 mm and residual stresses belonging to ECCS column curve b, since the flange is flame-cut in the shop (ECCS, 1976 and 1978).

Of course, the beneficial influence of the bending stiffness and of the torsional rigidity of the longitudinal stiffeners should be considered. In order to avoid the use of a spatial analysis (Fukumoto and Kubo, 1977), the planar simulation shown in Fig. 18 has been adopted.

The flange is continuous with supports at the cross bracings supposed rigid. The width and the thickness of the flange correspond to the shop drawings. In the axial forces, the contribution of the web and of the

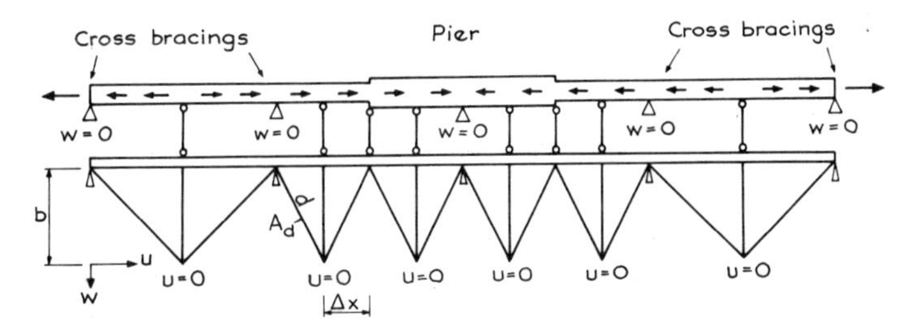

FIG. 18. Compressed flange with virtual truss simulating the reactions due to longitudinal hollow stiffeners.

longitudinal stiffeners are included by an equilibrium condition relating to the hinge line.

The stiffening truss in Fig. 18 simulates exactly the reaction of the longitudinal stiffeners on the flange: the bending stiffness of its top chord is equal to $\Sigma EJ_{st}{\cdot}(a/h)^2$, according to Fig. 17. The section area of the diagonals, on the other hand is chosen so that the corresponding support reactions substitute for the effect of the torsional stiffness. For a very high value of the section area of the top chord and with a condition of no horizontal displacement for the bottom nodes, only the elongation of the diagonals needs to be considered and simple calculations lead to the following diagonal area (Dubas, 1980), with the notations given in the figure

$$A_d = \frac{\sum GK}{Eh^2}\,\frac{d^3}{\Delta x{\cdot}b^2}$$

9.7.3 Numerical Results

Figure 19 recapitulates the main results of the computations. The lateral deflection w_m in the panel near the pier is non-linear from the beginning and increases very rapidly after plastification occurs. This influence appears also from the variation of the bending stiffness in an element in the vicinity of the pier and in an element near the middle of the panel. In the first case plasticity appears early whereas in the field, the load level at the onset of plasticity is higher. In the panel, however, plasticity involves direct collapse.

This difference in the behaviour becomes evident from Fig. 20, which exhibits typical strain distributions, residual strains included. Near the pier the flange axial force is very high so that, the compressive residual strains at the edges being taken into account, the exterior fibres show nearly symmetrical plastifications. The bending moment must therefore be sustained by the remaining elastic core of the flange and the stiffness decreases suddenly. The curvature $1/\rho$ therefore increases rapidly, even for a moderate increment in the bending moment.

At this stage the flange in the pier region will be relatively flexible and the corresponding bending moments increase only slowly. In the panel, on the other hand, the strains result principally from bending and in the vicinity of collapse, the value at the outer edge practically vanishes. In the frame of the first-order theory the limit distribution would correspond to the well-known plastic interaction between M and N. The nose shown in Fig. 20 and the area ΔA belong to the web contribution to

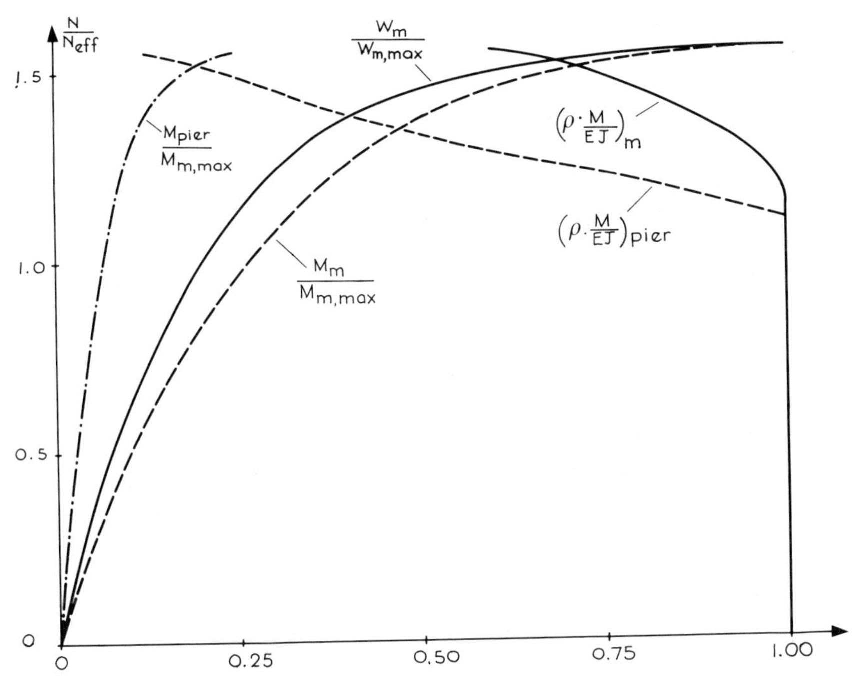

FIG. 19. Variation of bending moments, element stiffness, and deflections.

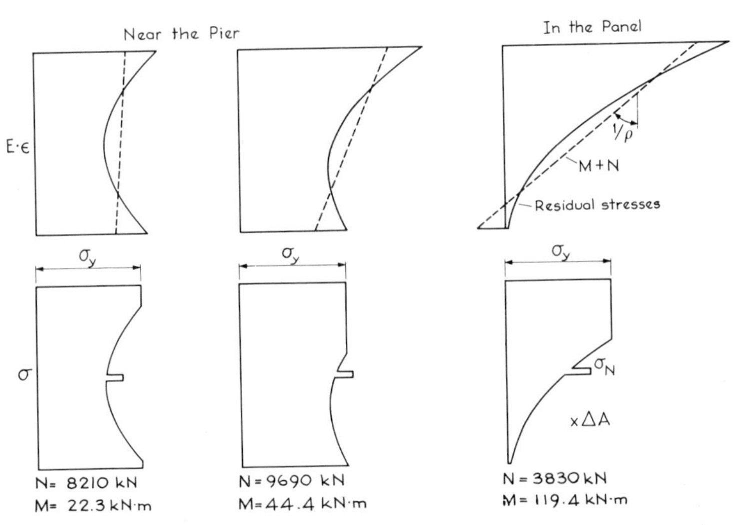

FIG. 20. Strain and stress distribution for two typical flange cross-sections.

the chord force, this girder part being assumed without residual stresses.

9.7.4 Comparison with the Approximate Method

Hand computations with a buckling modulus T, according to Section 9.4, for each flange element lead to a good agreement with the elasto-plastic calculations. In the present case, however, the iterations needed for the determination of T increase the amount of calculation to such an extent that this procedure can only be applied in special design cases.

9.8 CONCLUSIONS

The ultimate capacity of compression members with intermittent lateral supports can be calculated by the use of elasto-plastic second-order computations, taking into account the influence of geometrical imperfections, residual stresses, and other major effects. For current design procedure, however, such calculations are too expensive and will mainly be used to test simple design methods.

The examples shown earlier demonstrate the following conclusions:

(i) If the compressed chord in examination is restrained by elements which remain elastic till the chord collapses, the ultimate capacity computed from an effective length given by the bifurcation theory may be very conservative for stocky members, since the relative increase of the remaining elastic restraints is neglected. For this reason, this approach will often be uneconomical.

(ii) A better approximation is reached if the loss of stiffness of the stabilised compression member appears in the calculations, even in a simplified manner. The introduction of the buckling modulus defined in Section 9.4 leads to a satisfactory agreement with the results of elasto-plastic computations. Moreover, the increase of design work can generally be justified by a reduction in the structural costs.

9.9 ACKNOWLEDGEMENTS

I should like to express my thanks to Mr A. Piller, dipl. Ing. EPFZ, who was responsible for improvements of the programs and for all numerical computations.

REFERENCES

ARGYRIS, J. H., KELSEY, S., and KAMEL, H. (1964). Matrix methods of structural analysis—a précis of recent developments, *Matrix Methods of Structural Analysis*, AGARDograph 72, p. 1, Pergamon Press.

BAAR, S. (1968). Etude théorique et expérimentale du déversement des poutres à membrures tubulaires; Université de Liège, *Collection du Service de Résistance des Matériaux et de Stabilité des Constructions*, No 1.

CIDECT (1981). Monograph No 4, Paris.

CHWALLA, E. (1959). *Hilfstafeln zur Berechnung von Spannungsproblemen der Theorie zweiter Ordnung und von Knickproblemen*, Stahlbau-Verlag, Köln.

DIN 4114 (1952). *Stabilitätsfälle (Knickung, Kippung, Beulung)*.

DUBAS, P. (1977). Ultimate strength of compression members with intermittent rigid or flexible lateral supports, 2nd Intern. Coll. *Stability of Steel Structures, Preliminary Report*, p. 469, Liège.

DUBAS, P. (1980). Réflexions sur certains problèmes de sécurité et de stabilité en construction métallique, *Mémoires C.E.R.E.S*, No. 55, p. 1 Liège.

ENGESSER, F. (1884). Die Sicherung offener Brücken gegen Ausknicken, *Zentralblatt Bauverwaltung*, 415 and 1885, 93.

EUROPEAN CONVENTION FOR CONSTRUCTIONAL STEELWORK (1976). *2nd Intern. Coll. on Stability, Introductory Report*.

EUROPEAN CONVENTION FOR CONSTRUCTIONAL STEELWORK (1978). *European Recommendations for Steel Construction*.

FUKUMOTO, Y. and KUBO, M. (1977). Ultimate bending strength of plate girders with longitudinal stiffeners failed by lateral instability, *Der Stahlbau*, 365.

HILBERS, F. J. (1977). Tafeln zur näherungsweisen Bestimmung der Knicklast von Trogbrücken mit pfostenlosen Fachwerkhauptträgern, *Der Stahlbau*, 103.

HOLT, E. C. (1959). Buckling of a Pony Truss Bridge, Stability of Bridge Chords without Lateral Bracing, *Column Res. Counc. Rep.*, No. 2 (see also Rep. No. 3, 1956 and No. 4, 1957).

HRENNIKOFF, A. (1935). Elastic Stability of a Pony Truss, *Publ. IABSE*, Zürich, **3**, 192.

JOHNSTON, B. G. (1976). *Guide to Stability Design Criteria for Metal Structures*, 3rd Ed., p. 408, John Wiley & Sons.

KRISO, K. (1935). Die Knicksicherheit der Druckgurte offener Fachwerksbrücken; *Publ. IABSE*, Zürich, **3**, 271.

KRISO, K. (1940). Die Knickberechnung mehrfeldriger, in den Feldgrenzen beliebig gestützter Stäbe, *Publ. IABSE*, Zürich, **6**, 139.

OSTENFELD, A. (1916). Die Seitensteifigkeit offener Brücken, *Beton und Eisen*, 123.

PRINCE, C. and DELACOSTE, R. (1974). Equipment Storage Depot for the Telegraph and Telephone Department at Arlesheim (Switzerland), *Acier–Stahl–Steel*, 461.

ROUVÉ, B. (1976). Calcul du comportement postcritique des plaques raidies par la méthode des éléments finis, *Statique appliquée et Construction métallique*, 76–1, EPF Zürich.

SCHIBLER, W. (1946). Das Tragvermögen der Druckgurte offener Fachwerkbrücken mit parallelen Gurtungen, *Mitt. Inst. Baustatik* Zürich, No. 17.

SIA 161 (1979). Steel Structures (Swiss Code).

STAHLBAU-HANDBUCH 1949/50, W. Dorn 1950, p. 118 (see also Pflüger A. *Stabilitätsprobleme der Elastostatik*; 3rd Ed., Springer 1975).

STAHLBAU-KALENDER 1939, W. Ernst & Sohn, Berlin, p. 99.

TIMOSHENKO, S. (1936). *Theory of elastic stability*, 1st Ed., McGraw-Hill.

TURNER, M. J., MARTIN, H. C., and WEIKEL, B. C. (1964). Further development and applications of the stiffness method, *Matrix Methods of Structural Analysis*, AGARDograph 72, p. 203, Pergamon Press.

ZIMMERMANN, H. (1907). Der gerade Stab auf elastischen Einzelstützen mit Belastung durch längsgerichtete Kräfte. Sitzungsberichte der Kgl. Preussischen Akademie der Wissenschaften, 1907.

INDEX